Dimension Theory in Dynamical Systems

Chicago Lectures in Mathematics Series
Robert J. Zimmer, series editor
J. Peter May, Spencer J. Bloch, Norman R. Lebovitz,
    William Fulton, and Carlos Kenig, editors

Other *Chicago Lectures in Mathematics* titles available from the
University of Chicago Press:

Simplicial Objects in Algebraic Topology, by J. Peter May (1967)
Fields and Rings, Second Edition, by Irving Kaplansky (1969, 1972)
Lie Algebras and Locally Compact Groups, by Irving Kaplansky (1971)
Several Complex Variables, by Raghavan Narasimhan (1971)
Torsion-Free Modules, by Eben Matlis (1973)
Stable Homotopy and Generalised Homology, by J. F. Adams (1974)
Rings with Involution, by I. N. Herstein (1976)
Theory of Unitary Group Representation, by George V. Mackey (1976)
Commutative Semigroup Rings, by Robert Gilmer (1984)
Infinite-Dimensional Optimization and Convexity, by Ivar Ekeland and
    Thomas Turnbull (1983)
Navier-Stokes Equations, by Peter Constantin and Ciprian Foias (1988)
Essential Results of Functional Analysis, by Robert J. Zimmer (1990)
Fuchsian Groups, by Zvetlana Katok (1992)
*Unstable Modules over the Steenrod Algebra and Sullivan's Fixed Point
    Set Conjecture,* by Lionel Schwartz (1994)
Topological Classification of Stratified Spaces, by Shmuel Weinberger
    (1994)
Lectures on Exceptional Lie Groups, by J. F. Adams (1996)
Geometry of Nonpositively Curved Manifolds, by Patrick B. Eberlein (1996)

Yakov B. Pesin

DIMENSION THEORY IN DYNAMICAL SYSTEMS:
Contemporary Views and Applications

The University of Chicago Press
Chicago and London

Yakov B. Pesin is professor of mathematics at Pennsylvania State University, University Park. He is the author of *The General Theory of Smooth Hyperbolic Dynamical Systems* and co-editor of *Sinai's Moscow Seminar on Dynamical Systems.*

The University of Chicago Press, Chicago 60637
The University of Chicago Press, Ltd., London
© 1997 by The University of Chicago
All rights reserved. Published 1997
Printed in the United States of America
06 05 04 03 02 01 00 99 98 97   1 2 3 4 5

ISBN: 0-226-66221-7 (cloth)
ISBN: 0-226-66222-5 (paper)

Library of Congress Cataloging-in-Publication Data

Pesin, Ya. B.
    Dimension theory in dynamical systems : contemporary views and applications /
Yakov B. Pesin.
        p.   cm. — (Chicago lectures in mathematics series)
    Includes bibliographical references (p.   -) and index.
        ISBN 0-226-66221-7 (alk. paper). — ISBN 0-226-66222-5 (pbk.:   alk. paper)
        1. Dimension theory (Topology) 2. Differentiable dynamical systems. I. Title. II. Series:
Chicago lectures in mathematics.
    QA611.3.P47 1997
    515'.352--dc21   97-16686
                                                                                      CIP

For my wife, Natasha,
daughters, Ira and Lena,
and my father, Boris,
who make it all worthwhile

Contents

Preface

This is neither a book on dimension theory nor a book on the theory of dynamical systems. This book deals with a new direction of research that lies at the interface of these two theories. One would presumably start writing a book about a new area of research when this area is matured enough to be considered as an independent discipline. As indicators of the maturity, one can use, perhaps, the influence that the new discipline has on other areas and the presence of intriguing new ideas and profound methods of study, as well as exciting applications to other fields.

One can find all these features in the dimension theory of dynamical systems. Although this new discipline was formed only in the last 10–15 years, its impact on both "parents" — dimension theory and the theory of dynamical systems — is quite strong and fruitful and its concepts and results are widely used in many applied fields. The goal of the book is to lay down the mathematical foundation for the new area of research as well as to present the current stage in its development and systematic exposition of its most important accomplishments. I believe this will help to shape the new area as an independent discipline: establish its own language, isolate basic notions and methods of study, find its origin, and trace the history of its development. I also hope this will stimulate further active study.

I would like to point out another important circumstance: until recently physicists and applied mathematicians were the main creators of the dimension theory in dynamical systems. They developed many new concepts and posed a number of challenging problems. Although most "statements" were not rigorous but only intuitively clear and the "proofs" were based upon heuristic arguments, they essentially built up a new building and designed its architecture. Mathematicians came to lay up its foundation and to decorate the building. Their work paid off: they not only enjoyed some very interesting problems but also revealed some new and unexpected phenomena which did not fit in with the "physical" intuition and were not forestalled by physicists.

Hardly any book alone can cover such a diverse and broad area of research as the dimension theory in dynamical systems. This is why I had to select the material following my own interest and my (certainly subjective) point of view. I am mainly concerned with the general concept of characteristics of dimension type and with the "dimension" approach to the theory of dynamical systems.

Although this book is not designed as an introductory textbook on dimension theory or on the theory of dynamical systems some of its parts can be used for a special topics course since it contains all preliminary information from the "parent" disciplines. I suggest the following courses (they may also be considered as logically connected sequences of chapters which are recommended in the first reading):

(1) *Dimension of Cantor-like Sets and Symbolic Dynamics* — Chapters 1, 2, and 4, Appendix II (proofs are optional), and Chapter 5 (Section 15 is optional);

(2) *Dimension and Hyperbolic Dynamics* — Chapters 1, 2, and 4 (Sections 10 and 11), Appendix II (proofs are optional), Chapter 7 (Sections 20, 22, and 23) and Chapter 8 (Section 26 is optional);

(3) *Multifractal Analysis of Dynamical Systems* — Chapters 1, 2, and 3 (Section 9 and Appendix I are optional), Chapter 4 (Sections 10 and 11), Appendix II (proofs are optional), Appendix III (optional), Chapter 6 (Sections 17 and 18), and Chapter 7 (Sections 21 and 24 and Appendix IV is optional).

Let us point out that we number Theorems, Propositions, Examples, and Formulae in such a way that the numerals before the point indicate the number of the corresponding Section.

Throughout the book the reader will find numerous Remarks; they contain the material which is "aside" from the mainstream of the book but provide useful additional information.

There are five Appendices in the book. Appendices I–IV are brief (but sufficiently detailed) surveys on topics which are closely related to the main exposition and help extend reader's vision of the area. Appendix V provides the reader with some useful information from Analysis and Measure Theory and thus makes the book a little more self-contained.

Acknowledgments

I wish to acknowledge the invaluable assistance of several friends and collaborators including Valentine Afraimovich, Luis Barreira, Joerg Schmeling, and Howie Weiss, with whom I enjoyed numerous hours of fruitful discussions while this book was taking shape. They also reviewed the entire manuscript and made many cogent comments which helped me improve the exposition of the book significantly, crystallize its content, and avoid some mathematical and stylistical errors and misprints. I am particularly indebted to Luis Barreira for his devoted assistance in producing the figures and editing the text and to Howie Weiss for his help in polishing the text.

I sincerely thank Boris Hasselblatt who worked long hours modifying his TeX macros and thus enabling me to create a camera-ready copy of the manuscript.

In 1995, Valentine Afraimovich designed and taught a course on dimension theory in dynamical systems at Northwestern University. It was the first course of this type anywhere in the world that I am aware of. His comments helped me clarify the presentation of some parts of the book so that it can be used (at least partly) as a textbook for students, and for this I thank him.

In the Fall of 1996, I taught a graduate course on Dimension Theory and Dynamical Systems at The Pennsylvania State University based on a draft of this book. I would like to thank my students for their enthusiasm, patience, and assistance with numerous rough spots in my exposition. Our collaborative efforts that semester resulted in many additions and changes to the book. Special thanks go to Serge Ferleger, Boris Kalinin, Misha Guysinsky, and Serge Yaskolko.

My thanks also go to Kathy Wyland and Pat Snare at The Pennsylvania State University who typed the first draft of most chapters in TeX.

In the Spring of 1990, I enjoyed a wonderful and productive visit to the University of Chicago. While at Chicago, Robert Zimmer persuaded me to write a book for the Chicago Lectures in Mathematics Series, and this was a partial impetus for writing this book.

I had the great fortune to have top-notch editorial assistance at the Chicago University Press throughout the entire writing and editing process. Particular thanks go to Penelope Kaiserlian, Vicki Jennings, Michael Koplow, Margaret Mahan, and Dave Aftandilian.

Last but not least, I wish to thank Natasha Pesin, an experienced editor, who spent a great amount of time helping me edit and design the book in its present form. Most of all I thank her for her constant encouragement and inspiration.

State College, Pennsylvania
April, 1997   Yakov B. Pesin

Introduction

The dimension of invariant sets is among the most important characteristics of dynamical systems. The study of the dimension of these sets has recently spawned a new and exciting area in dynamical systems. This book presents a comprehensive and systematic treatment of dimension theory in dynamical systems.

The model for the notions of dimension we will consider is the Hausdorff dimension. Unlike the classical topological concept of dimension it is not of a purely topological nature. Carathéodory and Hausdorff originated this notion at the beginning of the twentieth century and later it has become a subject of intensive study by specialists in function theory, mainly by Besicovitch and his school. These investigators used the Hausdorff dimension as an appropriate quantitative characteristic of the complexity of topological structure of subsets in metric spaces; these sets are similar to the well-known Cantor set.

Soon after the discovery of strange attractors — invariant sets of a special type — specialists in dynamical systems became interested in studying the Hausdorff dimension of these attractors, and in relating the dimension to other invariants of dynamical systems. Local topological structure of a strange attractor is often the product of a submanifold and a Cantor-like set. The local submanifold is unstable with respect to the dynamics and corresponds to the directions "along" the attractor. The Cantor-like set lies in the directions "transverse" to the attractor that are stable with respect to the dynamics. The dimension can be interpreted as a quantitative characteristic of the complexity of the topological structure of the attractor in the transverse directions.

A classical example of a strange attractor is the Lorenz attractor. There are many other examples of invariant sets with "wild" topological structure. Among them are the well-known Smale–Williams solenoid and Smale horseshoe. The latter is an example of a hyperbolic invariant set whose local topological structure is the product of different Cantor-like sets. Strange attractors are always associated with trajectories having extremely irregular behavior and are thought of as the origin of "dynamical" chaos.

Specialists in dynamical systems strongly believe that there is a deep connection between the topology of the attractor and properties of the dynamics acting on it. This is a source for exciting relationships between dimension, as a characteristic of complexity of the topological structure, and invariants of the dynamics such as Lyapunov exponents and entropy which characterize instability and stochasticity of dynamical systems (see discussion in [GOY]). Revealing these interrelationships is one of the main goals in studying dimension in dynamical systems.

1

The great interest to this area among specialists in applied fields was inspired by the works of Mandelbrot and especially by his famous book *Fractal Geometry of Nature* [M1]. This book introduces the reader to a range of ideas connected with dimension. It also gives a convincing heuristic description of the way in which the complicated topological structure of invariant sets can influence the qualitative behavior of dynamical systems generating irregular "turbulent" regimes. Mandelbrot also revealed another important aspect of this phenomenon. Assume that a physical system admits a group of scale similarities, i.e., it "reproduces" itself on smaller scales. From a mathematical point of view this means that the dynamical system, which describes the physical phenomenon, possesses invariant sets of a special *self-similar* structure known as *fractals* — a word coined by Mandelbrot in 1975 (see Chapters 5 and 6, where various types of fractals are considered).

The works of Hausdorff, Besicovitch, and Mandelbrot shaped a new field in mathematics called *fractal geometry*. Many areas of science have adopted and widely used the methods and results of fractal geometry. The important feature of fractals is their independence of scaling. The rate of this scaling can be characterized quantitatively by a *fractal dimension*. Using many (often infinitely many) single fractals one can build a multifractal. Its topological structure is much more complicated and is the result of the "interaction" of topological structures of single fractals on different scales.

Multifractals are used in many applied sciences to cope with phenomena associated with intricate structures involving more than one scaling exponent (see Chapter 6). The important conclusion of fractal geometry, based on self-similar properties of multifractals, is that one can produce complex shapes with "highly unusual" properties starting from some simple ones and using simple iteration schemes (see [Ba]). During the past 10 years the concepts of fractal geometry have become enormously popular among specialists in most natural sciences (see [Sc]).

The "natural" popularity of fractal geometry has caused "natural" problems: the rigorous mathematical study was far behind applications. The "empty space" was immediately filled up with numerous "notions" and "results" obtained in studying fractals by a computer. Plausibility was the only criterion for immediate adoption of these notions and results into the theory. Unfortunately, Mandelbrot's book, being directed at specialists in applied fields, hardly contains any rigorous general definitions of dimension and related rigorous results. In particular, the book does not reflect the crucial fact that, in a variety of ways, characteristics of dimension type can be quite "treacherous" and have some "pathological" properties. These properties may not fit in with the intuition that physicists may have developed in working with other objects of research.

Undoubtedly, researchers were not quite satisfied with the notion of Hausdorff dimension which is not quite adopted to the dynamics. Moreover, in many cases the straightforward calculation of the Hausdorff dimension was very difficult. This prompted researchers to introduce other characteristics that, to a greater or lesser degree, aspire to be called dimensions. Among them are *correlation dimension, information dimension, similarity dimension*, etc. In many cases both the motivation for the introduction of these characteristics and their definitions were vague and could be understood in a variety of ways. Nevertheless, most researchers were convinced that in "sufficiently good cases", all these characteristics, if correctly interpreted, would coincide and determine what they called "the fractal dimension". Moreover, a number of conjectures connecting this dimension with other invariants of dynamics have been put forward. Although not all of these conjectures have been confirmed, on the whole, intuition did not lead the researchers astray.

The goal of this book is to study interrelations between dimension theory (and, in particular, fractal geometry) and the theory of dynamical systems. During the past 10–15 years these interrelations have grown from some isolated special results into a cohesive new area in the theory of dynamical systems with its own intrinsic structure. Its current state is characterized by an abundance of notions of dimension and exciting nontrivial relations between them and other invariants of dynamics, new promising methods of study, and growing interest from specialists in different fields. This book provides a rigorous mathematical description of the notions, methods, and results that shape the new area, and extends some concepts of fractal geometry by developing general ways to introduce various characteristics of dimension type.

Let us describe a general scheme for introducing the notion of dimension. Let X be a set and $m(\cdot, \alpha)$ a family of σ-sub-additive set functions on X depending on a real parameter α. Assume that for each $Z \subset X$ there exists a critical "overchanged" value α_0 of α such that $m(Z, \alpha) = 0$ for $\alpha > \alpha_0$ and $m(Z, \alpha) = \infty$ for $\alpha < \alpha_0$ while $m(Z, \alpha_0)$ can be any number in the interval $[0, \infty]$. The number α_0 is called the *dimension characteristic* of the set Z. Of course, from such a general standpoint, one cannot expect to obtain sufficiently meaningful results on dimension. Therefore, we propose a more restricted (but still sufficiently general) approach to construct a family of set functions with the above property. Our construction is an elaboration of the well-known *Carathéodory construction* in general measure theory [C]. We generalize it to include new phenomena associated with the use of dimension in dynamical systems.

Let us outline our approach. One can introduce the notion of dimension in a space X which is endowed with a special structure that we call a *C-structure*. The latter is given when one chooses a collection $\mathcal{F}$ of subsets in X and three non-negative set functions $\xi, \eta, \psi \colon \mathcal{F} \to \mathbb{R}$ that satisfy some conditions. The set function ψ is used to characterize the "size" of sets in $\mathcal{F}$: a set $U \in \mathcal{F}$ is "small" if $\psi(U)$ is small. The role of the set functions ξ and η can be understood using concepts of statistical physics. In this context the set X is viewed as a *configurations space* of a given physical system. To any cover $\mathcal{G} = \{U_1, \ldots, U_n\}$ of Z by sets $U_i \in \mathcal{F}$ one can associate the *free energy*

$$F(\mathcal{G}) = \sum_{U_i \in \mathcal{G}} \xi(U_i) \eta(U_i)^\alpha,$$

where ξ is the *weight function*, η is the *potential*, and $\frac{1}{\alpha}$ is the *temperature* of the physical system. One can now define the family of set functions $m(Z, \alpha)$ by

$$m(Z, \alpha) = \lim_{\epsilon \to 0} \inf_{\mathcal{G}} F(\mathcal{G}),$$

where the infimum is taken over all finite or countable covers $\mathcal{G} \subset \mathcal{F}$ of Z by sets U of "size" $\psi(U) \leq \epsilon$ $(U \in \mathcal{G})$.

In Chapter 1 we will show that the family of set functions $m(Z, \alpha)$, $Z \subset X$ has a critical value $\alpha_0 = \alpha_0(Z)$. We call it the *Carathéodory dimension of the set* Z and denote it by $\dim_C Z$. Another procedure, when one uses covers of Z by sets U with $\psi(U) = \epsilon$, leads to the definition of two other basic characteristics of dimension type — the *lower* and *upper Carathéodory capacities of the set* Z. We denote them by $\underline{\mathrm{Cap}}_C Z$ and $\overline{\mathrm{Cap}}_C Z$ respectively. The basic relationship between Carathéodory dimension and lower and upper Carathéodory capacities is the following inequality:

$$\dim_C Z \leq \underline{\mathrm{Cap}}_C Z \leq \overline{\mathrm{Cap}}_C Z. \tag{0.1}$$

One can generate a C-structure on X using other structures on X. For example, if X is a complete metric space then one can choose $\mathcal{F}$ to be the collection of open subsets in X, and set $\xi(U) = 1$, $\eta(U) = \psi(U) = \mathrm{diam}\, U$ for $U \in \mathcal{F}$. In this case the Carathéodory dimension of a set Z is its Hausdorff dimension, $\dim_H Z$, and the lower and upper Carathéodory capacities of Z coincide with the lower and upper box dimensions of Z, $\underline{\dim}_B Z$ and $\overline{\dim}_B Z$ (see Section 6).

We will be mostly interested in C-structures associated in one way or another with a dynamical system f acting on X. In this case the choice of the collection of subsets $\mathcal{F}$ and set functions ξ, η, ψ is determined by f in some "natural" way. The Carathéodory dimension and Carathéodory capacities of a set Z, which is invariant under f, turn out to be invariants of the dynamical system $f|Z$. Examples are:

(1) q-dimension of Z, $\dim_q Z$ $(q \geq -1$, $q \neq 0$ is a parameter) that is used to characterize the multifractal structure of Z generated by f (see Chapter 3);

(2) topological pressure of a function φ on Z, $P_Z(\varphi)$ and topological entropy on Z, $h_Z(f)$ (see Chapter 4); thus our approach exposes a "dimensional" nature of these well-known topological invariants of dynamical systems.

The study of C-structures generated by dynamical systems leads to another class of characteristics of dimension type specified by a measure μ on X. The formal definition does not involve any dynamics on X and is given in Chapter 1 (see Section 3). We call these characteristics the *Carathéodory dimension of* μ and *lower* and *upper Carathéodory capacities of* μ and denote them respectively by $\dim_C \mu$, $\underline{\mathrm{Cap}}_C \mu$, and $\overline{\mathrm{Cap}}_C \mu$. When μ is invariant under a dynamical system f these characteristics are invariants of f associated with μ. Among several examples given in Chapters 2, 3, and 4 let us mention the measure-theoretic entropy of f, $h_\mu(f)$ (see Section 11). Thus our approach provides a "dimensional" interpretation of this important metric invariant of dynamical systems.

The basic relationship between the Carathéodory dimension of μ and lower and upper Carathéodory capacities of μ is the following:

$$\dim_C \mu \leq \underline{\mathrm{Cap}}_C \mu \leq \overline{\mathrm{Cap}}_C \mu. \tag{0.2}$$

A challenging problem is to find sufficient conditions that would guarantee equalities in (0.1) and (0.2). We stress that the coincidence of the Carathéodory dimension and Carathéodory capacities relative to a set Z is a rare phenomenon and requires a special somewhat homogeneous structure of Z (see Chapter 5). As far as Carathéodory dimension and Carathéodory capacities relative to a measure μ are concerned we present a powerful criterion that guarantees their coincidence (see Section 4). We believe that in "good" cases the conditions of this criterion hold and, thus, the common value represents the *dimension* of μ defined by the C-structure on X.

Another aspect of the notions of dimension and capacity type characteristics relative to a measure has to do with the above-mentioned phenomenon of self-similarity of an invariant set. In "real" situations, self-similarity is hardly ever exact. However, the invariant set sometimes can be "broken into pieces" each of which turns out to be "asymptotically self-similar" in a way. Such pieces are the supports of invariant ergodic measures, and "self-similarity scales" can be expressed in terms of the Lyapunov exponents of these measures. This is the clue to reveal the fundamental relation between dimension, Lyapunov exponents, and metric entropy (see Chapter 8).

The book consists of two parts. In the first part we develop the general theory of Carathéodory dimension. In Chapter 1 we describe a generalized version of the classical Carathéodory construction in a space X and introduce the notions of C-structure and Carathéodory dimension characteristics: Carathéodory dimension and lower and upper Carathéodory capacities of subsets of X and measures on X.

In Chapter 2 we study C-structures generated by metrics on Euclidean spaces. We introduce the notions of Hausdorff dimension and lower and upper box dimensions of sets and measures and describe their basic properties. In Chapter 3 we deal with C-structures generated by both metrics and measures and introduce the notions of q-dimension and lower and upper q-box dimensions of sets and measures. This leads to an important application of the general Carathéodory construction developed in Chapter 1: the q-box dimension is closely related to dimension spectra of dynamical systems which are widely used in numerical study of dynamical systems. We describe these spectra in detail in Chapter 6.

In Appendix I we extend results of Chapters 2 and 3 to arbitrary complete separable metric spaces.

Our main example of a C-structure is given in Chapter 4, where we consider C-structures generated by dynamical systems acting on compact metric spaces and continuous functions. This example is one of the main manifestations of the general Carathéodory construction: we demonstrate how the "dimension" approach can be used to introduce a general concept of topological pressure and topological entropy for arbitrary subsets of the space as well as a concept of measure-theoretic entropy. In Appendix II we use the "dimension" approach to discuss various versions of the thermodynamic formalism (including the classical thermodynamic formalism of dynamical systems created by Bowen, Ruelle, Sinai, and Walters). Although this lies not strictly along the line of the main exposition, it is an important addition to Chapter 4 and is crucial for results in the second part of the book. In Appendix III we describe an example of C-structure which plays a crucial role in studying some "weird" sets for dynamical systems.

The second part of the book is devoted to applications of results in Part I to dimension theory and the theory of dynamical systems. In Chapter 5 we describe various geometric constructions — one of the most popular subjects in dimension theory. We demonstrate that the theory of dynamical systems grants powerful methods to study geometric constructions with complicated geometry of basic sets and essentially arbitrarily symbolic representation. Our main tool is the thermodynamic formalism developed in Chapter 4 that we apply to the symbolic dynamical system, associated with the geometric construction.

In Chapter 6 we study another popular subject of dimension theory intimately connected with fractals and multifractals. We introduce various dimension spectra (the Hentschel–Procaccia spectrum for dimensions, Rényi spectrum for dimensions, and the spectrum of so-called pointwise dimensions) and describe their relations to some well-known dimension characteristics of dynamical systems such as the correlation dimension and information dimension. We also use dimension spectra to discuss the mathematical content of the notion of multifractality, and we effect a complete multifractal analysis of Gibbs measures supported on the limit sets of Moran geometric constructions.

The interrelation between dimension theory and the theory of dynamical systems is of benefit to both sides. In the last two chapters we show how methods of dimension theory can be applied to study various characteristics of dimension type of sets and measures invariant under hyperbolic dynamical systems. We consider repellers for smooth expanding maps (including hyperbolic Julia sets, repellers for one-dimensional Markov maps, and limit sets for Schottky groups), basic sets of Axiom A diffeomorphisms (including Smale horseshoes), and Smale–Williams solenoids. We obtain most definite results in the case when dynamics is conformal and sharp "dimension estimates" in the non-conformal case. The approach we use demonstrates the power of the general Carathéodory construction which allows one to extend and unify many results on dimension of invariant sets for dynamical systems with hyperbolic behavior.

A significant part of Chapter 7 is to develop a complete multifractal analysis of Gibbs measures for smooth conformal dynamical systems. In particular, we obtain a complete and surprisingly simple description of a highly non-trivial and intricate multifractal structure of conformal repellers and conformal hyperbolic sets associated with the pointwise dimension, local entropy, and Lyapunov exponent. The approach is built upon the general concept of multifractal spectra — a recent new direction of research in the theory of dynamical systems which we sketch in Appendix IV. Multifractal spectra provide refined information on some ergodic properties of dynamical systems. For example, multifractal spectra for local entropies describe their deviation from the mean value provided by the Shannon–McMillan–Breiman Theorem while multifractal spectra for Lyapunov exponents describe their distribution around the mean value given by the Multiplicative Ergodic Theorem.

In Chapter 8 we deal with the dimension of invariant measures. In particular, we discuss the recent achievement in dimension theory of smooth hyperbolic dynamical systems — the affirmative solution of the Eckmann–Ruelle conjecture obtained in [BPS1]. It establishes the existence of pointwise dimension for almost every point with respect to a hyperbolic invariant measure. This implies that all characteristics of dimension type of the measure coincide and thus, it justifies the strong opinion among experts that in "good cases" (and hyperbolic measures are "good" ones) all known methods of computing the dimension of a measure lead to the same quantity. Since hyperbolic measures are "responsible" for chaotic regimes generated by dynamical systems, this quantity stands in a row of most fundamental characteristics of such complicated motions.

Part I

Carathéodory Dimension Characteristics

Chapter 1

General Carathéodory Construction

The classical Carathéodory construction in the general measure theory was originated by Carathéodory in [C] (a contemporary exposition can be found in [Fe]). It was designed to produce a family of α-measures on a metric space X given by

$$m(Z, \alpha, \eta) = \liminf_{\varepsilon \to 0} \left\{ \sum_{U_i \in \mathcal{G}} \eta(U_i)^\alpha \right\},$$

where the infimum is taken over all finite or countable covers $\mathcal{G} = \{U_i\}$ of Z by open sets U_i with $\operatorname{diam} U_i \leq \varepsilon$ (one can easily see that the limit exists). Here η is a positive set function. One can also use an arbitrarily chosen collection of subsets of X to make up covers of Z.

In this chapter we introduce a construction which is a generalization of the classical Carathéodory construction. It was elaborated by Pesin in [P2] to produce various characteristics of dimension type. The starting point for the construction is a space X which is endowed with a special structure. We introduce this structure axiomatically by describing its basic elements and relations between them, and we call it the Carathéodory dimension structure (or, briefly, C-structure). This structure enables us to yield two types of quantities that are the dimension of subsets of X and the dimension of measures on X. We call these quantities the Carathéodory dimension characteristics (of sets and measures). They include the Carathéodory dimension and lower and upper Carathéodory capacities. We study some of their fundamental properties which will be widely used in the book.

C-structures can be generated by some other structures on X associated with metrics, measures, functions, etc. In Chapters 2 and 3, we will give some examples of C-structures and will illustrate general properties of Carathéodory dimension characteristics established in this chapter. We will be mainly interested in C-structures generated by dynamical systems acting on X.

Let us notice that Carathéodory dimension characteristics are invariants of an isomorphism which preserves the C-structure. In the case when the C-structure is generated by a dynamical system on X, the Carathéodory dimension and lower and upper Carathéodory capacities become invariants of the dynamics and can be used to characterize invariant sets and invariant measures. There are deep relations between them and other important invariants of dynamics which we will consider in the second part of the book.

The original Carathéodory construction was created within the framework of the classical function theory where the dimension of subsets was a "natural" subject of study. The dimension of measures was introduced within the theory of dynamical systems in order to characterize invariant subsets and measures concentrated on them. The most exciting applications of the general Carathéodory construction can be obtained in this case and are related to the dimension characteristics of measures. It is worth emphasizing that the general Carathéodory construction does not require any dynamics on X and thus, its applications go far beyond the theory of dynamical systems and include such fields as function theory, geometry, etc.

1. Carathéodory Dimension of Sets

Let X be a set and $\mathcal{F}$ a collection of subsets of X. Assume that there exist two set functions $\eta, \psi \colon \mathcal{F} \to \mathbb{R}^+ (= [0, \infty))$ satisfying the following conditions:

A1. $\varnothing \in \mathcal{F}$; $\eta(\varnothing) = 0$ and $\psi(\varnothing) = 0$; $\eta(U) > 0$ and $\psi(U) > 0$ for any $U \in \mathcal{F}$, $U \neq \varnothing$;

A2. for any $\delta > 0$ one can find $\varepsilon > 0$ such that $\eta(U) \leq \delta$ for any $U \in \mathcal{F}$ with $\psi(U) \leq \varepsilon$.

A3. for any $\varepsilon > 0$ there exists a finite or countable subcollection $\mathcal{G} \subset \mathcal{F}$ which covers X (i.e., $\cup_{U \in \mathcal{G}} U \supset X$) and $\psi(\mathcal{G}) \stackrel{\text{def}}{=} \sup\{\psi(U) : U \in \mathcal{G}\} \leq \varepsilon$.

Let $\xi \colon \mathcal{F} \to \mathbb{R}^+$ be a set function. We say that the collection of subsets $\mathcal{F}$ and the set functions ξ, η, ψ, satisfying Conditions $A1, A2$, and $A3$, introduce the **Carathéodory dimension structure** or **C-structure** τ on X and we write $\tau = (\mathcal{F}, \xi, \eta, \psi)$. C-structures on X can be generated by other structures associated with metrics and measures on X, maps acting on X, etc. We illustrate this by the following examples; more general setups are given in Chapters 2, 3, and 4.

Examples.

(1) Define the C-structure on the Euclidean space $\mathbb{R}^m$ as follows. Let $\mathcal{F}$ be the collection of open sets, $\xi(U) = 1$, $\eta(U) = \operatorname{diam} U$, $\psi(U) = \operatorname{diam} U$ for $U \in \mathcal{F}$. It is easy to see that the collection of subsets $\mathcal{F}$ and set functions ξ, η, ψ satisfy Conditions $A1, A2$, and $A3$ and hence introduce the C-structure $\tau = (\mathcal{F}, \xi, \eta, \psi)$ on X.

(2) Let μ be a Borel probability measure on the Euclidean space $\mathbb{R}^m$. Fix any $\gamma > 0$ and $q \geq 0$. Define $\mathcal{F}$ to be the collection of open balls, $B(x, \varepsilon)$ ($x \in X$, $\varepsilon > 0$) and set $\xi(B(x, \varepsilon)) = \mu(B(x, \gamma\varepsilon))^q$, $\eta(B(x, \varepsilon)) = \psi(B(x, \varepsilon)) = \varepsilon$. One can easily check that the structure $\tau_{q,\gamma} = (\mathcal{F}, \xi, \eta, \psi)$ is a C-structure on X.

(3) Let X be a compact metric space endowed with a metric ρ and $f \colon X \to X$ a continuous map. Given $\delta > 0$, $n \geq 0$, and $x \in X$ we denote

$$B_n(x, \delta) = \{y \in X : \rho(f^i(x), f^i(y)) \leq \delta \text{ for } 0 \leq i \leq n\}.$$

We define the C-structure $\tau_\delta = (\mathcal{F}_\delta, \xi, \eta, \psi)$ on X by setting $\mathcal{F}_\delta = \{B_n(x, \delta) : x \in X, n \geq 0\}$, $\xi(B_n(x, \delta)) = 1$, $\eta(B_n(x, \delta)) = e^{-n}$, and $\psi(B_n(x, \delta)) = \frac{1}{n}$. (Remark: we assume, for simplicity, that the map f is such that $B_n(x, \delta) \neq B_m(y, \delta)$ if $n \neq m$; the general case is considered in Section 11).

Consider a set X endowed with a C-structure $\tau = (\mathcal{F}, \xi, \eta, \psi)$. Given a set $Z \subset X$ and numbers $\alpha \in \mathbb{R}$, $\varepsilon > 0$, we define

$$M_C(Z, \alpha, \varepsilon) = \inf_{\mathcal{G}}\{\sum_{U \in \mathcal{G}} \xi(U)\eta(U)^\alpha\}, \qquad (1.1)$$

where the infimum is taken over all finite or countable subcollections $\mathcal{G} \subset \mathcal{F}$ covering Z with $\psi(\mathcal{G}) \le \varepsilon$. By Condition $A3$ the function $M_C(Z, \alpha, \varepsilon)$ is correctly defined. It is non-decreasing as ε decreases. Therefore, there exists the limit

$$m_C(Z, \alpha) = \lim_{\varepsilon \to 0} M_C(Z, \alpha, \varepsilon). \qquad (1.2)$$

We shall study the function $m_C(Z, \alpha)$.

Proposition 1.1. *For any $\alpha \in \mathbb{R}$ the set function $m_C(\cdot, \alpha)$ satisfies the following properties:*

(1) $m_C(\varnothing, \alpha) = 0$ for $\alpha > 0$;
(2) $m_C(Z_1, \alpha) \le m_C(Z_2, \alpha)$ if $Z_1 \subset Z_2 \subset X$;
(3) $m_C(\bigcup_{i \ge 0} Z_i, \alpha) \le \sum_{i \ge 0} m_C(Z_i, \alpha)$, where $Z_i \subset X, i = 0, 1, 2, \ldots$.

Proof. The first two statements follow directly from the definitions. We shall prove the third one. Given $\delta > 0$, $\varepsilon > 0$, and $i \ge 0$ one can find ε_i, $0 < \varepsilon_i \le \varepsilon$ and a cover $\mathcal{G}_i = \{U_{ij} \in \mathcal{F}, \ j \ge 0\}$ of the set Z_i with $\psi(\mathcal{G}_i) \le \varepsilon_i$ such that

$$\left| m_C(Z_i, \alpha) - \sum_{j \ge 0} \xi(U_{ij})\eta(U_{ij})^\alpha \right| \le \frac{\delta}{2^i}.$$

The collection $\mathcal{G}$ of sets $\{U_{ij}, i \ge 0, j \ge 0\}$ covers $Z = \cup_{i \ge 0} Z_i$ and satisfies $\psi(\mathcal{G}) \le \varepsilon$. Now we have that

$$M_C(Z, \alpha, \varepsilon) \le \sum_{U_{ij} \in \mathcal{G}} \xi(U_{ij})\eta(U_{ij})^\alpha \le 2\delta + \sum_{i \ge 0} m_C(Z_i, \alpha).$$

Since ε and δ can be chosen arbitrarily small this implies the desired result. ∎

If $m_C(\varnothing, \alpha) = 0$ (this holds true for $\alpha > 0$ but can also happen for some negative α) the set function $m_C(\cdot, \alpha)$ becomes an outer measure on X (see definition of outer measures and other relevant information in Appendix V). We call it the α-**Carathéodory outer measure** (specified by the collection of subsets $\mathcal{F}$ and the set functions ξ, η, ψ). According to the general measure theory the outer measure induces a σ-additive measure on the σ-field of measurable sets in X (see for example, [Fe] and also Appendix V). We call this measure the α-**Carathéodory measure**. Note that this measure is not necessarily σ-finite.

We shall now describe a crucial property of the function $m_C(Z, \cdot)$ for a fixed set Z.

Proposition 1.2. *There exists a critical value α_C, $-\infty \le \alpha_C \le +\infty$ such that $m_C(Z, \alpha) = \infty$ for $\alpha < \alpha_C$ and $m_C(Z, \alpha) = 0$ for $\alpha > \alpha_C$.*

Proof. It follows from $A2$ that if $\infty > m_C(Z, \alpha) \ge 0$ for some $\alpha \in \mathbb{R}$ then $m_C(Z, \beta) = 0$ for any $\beta > \alpha$ and if $m_C(Z, \alpha) = \infty$ for some $\alpha \in \mathbb{R}$ then $m_C(Z, \beta) = \infty$ for any $\beta < \alpha$. This proves the desired statement. ∎

Let us remark that $m_C(Z, \alpha_C)$ can be 0, ∞, or a finite positive number.

We define the **Carathéodory dimension of a set** $Z \subset X$ by

$$\dim_C Z = \alpha_C = \inf\{\alpha : m_C(Z, \alpha) = 0\} = \sup\{\alpha : m_C(Z, \alpha) = \infty\}. \qquad (1.3)$$

The Carathéodory dimension clearly depends on the choice of C-structure $\tau = (\mathcal{F}, \xi, \eta, \psi)$ on X. We will use the more explicit notation $\dim_{C,\tau} X$ when we want to emphasize the C-structure we are dealing with.

Let us point out that the definition of the Carathéodory dimension does not require any assumptions on the set function ξ. The use of this function allows one to broaden remarkably applications of the above construction.

One can obtain different examples of Carathéodory dimension of sets by using the C-structures in Examples 1, 2, and 3 above. Thus, we have respectively:

(1) the Hausdorff dimension of the set Z which we denote by $\dim_H Z$ (see Section 6);

(2) the (q, γ)-dimension of the set Z which we denote by $\dim_{q,\gamma} Z$; we define the q-dimension of the set Z by $\dim_q Z = \sup_{\gamma > 1} \dim_{q,\gamma} Z$ (see Section 8);

(3) the δ-topological entropy of the map f on the set Z which we denote by $h_Z(f, \delta)$; we define the topological entropy of f on Z by $h_Z(f) = \varlimsup_{\delta \to 0} h_Z(f, \delta)$; (we show in Section 11 that if Z is f-invariant and compact then this definition is equivalent to the well-known definition of topological entropy).

We now state some basic properties of the Carathéodory dimension.

Theorem 1.1.

(1) $\dim_C \varnothing \leq 0$.

(2) $\dim_C Z_1 \leq \dim_C Z_2$ if $Z_1 \subset Z_2 \subset X$.

(3) $\dim_C \left(\bigcup_{i \geq 0} Z_i \right) = \sup_{i \geq 0} \dim_C Z_i$, where $Z_i \subset X, i = 0, 1, 2, \dots$.

Proof. The first two statements follow directly from Proposition 1.1. In order to prove the third statement assume that $\dim_C Z_i < \alpha$ for all $i = 0, 1, 2, \dots$. It follows that $m_C(Z_i, \alpha) = 0$ and hence by Proposition 1.1, $m_C(\cup_{i \geq 0} Z_i) = 0$. Therefore, $\dim_C(\cup_{i \geq 0} Z_i) \leq \alpha$. This implies that $\dim_C(\cup_{i \geq 0} Z_i) \leq \sup_{i \geq 0} \dim_C Z_i$. The opposite inequality immediately follows from the second statement of the theorem. ∎

Let us notice that, in general, there may exist sets with *negative* Carathéodory dimension. Whether this is the case depends on the set function ξ. If $\xi(U) \geq$ constant for any $U \in \mathcal{F}$ then $\dim_C Z \geq 0$ for any $Z \subset X$. However, if the set function $\xi(U)$ decays "sufficiently fast" towards zero as $\psi(U) \to 0$, the Carathéodory dimension of the space X (and hence every subset in X) may become negative or even $-\infty$. On the other hand, if $\xi(U)$ increases to infinity "sufficiently fast" as $\psi(U) \to 0$ the Carathéodory dimension of X may become equal to $+\infty$ (see Example 6.3).

Let $\tau = (\mathcal{F}, \xi, \eta, \psi)$ and $\tau' = (\mathcal{F}', \xi', \eta', \psi')$ be two C-structures on X. We say that τ is **subordinated** to τ' and write $\tau < \tau'$ if the following conditions hold:

(a) for any $U \in \mathcal{F}$ there exists $U' \in \mathcal{F}'$ such that $U \subset U'$;

(b) there exists a constant $K > 0$ such that for any $U \in \mathcal{F}$,

$$\xi'(U') \le K\,\xi(U),\ \eta'(U') \le K\,\eta(U),\ \psi'(U') \le K\psi(U),$$

where $U' \in \mathcal{F}'$ and $U' \supset U$.

Theorem 1.2. *If $\tau < \tau'$ then $\dim_{C,\tau'} Z \le \dim_{C,\tau} Z$ for any $Z \subset X$.*

Proof. Given $Z \subset X$, $\alpha \in \mathbb{R}$, and $\delta > 0$, there exists $\varepsilon_0 > 0$ such that for any ε, $0 < \varepsilon \le \varepsilon_0$ one can find a cover $\mathcal{G}$ of Z with $\psi(\mathcal{G}) \le \varepsilon$ and

$$\sum_{U \in \mathcal{G}} \xi(U)\eta(U)^\alpha \le M_{C,\tau}(Z, \alpha, \varepsilon) + \delta.$$

For each $U \in \mathcal{G}$ we choose a set U' which satisfies Conditions (a) and (b) above. The sets U' comprise the cover $\mathcal{G}'$ of Z with $\psi'(\mathcal{G}') \le K\varepsilon$. We have now

$$M_{C,\tau'}(Z, \alpha, K\varepsilon) \le \sum_{U' \in \mathcal{G}'} \xi'(U')\eta'(U')^\alpha$$

$$\le K^{1+\alpha} \sum_{U \in \mathcal{G}} \xi(U)\eta(U)^\alpha \le K^{1+\alpha}(M_{C,\tau}(Z, \alpha, \varepsilon) + \delta).$$

Taking the limit as $\varepsilon \to 0$ we obtain from here that

$$m_{C,\tau'}(Z, \alpha) \le K^{1+\alpha}(m_{C,\tau}(Z, \alpha) + \delta).$$

Since δ can be chosen arbitrarily small this implies that

$$m_{C,\tau'}(Z, \alpha) \le K^{1+\alpha} m_{C,\tau}(Z, \alpha).$$

The desired result follows immediately.   ∎

We show that the Carathéodory dimension is invariant under an isomorphism which preserves the C-structure.

Let X and X' be sets endowed respectively with C-structures $\tau = (\mathcal{F}, \xi, \eta, \psi)$ and $\tau' = (\mathcal{F}', \xi', \eta', \psi')$. Let also $\chi: X \to X'$ be a bijective map.

Theorem 1.3. *Assume that there exists a constant $K \ge 1$ such that for any $U \in \mathcal{F}$ one can find sets $U_1', U_2' \in \mathcal{F}'$ satisfying*

(1) $U_2' \subset \chi(U) \subset U_1'$;

(2) $K^{-1}\xi'(U_1') \le \xi(U) \le K\xi'(U_2')$, $K^{-1}\eta'(U_1') \le \eta(U) \le K\eta'(U_2')$, and $K^{-1}\psi'(U_1') \le \psi(U) \le K\psi'(U_2')$.

Then $\dim_{C,\tau'} \chi(Z) = \dim_{C,\tau} Z$ for any $Z \subset X$.

Proof. Define

$$\mathcal{F}'' = \{\chi(U) : U \in \mathcal{F}\}, \quad \xi'' = \xi \circ \chi^{-1}, \quad \eta'' = \eta \circ \chi^{-1}, \quad \psi'' = \psi \circ \chi^{-1}.$$

It is easy to see that $\tau'' = (\mathcal{F}'', \xi'', \eta'', \psi'')$ is a C-structure on X' and that $\tau' < \tau''$ and $\tau'' < \tau'$. The desired result follows from Theorem 1.2.   ∎

2. Carathéodory Capacity of Sets

Let X be a set endowed with a C-structure $\tau = (\mathcal{F}, \xi, \eta, \psi)$. We now modify the above construction to produce another type of Carathéodory dimension characteristics.

We shall assume that the following Condition $A3'$ holds which is stronger than Condition $A3$:

A3′. there exists $\epsilon > 0$ such that for any $\epsilon \geq \varepsilon > 0$, one can find a finite subcollection $\mathcal{G} \subset \mathcal{F}$ covering X such that $\psi(U) = \varepsilon$ for any $U \in \mathcal{G}$.

Given $\alpha \in \mathbb{R}$, $\varepsilon > 0$, and a set $Z \subset X$, define

$$R_C(Z, \alpha, \varepsilon) = \inf_{\mathcal{G}} \left\{ \sum_{U \in \mathcal{G}} \xi(U) \eta(U)^\alpha \right\},$$

where the infimum is taken over all finite or countable subcollections $\mathcal{G} \subset \mathcal{F}$ covering Z such that $\psi(U) = \varepsilon$ for any $U \in \mathcal{G}$. According to $A3'$, $R_C(Z, \alpha, \varepsilon)$ is correctly defined. We set

$$\underline{r}_C(Z, \alpha) = \varliminf_{\varepsilon \to 0} R_C(Z, \alpha, \varepsilon), \quad \overline{r}_C(Z, \alpha) = \varlimsup_{\varepsilon \to 0} R_C(Z, \alpha, \varepsilon).$$

The following statement describes the behavior of the functions $\underline{r}_C(Z, \cdot)$ and $\overline{r}_C(Z, \cdot)$. The proof is analogous to the proof of Proposition 1.2.

Proposition 2.1. *For any $Z \subset X$, there exist $\underline{\alpha}_C, \overline{\alpha}_C \in \mathbb{R}$ such that*

(1) $\underline{r}_C(Z, \alpha) = \infty$ *for $\alpha < \underline{\alpha}_C$ and $\underline{r}_C(Z, \alpha) = 0$ for $\alpha > \underline{\alpha}_C$ (while $\underline{r}_C(Z, \underline{\alpha}_C)$ can be 0, ∞, or a finite positive number);*

(2) $\overline{r}_C(Z, \alpha) = \infty$ *for $\alpha < \overline{\alpha}_C$ and $\overline{r}_C(Z, \alpha) = 0$ for $\alpha > \overline{\alpha}_C$ (while $\overline{r}_C(Z, \overline{\alpha}_C)$ can be 0, ∞, or a finite positive number).*

Given $Z \subset X$, we define the **lower** and **upper Carathéodory capacities of the set** Z by

$$\begin{aligned} \underline{\mathrm{Cap}}_C Z &= \underline{\alpha}_C = \inf\{\alpha : \underline{r}_C(Z, \alpha) = 0\} = \sup\{\alpha : \underline{r}_C(Z, \alpha) = \infty\}, \\ \overline{\mathrm{Cap}}_C Z &= \overline{\alpha}_C = \inf\{\alpha : \overline{r}_C(Z, \alpha) = 0\} = \sup\{\alpha : \underline{r}_C(Z, \alpha) = \infty\}. \end{aligned} \tag{2.1}$$

These quantities clearly depend on the choice of the collection $\mathcal{F}$ and the set functions ξ, η, ψ. We will use the more explicit notation $\underline{\mathrm{Cap}}_{C,\tau} Z$ and $\overline{\mathrm{Cap}}_{C,\tau} Z$ when we want to emphasize the structure τ we are dealing with.

One can obtain examples of the lower and upper Carathéodory capacities by exploiting the C-structures in Examples 1, 2, and 3 in Section 1 (these C-structures clearly satisfy Condition $A3'$). Thus, we have respectively:

(1) the lower and upper box dimensions of the set Z which we denote by $\underline{\dim}_B Z$, $\overline{\dim}_B Z$ (see Section 6);

(2) the lower and upper (q, γ)-box dimensions of the set Z which we denote by $\underline{\dim}_{q,\gamma} Z$ and $\overline{\dim}_{q,\gamma} Z$; we define the lower and upper q-box dimensions

of Z by $\underline{\dim}_q Z = \sup_{\gamma>1} \underline{\dim}_{q,\gamma} Z$ and $\overline{\dim}_q Z = \sup_{\gamma>1} \overline{\dim}_{q,\gamma} Z$ (see Section 8); in Section 18 we establish relations between the lower and upper q-box dimensions and dimension spectra of dynamical systems;

(3) the lower and upper δ-capacity topological entropies of f on Z which we denote by $\underline{Ch}_Z(f,\delta)$ and $\overline{Ch}_Z(f,\delta)$; we define the lower and upper capacity topological entropies of f on Z by setting $\underline{Ch}_Z(f) = \underline{\lim}_{\delta\to 0} \underline{Ch}_Z(f,\delta)$ and $\overline{Ch}_Z(f) = \underline{\lim}_{\delta\to 0} \overline{Ch}_Z(f,\delta)$; we show in Section 11 that if Z is invariant under f then $\underline{Ch}_Z(f) = \overline{Ch}_Z(f)$ and if Z is invariant and compact then $\underline{Ch}_Z(f) = \overline{Ch}_Z(f) = h_Z(f)$.

Remark.

The set functions $\underline{r}_C(\cdot,\alpha)$ and $\overline{r}_C(\cdot,\alpha)$ satisfy:

(a) $\underline{r}_C(\varnothing,\alpha) \geq 0$ and $\overline{r}_C(\varnothing,\alpha) \geq 0$ for $\alpha > 0$;

(b) $\underline{r}_C(Z_1,\alpha) \leq \underline{r}_C(Z_2,\alpha)$, and $\overline{r}_C(Z_1,\alpha) \leq \overline{r}_C(Z_2,\alpha)$ if $Z_1 \subset Z_2 \subset X$.

In general, the set function $\underline{r}_C(\cdot,\alpha)$ is not finitely sub-additive while the set function $\overline{r}_C(\cdot,\alpha)$ does have this property: one can show that for any finite number of sets $Z_i \subset X$, $i = 1,\ldots,n$

$$\overline{r}_C\left(\bigcup_{i=1}^n Z_i, \alpha\right) \leq \sum_{i=1}^n \overline{r}_C(Z_i, \alpha).$$

Note that if $\underline{r}_C(Z,\alpha) = \overline{r}_C(Z,\alpha) \overset{\text{def}}{=} r_C(Z,\alpha)$ for any $Z \subset X$ then the set function $r_C(\cdot,\alpha)$ is a finite sub-additive outer measure on X provided $r_C(\varnothing,\alpha) = 0$.

We now state some basic properties of the lower and upper Carathéodory capacities of sets. The proofs follow directly from the definitions.

Theorem 2.1.

(1) $\dim_C Z \leq \underline{\text{Cap}}_C Z \leq \overline{\text{Cap}}_C Z$ for any $Z \subset X$.

(2) $\underline{\text{Cap}}_C Z_1 \leq \underline{\text{Cap}}_C Z_2$ and $\overline{\text{Cap}}_C Z_1 \leq \overline{\text{Cap}}_C Z_2$ for any $Z_1 \subset Z_2 \subset X$.

(3) For any sets $Z_i \subset X$, $i = 1,2,\ldots$

$$\underline{\text{Cap}}_C\left(\bigcup_{i\geq 0} Z_i\right) \geq \sup_{i\geq 0} \underline{\text{Cap}}_C Z_i, \quad \overline{\text{Cap}}_C\left(\bigcup_{i\geq 0} Z_i\right) \geq \sup_{i\geq 0} \overline{\text{Cap}}_C Z_i. \qquad (2.2)$$

For any $\varepsilon > 0$ and any set $Z \subset X$ let us set

$$\Lambda(Z,\varepsilon) = \inf_{\mathcal{G}}\left\{\sum_{U\in\mathcal{G}} \xi(U)\right\}, \qquad (2.3)$$

where the infimum is taken over all finite or countable subcollections $\mathcal{G} \subset \mathcal{F}$ covering Z for which $\psi(U) = \varepsilon$ for all $U \in \mathcal{G}$.

Let us assume that the set function η satisfies the following condition:

A4. $\eta(U_1) = \eta(U_2)$ for any $U_1, U_2 \in \mathcal{F}$ for which $\psi(U_1) = \psi(U_2)$.

Provided this condition holds the function $\eta(\varepsilon) = \eta(U)$ if $\psi(U) = \varepsilon$ is correctly defined and the lower and upper Carathéodory capacities admit the following description.

Theorem 2.2. *If the set function η satisfies Condition A4 then for any $Z \subset X$,*

$$\underline{\text{Cap}}_C Z = \lim_{\varepsilon \to 0} \frac{\log \Lambda(Z, \varepsilon)}{\log(1/\eta(\varepsilon))}, \quad \overline{\text{Cap}}_C Z = \overline{\lim_{\varepsilon \to 0}} \frac{\log \Lambda(Z, \varepsilon)}{\log(1/\eta(\varepsilon))}.$$

Proof. We will prove the first equality; the second one can be proved in a similar fashion. Let us put

$$\alpha = \underline{\text{Cap}}_C Z, \quad \beta = \underline{\lim_{\varepsilon \to 0}} \frac{\log \Lambda(Z, \varepsilon)}{\log(1/\eta(\varepsilon))}.$$

Given $\gamma > 0$, one can choose a sequence $\varepsilon_n \to 0$ such that

$$0 = \underline{r}_C(Z, \alpha + \gamma) = \lim_{n \to \infty} R_C(Z, \alpha + \gamma, \varepsilon_n).$$

It follows that $R_C(Z, \alpha + \gamma, \varepsilon_n) \leq 1$ for all sufficiently large n. Therefore, for such numbers n,

$$\Lambda(Z, \varepsilon_n)\eta(\varepsilon_n)^{\alpha+\gamma} \leq 1. \tag{2.4}$$

By virtue of Condition $A2$ one can assume that $\eta(\varepsilon_n) < 1$ for all sufficiently large n. For such n we obtain from (2.4) that

$$\alpha + \gamma \geq \frac{\log \Lambda(Z, \varepsilon_n)}{\log(1/\eta(\varepsilon_n))}.$$

Therefore,

$$\alpha + \gamma \geq \underline{\lim_{n \to \infty}} \frac{\log \Lambda(Z, \varepsilon_n)}{\log(1/\eta(\varepsilon_n))} \geq \beta.$$

Hence,

$$\alpha \geq \beta - \gamma. \tag{2.5}$$

Let us now choose a sequence ε_n' such that

$$\beta = \lim_{n \to \infty} \frac{\log \Lambda(Z, \varepsilon_n')}{\log(1/\eta(\varepsilon_n'))}.$$

We have that

$$\lim_{n \to \infty} R_C(Z, \alpha - \gamma, \varepsilon_n') \geq \underline{r}_C(Z, \alpha - \gamma) = \infty.$$

This implies that $R_C(Z, \alpha - \gamma, \varepsilon_n') \geq 1$ for all sufficiently large n. Therefore, for such n,

$$\Lambda(Z, \varepsilon_n')\eta(\varepsilon_n')^{\alpha-\gamma} \geq 1,$$

and hence,

$$\alpha - \gamma \leq \frac{\log \Lambda(Z, \varepsilon_n')}{\log(1/\eta(\varepsilon_n'))}.$$

Taking the limit as $n \to \infty$ we obtain that

$$\alpha - \gamma \leq \lim_{n \to \infty} \frac{\log \Lambda(Z, \varepsilon_n')}{\log(1/\eta(\varepsilon_n'))} = \beta,$$

and consequently,

$$\alpha \leq \beta + \gamma. \tag{2.6}$$

Since γ can be chosen arbitrarily small the inequalities (2.5) and (2.6) imply that $\alpha = \beta$. ∎

Using Theorem 2.2 one can obtain a simple but useful criterion that guarantees the coincidence of the lower and upper Carathéodory capacities of a set $Z \subset X$.

Theorem 2.3. *Assume that the set functions ξ and η satisfy Condition A4 and the following conditions:*

(1) *for any set $Z \subset X$ the function $\Lambda(Z, \varepsilon)$ increases monotonically while ε decreases;*

(2) *$\eta(\varepsilon)$ decreases monotonically with ε;*

(3) *there exists a monotonically decreasing sequence $\varepsilon_n \to 0$ and a number $C > 0$ such that $\eta(\varepsilon_{n+1}) \geq C\eta(\varepsilon_n)$;*

(4) *there exists the limit*

$$\lim_{n \to \infty} \frac{\log \Lambda(Z, \varepsilon_n)}{\log(1/\eta(\varepsilon_n))} \overset{\text{def}}{=} d.$$

Then, $\underline{\operatorname{Cap}}_C Z = \overline{\operatorname{Cap}}_C = d.$

Proof. Given $\varepsilon > 0$, choose n such that $\eta(\varepsilon_{n+1}) \leq \eta(\varepsilon) < \eta(\varepsilon_n)$. Then by the conditions of the theorem,

$$\frac{\log \Lambda(Z, \varepsilon)}{\log(1/\eta(\varepsilon))} \leq \frac{\log \Lambda(Z, \varepsilon_{n+1})}{\log(1/\eta(\varepsilon_n))} \leq \frac{\log \Lambda(Z, \varepsilon_{n+1})}{\log(1/\eta(\varepsilon_{n+1})) + \log C}.$$

Therefore,

$$\varlimsup_{\varepsilon \to 0} \frac{\log \Lambda(Z, \varepsilon)}{\log(1/\eta(\varepsilon))} \leq \varlimsup_{n \to \infty} \frac{\log \Lambda(Z, \varepsilon_n)}{\log(1/\eta(\varepsilon_n))}.$$

The opposite inequality is obvious. The case of lower limits can be considered in a similar fashion. ∎

In general, we cannot expect equalities in (2.2) (see Examples in Section 6) unless we consider the union of a finite number of sets Z_i. We shall see that regarding to this the lower and upper Carathéodory capacities expose different types of behavior.

Theorem 2.4. *Under Condition A4 we have that*

(1) *for any sets $Z_i \subset X$, $i = 1, \ldots, n$*

$$\overline{\operatorname{Cap}}_C \left(\bigcup_{i=1}^{n} Z_i \right) = \max_{1 \leq i \leq n} \left\{ \overline{\operatorname{Cap}}_C Z_i \right\};$$

(2) *if $Z_i \subset X$, $i = 1, \ldots, n$ are sets such that $\underline{\operatorname{Cap}}_C Z_i = \overline{\operatorname{Cap}}_C Z_i$ then*

$$\underline{\operatorname{Cap}}_C \left(\bigcup_{i=1}^{n} Z_i \right) = \max_{1 \leq i \leq n} \left\{ \underline{\operatorname{Cap}}_C Z_i \right\}.$$

Proof. We begin with the proof of the first statement. It is sufficient to consider only the case $n = 2$ since the general case can be easily derived by induction. Let $Z = Z_1 \cup Z_2$. We shall first show that for each $\varepsilon > 0$,

$$\Lambda(Z, \varepsilon) \leq \Lambda(Z_1, \varepsilon) + \Lambda(Z_2, \varepsilon), \tag{2.7}$$

where $\Lambda(\cdot, \varepsilon)$ is defined by (2.3). In fact, given $\delta > 0$, there exist covers $\mathcal{G}_1 \subset \mathcal{F}$ of Z_1 and $\mathcal{G}_2 \subset \mathcal{F}$ of Z_2 such that $\psi(U) = \varepsilon$ for any $U \in \mathcal{G}_1$ or $U \in \mathcal{G}_2$ and

$$\sum_{U \in \mathcal{G}_1} \xi(U) \leq \Lambda(Z_1, \varepsilon) + \delta, \quad \sum_{U \in \mathcal{G}_2} \xi(U) \leq \Lambda(Z_2, \varepsilon) + \delta.$$

It follows that

$$\Lambda(Z, \varepsilon) \leq \sum_{U \in \mathcal{G}_1} \xi(U) + \sum_{U \in \mathcal{G}_2} \xi(U) \leq \Lambda(Z_1, \varepsilon) + \Lambda(Z_2, \varepsilon) + 2\delta.$$

Since δ can be chosen arbitrarily small this implies (2.7). By Theorem 2.2 there is a sequence $\varepsilon_n \to 0$ such that

$$\overline{\mathrm{Cap}}_C Z = \lim_{n \to \infty} \frac{\log \Lambda(Z, \varepsilon_n)}{\log(1/\eta(\varepsilon_n))}.$$

Passing to a subsequence if necessary we may assume that for $i = 1, 2$ the limits

$$\alpha_i = \lim_{n \to \infty} \frac{\log \Lambda(Z_i, \varepsilon_n)}{\log(1/\eta(\varepsilon_n))} \leq \overline{\mathrm{Cap}}_C Z_i$$

exist. Moreover, we may also assume that $\Lambda(Z_1, \varepsilon_n) \neq 0$ for all $n \geq 0$ (in the case $\Lambda(Z_1, \varepsilon_n) = \Lambda(Z_2, \varepsilon_n) = 0$ for all sufficiently large n, the result is obvious). We have

$$\begin{aligned}\log \Lambda(Z, \varepsilon_n) &\leq \log \Lambda(Z_1, \varepsilon_n) + \log\Big(1 + \frac{\Lambda(Z_2, \varepsilon_n)}{\Lambda(Z_1, \varepsilon_n)}\Big)\\ &= \log \Lambda(Z_1, \varepsilon_n) + \log(1 + \eta(\varepsilon_n)^{a_n}),\end{aligned} \tag{2.8}$$

where $a_n = \log \Lambda(Z_2, \varepsilon_n)/\log \eta(\varepsilon_n) - \log \Lambda(Z_1, \varepsilon_n)/\log \eta(\varepsilon_n)$.

Let us consider the following cases:

a) $\alpha_1 > \alpha_2$. There is a number $a > 0$ such that $a_n \geq a$ for all sufficiently large n. For such n we obtain from (2.8) that

$$\log \Lambda(Z, \varepsilon_n) \leq \log \Lambda(Z_1, \varepsilon_n) + C_1 \eta(\varepsilon_n)^a,$$

where $C_1 > 0$ is a constant. Dividing this inequality by $\log(1/\eta(\varepsilon_n))$ and passing to the limit as $n \to \infty$ we have

$$\overline{\mathrm{Cap}}_C Z \leq \alpha_1 \leq \max\{\overline{\mathrm{Cap}}_C Z_1, \ \overline{\mathrm{Cap}}_C Z_2\}. \tag{2.9}$$

b) $\alpha_1 = \alpha_2$. For any $\delta > 0$ and any sufficiently large $n = n(\delta)$,

$$\log(1 + \eta(\varepsilon_n)^{a_n}) \leq \delta \log(1/\eta(\varepsilon_n)).$$

Therefore, using (2.8), we have $\overline{\mathrm{Cap}}_C Z \leq \alpha_1 + \delta$. Since δ is arbitrary this leads to (2.9).

c) $\alpha_1 < \alpha_2$. This case is analogous to the case 1 with $\Lambda(Z_1, \varepsilon_n)$ replaced by $\Lambda(Z_2, \varepsilon_n)$. (Let us notice that in this case $\Lambda(Z_2, \varepsilon_n) \neq 0$ for all sufficiently large n since $\Lambda(Z_2, \varepsilon_n) \geq \Lambda(Z_1, \varepsilon_n) > 0$).

Thus, in all the cases (2.9) holds. Therefore, the result follows from Statement 3 of Theorem 2.1.

We now proceed with the proof of the second statement. It follows directly from the first statement that

$$\underline{\mathrm{Cap}}_C \left(\bigcup_{i=1}^{n} Z_i \right) \leq \overline{\mathrm{Cap}}_C \left(\bigcup_{i=1}^{n} Z_i \right) = \max_{1 \leq i \leq n} \left\{ \overline{\mathrm{Cap}}_C Z_i \right\}$$

$$= \max_{1 \leq i \leq n} \left\{ \underline{\mathrm{Cap}}_C Z_i \right\} \leq \underline{\mathrm{Cap}}_C \left(\bigcup_{i=1}^{n} Z_i \right).$$

The opposite inequality follows from Statement 3 of Theorem 2.1. This completes the proof of the theorem. ∎

Let X and X' be sets endowed with C-structures $\tau = (\mathcal{F}, \xi, \eta, \psi)$ and $\tau' = (\mathcal{F}', \xi', \eta', \varphi')$ satisfying Conditions $A1, A2$, and $A3'$. The following statement shows that the lower and upper Carathéodory capacities are invariant under a bijective map $\chi: X \to X'$ which preserves the C-structures. Its proof is similar to the proof of Theorem 1.3.

Theorem 2.5. *Assume that there are positive constants K, K_1, and K_2 such that for any $U \in \mathcal{F}$ one can find sets $U_1', U_2' \in \mathcal{F}'$ satisfying*

(1) $U_1' \subset \chi(U) \subset U_2'$;
(2) $K^{-1}\xi'(U_1') \leq \xi(U) \leq K\xi'(U_2')$; $K^{-1}\eta'(U_1') \leq \eta(U) \leq K\eta'(U_2')$;
$\psi'(U_1') = K_1\psi(U)$, $\psi'(U_2') = K_2\psi(U)$.

Then for any $Z \subset X$

$$\underline{\mathrm{Cap}}_{C,\tau'}\chi(Z) = \underline{\mathrm{Cap}}_{C,\tau} Z, \quad \overline{\mathrm{Cap}}_{C,\tau'}\chi(Z) = \overline{\mathrm{Cap}}_{C,\tau} Z.$$

3. Carathéodory Dimension and Capacity of Measures

Let (X, μ) be a Lebesgue space with a finite measure μ. Let also $\tau = (\mathcal{F}, \xi, \eta, \psi)$ be a C-structure on X, i.e., the functions ξ and η satisfy Conditions $A1$, $A2$, and $A3$. We assume that any set $U \in \mathcal{F}$ is measurable.

The measure μ may actually "occupy" only a "small" part of the space. In other words, if A is a subset of measure zero then it is negligible regarding to μ but its Carathéodory dimension may exceed the Carathéodory dimension of the set $X \setminus A$ of full measure (i.e., the set A is "bigger" than the set $X \setminus A$). This motivates the introduction of the following notions. The quantity

$$\dim_C \mu = \inf\{\dim_C Z : \mu(Z) = 1\} \tag{3.1}$$

is called the **Carathéodory dimension of the measure μ**. Let us stress that *a priori* there may be no sets that support the measure μ (i.e., of full measure) and carries its dimension (i.e., the infimum may not be reached).

If we assume the stronger Condition $A3'$, then one can define the quantities

$$\underline{\mathrm{Cap}}_C\mu = \lim_{\delta\to0}\inf\{\underline{\mathrm{Cap}}_C Z : \mu(Z) \geq 1-\delta\},$$
$$\overline{\mathrm{Cap}}_C\mu = \lim_{\delta\to0}\inf\{\overline{\mathrm{Cap}}_C Z : \mu(Z) \geq 1-\delta\} \tag{3.2}$$

that are called respectively the **lower** and **upper Carathéodory capacities of the measure** μ.

We have, therefore, two classes of Carathéodory dimension characteristics; they are Carathéodory dimension and (lower and upper) Carathéodory capacities of sets and measures. We collect them in the following table:

Carathéodory Dimension Characteristics of Sets	Carathéodory Dimension Characteristics of Measures
$\dim_C Z$	$\dim_C \mu$
$\underline{\mathrm{Cap}}_C Z$	$\underline{\mathrm{Cap}}_C \mu$
$\overline{\mathrm{Cap}}_C Z$	$\overline{\mathrm{Cap}}_C \mu$

The study of these characteristics constitutes two main branches of the general Carathéodory construction.

In the following chapters of the book we will demonstrate an essential role played by Carathéodory dimension characteristics of measures in the theory of dynamical systems. One of the main reasons for this is that for a broad class of measures invariant under dynamical systems their Carathéodory dimension and lower and upper Carathéodory capacities coincide. The common value turns out to be a fundamental characteristic of measures and is intimately related to other characteristics of dynamical systems.

Using the C-structures in Examples 1, 2, and 3 one can obtain examples of Carathéodory dimension characteristics of measures. We consider these examples in detail in Chapters 2, 3, and 4. For instance, in Chapter 4 we will show that the measure-theoretic entropy, introduced by Kolmogorov and Sinai within the framework of general measure theory, is one of the Carathéodory dimension characteristics of the measure. This reveals the "dimension" nature of this famous invariant of dynamical systems and allows one to establish new properties of measure-theoretic entropy (see Section 11).

Remarks.

(1) It is not difficult to verify that

$$\dim_C \mu = \lim_{\delta\to0}\inf\{\dim_C Z : \mu(Z) \geq 1-\delta\}. \tag{3.3}$$

Indeed, let $a = \lim_{\delta\to0}\inf\{\dim_C Z : \mu(Z) \geq 1-\delta\}$. For every $\delta > 0$, we have that $\dim_C \mu \geq \inf\{\dim_C Z : \mu(Z) \geq 1-\delta\}$ and hence $\dim_C \mu \geq a$. On the other hand, there is $\delta > 0$ and a sequence of sets Z_n, $n \geq 2$ such that $\mu(Z_n) \geq 1 - \frac{1}{n}$, $Z_n \subset Z_{n+1}$, and $a = \lim\limits_{n \to \infty} \dim_C Z_n$. If $Z = \cup_n Z_n$ then

$$\dim_C \mu \leq \dim_C Z = \sup_{n \geq 2} \dim_C Z_n = a.$$

It also follows from the definition that

$$\underline{\mathrm{Cap}}_C \mu \leq \inf\{\underline{\mathrm{Cap}}_C Z : \mu(Z) = 1\},$$
$$\overline{\mathrm{Cap}}_C \mu \leq \inf\{\overline{\mathrm{Cap}}_C Z : \mu(Z) = 1\}. \tag{3.4}$$

Example 7.1 in Section 7 illustrates that strict inequalities in (3.4) can occur.

(2) Let us set

$$m_C(\mu, \alpha) = \inf\{m_C(Z, \alpha) : \mu(Z) = 1\},$$
$$\underline{r}_C(\mu, \alpha) = \lim_{\delta \to 0} \inf\{\underline{r}_C(Z, \alpha) : \mu(Z) \geq 1 - \delta\},$$
$$\overline{r}_C(\mu, \alpha) = \lim_{\delta \to 0} \inf\{\overline{r}_C(Z, \alpha) : \mu(Z) \geq 1 - \delta\}.$$

It is not hard to show that the functions $m_C(\mu, \cdot)$, $\underline{r}_C(\mu, \cdot)$, and $\overline{r}_C(\mu, \cdot)$ satisfy Propositions 1.2 and 2.1 and define the Carathéodory dimension and lower and upper Carathéodory capacities which coincide respectively with $\dim_C \mu$, $\underline{\mathrm{Cap}}_C \mu$, and $\overline{\mathrm{Cap}}_C \mu$. Indeed, if $m(s, \cdot)$, $s \in S$ is a family of functions, depending on a parameter s and satisfying Proposition 1.2 with the corresponding critical values α_s, then the functions $\underline{m}(\cdot) = \inf_{s \in S} m(s, \cdot)$ and $\overline{m}(\cdot) = \sup_{s \in S} m(s, \cdot)$ satisfy Proposition 1.2 with the critical values $\underline{\alpha} = \inf_{s \in S} \alpha_s$ and $\overline{\alpha} = \sup_{s \in S} \alpha_s$ respectively.

(3) Let X be a separable topological space and μ a Borel probability measure on X. The Carathéodory dimension characteristics of μ can be used to estimate the Carathéodory dimension characteristics of its support $S(\mu)$ (the smallest closed subset of full measure). Namely,

$$\dim_C \mu \leq \dim_C S(\mu), \quad \underline{\mathrm{Cap}}_C \mu \leq \underline{\mathrm{Cap}}_C S(\mu), \quad \overline{\mathrm{Cap}}_C \mu \leq \overline{\mathrm{Cap}}_C S(\mu).$$

In general, one can expect strict inequalities (see Remark 3 in Section 7 and baker's transformation in Section 23).

We establish relations between the Carathéodory dimension and lower and upper Carathéodory capacities of measures. It follows immediately from the definitions that

$$\dim_C \mu \leq \underline{\mathrm{Cap}}_C \mu \leq \overline{\mathrm{Cap}}_C \mu.$$

Below we shall give some general sufficient conditions to produce lower and upper estimates for these quantities and to study the problem of their coincidence.

Let us fix $x \in X$. By Condition A3 for any $\varepsilon > 0$ there is a set $U \in \mathcal{F}, U \ni x$ with $\psi(U) \leq \varepsilon$. Furthermore, by Condition A1, we also have that $\eta(U) > 0$. We shall now assume that the following condition holds:

A5. for μ-almost every $x \in X$ and any $\varepsilon > 0$, if $U \in \mathcal{F}$, $U \ni x$ is a set with $\psi(U) \leq \varepsilon$ then $\mu(U) > 0$ and $\xi(U) > 0$.

By virtue of Condition $A3'$ given $x \in X$ and ε, $0 < \varepsilon \leq \epsilon$, there exists a set $U(x,\varepsilon) \in \mathcal{F}$ containing x with $\psi(U(x,\varepsilon)) = \varepsilon$.

Once a choice of a set $U(x,\varepsilon)$ is made for each $x \in X$ and each $0 < \varepsilon \leq \epsilon$ we obtain the subcollection

$$\mathcal{F}' = \{U(x,\varepsilon) \in \mathcal{F} : x \in X, \, 0 < \varepsilon \leq \epsilon\}.$$

Let us now set for $\alpha \in \mathbb{R}$,

$$\begin{aligned}
\underline{d}_{C,\mu,\alpha}(x) &= \varliminf_{\varepsilon \to 0} \frac{\alpha \log \mu(U(x,\varepsilon))}{\log\left(\xi(U(x,\varepsilon))\eta(U(x,\varepsilon))^\alpha\right)}, \\
\overline{d}_{C,\mu,\alpha}(x) &= \varlimsup_{\varepsilon \to 0} \frac{\alpha \log \mu(U(x,\varepsilon))}{\log\left(\xi(U(x,\varepsilon))\eta(U(x,\varepsilon))^\alpha\right)}.
\end{aligned} \tag{3.5}$$

According to Condition $A5$ these quantities are correctly defined at least for μ-almost every $x \in X$. They are called respectively the **lower** and **upper** α-**Carathéodory pointwise dimensions** of the measure μ at the point x specified by the subcollection $\mathcal{F}'$.

It turns out that these characteristics can be effectively used to produce sharp lower bounds for the Carathéodory dimension of the measure and sharp upper bounds for the upper Carathéodory capacity of the measure. In particular, this gives a powerful criterion for coincidence of the Carathéodory dimension characteristics of measures (see Section 4).

In general, the lower and upper bounds depend on the choice of the subcollection $\mathcal{F}'$. One can try to achieve optimal estimates by an appropriate choice of this subcollection: in practical situations such a subcollection can be often found in a "natural" way. In what follows $\mathcal{F}'$ is assumed to be fixed.

For any $x \in X$ for which Condition $A5$ holds,

$$\underline{d}_{C,\mu,\alpha}(x) \leq \overline{d}_{C,\mu,\alpha}(x).$$

If $\xi(U) = 1$ for any $U \in \mathcal{F}$ then $\underline{d}_{C,\mu,\alpha}(x) \geq 0$. In general, the lower and upper α-Carathéodory pointwise dimensions of μ can take on both positive and negative values. We shall see below that the two cases should be treated separately when: a) the "essential" lower bound for $\underline{d}_{C,\mu,\alpha}(x)$ is positive and b) the "essential" upper bound for $\overline{d}_{C,\mu,\alpha}(x)$ is negative. In the first case the "essential" lower bound for $\underline{d}_{C,\mu,\alpha}(x)$ and the "essential" upper bound for $\overline{d}_{C,\mu,\alpha}(x)$ produce respectively the lower bound for $\dim_C \mu$ and upper bound for $\overline{\mathrm{Cap}}_C \mu$ and vice-versa in the second case. No efficient estimates for $\dim_C \mu$ and $\overline{\mathrm{Cap}}_C \mu$ can be obtained, in general, if $\underline{d}_{C,\mu,\alpha}(x)$ admits only a non-positive "essential" lower bound while $\overline{d}_{C,\mu,\alpha}(x)$ admits only a non-negative one.

From now on we shall assume that in the statements dealing with $\dim_C \mu$ Conditions $A1$, $A2$, and $A3$ hold while in the statements involving $\overline{\mathrm{Cap}}_C \mu$ (or both $\dim_C \mu$ and $\overline{\mathrm{Cap}}_C \mu$) Conditions $A1$, $A2$, and $A3'$ hold. We shall also assume that the measures we consider satisfy Condition $A5$.

We first obtain a lower estimate for $\dim_C \mu$. We say that the subcollection $\mathcal{F}'$ is *sufficient* if for any $Z \subset X$, $\alpha \neq \alpha_C$, and $\varepsilon > 0$ the quantity $M_C(Z, \alpha, \varepsilon)$ can be computed while taking the infimum over $\mathcal{G} \subset \mathcal{F}'$.

Theorem 3.1. *Assume that there are a number $\beta \neq 0$ and an interval $[\beta_1, \beta_2]$ such that $\beta \in (\beta_1, \beta_2)$ and for μ-almost every $x \in X$ and any $\alpha \in [\beta_1, \beta_2]$*

(1) *if $\beta > 0$ then $\underline{d}_{C,\mu,\alpha}(x) \geq \beta$ and if $\beta < 0$ then $\overline{d}_{C,\mu,\alpha}(x) \leq \beta$;*

(2) *there exists $\varepsilon(x) > 0$ such that $\xi(U(x,\varepsilon))\eta(U(x,\varepsilon))^\alpha < 1$ for any set $U(x,\varepsilon) \in \mathcal{F}'$; moreover, the function $\varepsilon(x)$ is measurable;*

(3) *$\mathcal{F}'$ is sufficient.*

Then $\dim_C \mu \geq \beta$.

Proof. We consider the case $\beta > 0$; the case, $\beta < 0$ is similar. Without loss of generality we can assume $\beta_1 > 0$. Fix any $\gamma \in (0, \beta - \beta_1)$ and set $\alpha = \beta - \gamma$. Let Λ be the set of points $x \in X$ satisfying Condition $A5$ and Conditions 1 and 2 of the theorem. It follows from (3.5) and Condition 1 of the theorem that for any $x \in \Lambda$ one can find $\varepsilon_1(x)$, $0 < \varepsilon_1(x) \leq \varepsilon(x)$ such that if $0 < \varepsilon \leq \varepsilon_1(x)$ then

$$\frac{\log \mu(U(x,\varepsilon))}{\log(\xi(U(x,\varepsilon))\eta(U(x,\varepsilon))^{\beta-\gamma})} \geq 1.$$

Using the second condition of the theorem we conclude that

$$\mu(U(x,\varepsilon)) \leq \xi(U(x,\varepsilon))\eta(U(x,\varepsilon))^{\beta-\gamma}. \tag{3.6}$$

Given $\rho > 0$, set $\Lambda_\rho = \{x \in \Lambda : \varepsilon_1(x) \geq \rho\}$. It is easy to see that $\Lambda_{\rho_1} \subset \Lambda_{\rho_2}$ if $\rho_1 \geq \rho_2$ and $\Lambda = \cup_{\rho>0} \Lambda_\rho$. Therefore, there exists $\rho > 0$ such that $\mu(\Lambda_\rho) \geq \frac{1}{2}$. Let $Z \subset \Lambda$ be an arbitrary set of full measure. Let us fix $0 < \varepsilon \leq \rho$ and choose a cover $\mathcal{G} \subset \mathcal{F}'$ of the set $\Lambda_\rho \cap Z$ with $\psi(\mathcal{G}) \leq \varepsilon$. Now, using (3.6) we obtain

$$\sum_{U(x,\varepsilon)\in\mathcal{G}} \xi(U(x,\varepsilon))\eta(U(x,\varepsilon))^{\beta-\gamma} \geq \sum_{U(x,\varepsilon)\in\mathcal{G}} \mu(U(x,\varepsilon)) \geq \mu(\Lambda_\rho) \geq \frac{1}{2}.$$

Since $\mathcal{G}$ is arbitrary and $\mathcal{F}'$ is sufficient it follows that $M_C(\Lambda_\rho \cap Z, \beta-\gamma, \varepsilon) \geq 1/2$. Taking the limit as $\varepsilon \to 0$, we have $m_C(\Lambda_\rho \cap Z, \beta - \gamma) \geq 1/2$. This implies that $\dim_C Z \geq \dim_C \Lambda_\rho \cap Z \geq \beta - \gamma$. It remains to note that since γ can be chosen arbitrarily small we obtain that $\dim_C Z \geq \beta$ and hence $\dim_C \mu \geq \beta$. ∎

We now obtain an upper estimate for the upper Carathéodory capacity of the measure.

Theorem 3.2. *Assume that there are a number $\beta \neq 0$ and an interval $[\beta_1, \beta_2]$ such that $\beta \in (\beta_1, \beta_2)$ and for μ-almost every $x \in X$ and any $\alpha \in [\beta_1, \beta_2]$*

(1) *if $\beta > 0$ then $\overline{d}_{C,\mu,\alpha}(x) \leq \beta$ and if $\beta < 0$ then $\underline{d}_{C,\mu,\alpha}(x) \geq \beta$;*

(2) *there exists $\varepsilon(x) > 0$ such that $\xi(U(x,\varepsilon))\eta(U(x,\varepsilon))^\alpha < 1$ for any set $U(x,\varepsilon) \in \mathcal{F}'$; moreover, the function $\varepsilon(x)$ is measurable;*

(3) *there exist $K > 0$ and $\varepsilon_0 > 0$ such that for any measurable set $Z \subset X$ of positive measure and any $0 < \varepsilon \leq \varepsilon_0$ one can find a cover $\mathcal{G} = \{U(x,\varepsilon)\} \subset \mathcal{F}'$ of Z for which*

$$\sum_{U(x,\varepsilon)\in\mathcal{G}} \mu(U(x,\varepsilon)) \leq K. \tag{3.7}$$

Then $\overline{\mathrm{Cap}}_C \mu \leq \beta$.

Remark.

Condition (3.7) establishes the relation between the C-structure in X and the structure induced by the measure μ. Roughly speaking it means that sets of positive measure in X admit "finite multiplicity covers" comprised from sets $U(x, \varepsilon) \in \mathcal{F}'$. In Chapters 2–4 we will show that this condition holds for a broad class of measures.

Proof of the theorem. We consider again only the case $\beta > 0$. Fix any $\gamma \in (0, \beta_2 - \beta)$ and set $\alpha = \beta + \gamma$. Let Λ be the set of points $x \in X$ for which Condition $A5$ and Conditions 1 and 2 of the theorem hold. It follows from (3.5) and the first condition of the theorem that for any $x \in \Lambda$ there exists $\varepsilon_1(x)$, $0 < \varepsilon_1(x) \leq \varepsilon(x)$ such that if $0 < \varepsilon \leq \varepsilon_1(x)$ then

$$\frac{\log \mu(U(x, \varepsilon))}{\log(\xi(U(x, \varepsilon))\eta(U(x, \varepsilon))^{\beta+\gamma})} \leq 1.$$

Using the second condition of the theorem we obtain that

$$\xi(U(x, \varepsilon))\eta(U(x, \varepsilon))^{\beta+\gamma} \leq \mu(U(x, \varepsilon)) \tag{3.8}$$

Given $\rho > 0$, define $\Lambda_\rho = \{x \in \Lambda : \varepsilon_1(x) \geq \rho\}$. It is easy to see that $\Lambda_{\rho_1} \subset \Lambda_{\rho_2}$ if $\rho_1 \geq \rho_2$ and $\Lambda = \cup_{\rho > 0} \Lambda_\rho$. Therefore, given $\delta > 0$ there exists $\rho_0 > 0$ such that $\mu(\Lambda_\rho) \geq 1 - \delta$ for any $0 < \rho \leq \rho_0$. Let us fix $0 < \rho \leq \rho_0$ and choose K and ε_0 in accordance with Condition 3 of the theorem. Furthermore, for ε such that $0 < \varepsilon \leq \min\{\varepsilon_0, \rho\}$ let us choose a cover $\mathcal{G} = \{U(x, \varepsilon)\} \subset \mathcal{F}'$ of the set Λ_ρ which satisfies (3.7). Since for each element of this cover condition (3.8) holds we obtain

$$\sum_{U(x,\varepsilon) \in \mathcal{G}} \xi(U(x, \varepsilon))\eta(U(x, \varepsilon))^{\beta+\gamma} \leq \sum_{U(x,\varepsilon) \in \mathcal{G}} \mu(U(x, \varepsilon)) \leq K.$$

It follows that $R_C(\Lambda_\rho, \beta + \gamma, \varepsilon) \leq K$. Taking the upper limit as $\varepsilon \to 0$ yields $\bar{r}_C(\Lambda_\rho, \beta + \gamma) \leq K$. This implies that $\overline{\mathrm{Cap}}_C \Lambda_\rho \leq \beta + \gamma$. Remembering that $\mu(\Lambda_\rho) \geq 1 - \delta$ and that δ was arbitrarily chosen (with $\rho = \rho(\delta) \to 0$ as $\delta \to 0$) we see that

$$\overline{\mathrm{Cap}}_C \mu \leq \lim_{\delta \to 0} \overline{\mathrm{Cap}}_C \Lambda_\rho \leq \beta + \gamma.$$

Since γ was also arbitrarily chosen it follows that $\overline{\mathrm{Cap}}_C \mu \leq \beta$. ∎

We consider the particular case $\xi(U) = 1$ for any $U \in \mathcal{F}$. Obviously, $\underline{d}_{C,\mu,\alpha}(x)$ and $\overline{d}_{C,\mu,\alpha}(x)$ are independent of α, i.e.,

$$\underline{d}_{C,\mu,\alpha}(x) \overset{\mathrm{def}}{=} \underline{d}_{C,\mu}(x), \quad \overline{d}_{C,\mu,\alpha}(x) \overset{\mathrm{def}}{=} \overline{d}_{C,\mu}(x).$$

Moreover,

$$0 \leq \underline{d}_{C,\mu}(x) \leq \overline{d}_{C,\mu}(x).$$

One can now reformulate Theorems 3.1 and 3.2 in the following way.

Theorem 3.3. *Assume that $\xi(U) = 1$ for any $U \in \mathcal{F}$. We have*

(1) *if there exists $\beta \geq 0$ such that $\underline{d}_{C,\mu}(x) \geq \beta$ for μ-almost every $x \in X$ and Condition 3 of Theorem 3.1 holds then $\dim_C \mu \geq \beta$;*

(2) *if there exists $\beta \geq 0$ such that $\overline{d}_{C,\mu}(x) \leq \beta$ for μ-almost every $x \in X$ and Condition 3 of Theorem 3.2 holds then $\overline{\mathrm{Cap}}_C \mu \leq \beta$.*

Proof. We remark that in the case we consider, the second conditions of Theorems 3.1 and 3.2 hold in view of Condition A2. Thus, if $\beta > 0$ the result follows from these theorems. If $\beta = 0$ the first statement is obvious since $\dim_C \mu \geq 0$. The second statement follows from the observation that for any $\varepsilon > 0$ we have $\overline{d}_{C,\mu}(x) \leq \varepsilon$. ∎

We now obtain lower bounds for the Carathéodory dimension of the measure and upper bounds for the upper Carathéodory capacity of the measure regardless of the choice of the subcollection $\mathcal{F}'$.

Given $\alpha \in \mathbb{R}$, set

$$
\begin{aligned}
\underline{\mathcal{D}}_{C,\mu,\alpha}(x) &= \varliminf_{\varepsilon \to 0} \inf_{\substack{U \ni x \\ \psi(U)=\varepsilon}} \frac{\alpha \log \mu(U)}{\log\left(\xi(U)\eta(U)^{\alpha}\right)}, \\
\overline{\mathcal{D}}_{C,\mu,\alpha}(x) &= \varlimsup_{\varepsilon \to 0} \sup_{\substack{U \ni x \\ \psi(U)=\varepsilon}} \frac{\alpha \log \mu(U)}{\log\left(\xi(U)\eta(U)^{\alpha}\right)}.
\end{aligned}
\tag{3.9}
$$

According to Condition A5 these quantities are correctly defined at least for μ-almost every $x \in X$. If $\mathcal{F}' = \{U(x,\varepsilon)\}$ is a subcollection of $\mathcal{F}$ then

$$
\underline{\mathcal{D}}_{C,\mu,\alpha}(x) \leq \underline{d}_{C,\mu,\alpha}(x) \leq \overline{d}_{C,\mu,\alpha}(x) \leq \overline{\mathcal{D}}_{C,\mu,\alpha}(x).
$$

Therefore, the following theorems are immediate consequences of Theorems 3.1 and 3.2.

Theorem 3.4. *Assume that there are a number $\beta \neq 0$ and an interval $[\beta_1, \beta_2]$ such that $\beta \in (\beta_1, \beta_2)$ and for μ-almost every $x \in X$ and any $\alpha \in [\beta_1, \beta_2]$*

(1) *if $\beta > 0$ then $\underline{\mathcal{D}}_{C,\mu,\alpha}(x) \geq \beta$ and if $\beta < 0$ then $\overline{\mathcal{D}}_{C,\mu,\alpha}(x) \leq \beta$;*

(2) *the second condition of Theorem 3.1 holds.*

Then $\dim_C \mu \geq \beta$.

Theorem 3.5. *Assume that there are a number $\beta \neq 0$ and an interval $[\beta_1, \beta_2]$ such that $\beta \in (\beta_1, \beta_2)$ and for μ-almost every $x \in X$ and any $\alpha \in [\beta_1, \beta_2]$*

(1) *if $\beta > 0$ then $\overline{\mathcal{D}}_{C,\mu,\alpha}(x) \leq \beta$ and if $\beta < 0$ then $\underline{\mathcal{D}}_{C,\mu,\alpha}(x) \geq \beta$;*

(2) *the second and third conditions of Theorem 3.2 hold.*

Then $\overline{\mathrm{Cap}}_C \mu \leq \beta$.

4. Coincidence of Carathéodory Dimension and Carathéodory Capacity of Measures

Using results of the previous section we obtain now some sufficient conditions that guarantee the coincidence of the Carathéodory dimension characteristics relative to measures. Let us point out that the Carathéodory dimension characteristics of sets are more "sensitive": the coincidence of them is a rare phenomenon although it can happen in some specific *rigid* situations (see Sections 13, 14, and 15).

We fix a subcollection $\mathcal{F}' = \{U(x,\varepsilon) \in \mathcal{F} : x \in X, 0 < \varepsilon \leq \epsilon\}$ as in Section 3. Let μ be a probability measure on X satisfying Condition $A5$. Consider the lower and upper α-pointwise Carathéodory dimensions of μ specified by the subcollection $\mathcal{F}'$.

We first consider the case when the set function ξ is trivial, i.e., $\xi(U) = 1$ for any $U \in \mathcal{F}$. Then according to (3.5) the quantities $\underline{d}_{C,\mu,\alpha}(x)$ and $\overline{d}_{C,\mu,\alpha}(x)$ do not depend on $\alpha \in \mathbb{R}$ but may depend on $x \in X$ and may also be different. They are also non-negative. If they are "essentially" the same and are constant, Theorem 3.3 gives us sufficient conditions for coincidence of the Carathéodory dimension characteristics of measures.

Theorem 4.1. *Assume that $\xi(U) = 1$ for any $U \in \mathcal{F}$. Assume also that there exists $\beta \geq 0$ such that for μ-almost every $x \in X$*

(1) $\underline{d}_{C,\mu,\alpha}(x) = \overline{d}_{C,\mu,\alpha}(x) = \beta$;
(2) *Condition 3 of Theorem 3.1 and Condition 3 of Theorem 3.2 hold.*

Then $\dim_C \mu = \underline{\mathrm{Cap}}_C \mu = \overline{\mathrm{Cap}}_C \mu = \beta$.

We now turn to the case of non-trivial set function ξ. The quantities $\underline{d}_{C,\mu,\alpha}(x)$ and $\overline{d}_{C,\mu,\alpha}(x)$ "essentially" depend on $\alpha \in \mathbb{R}$ and Condition 1 of Theorem 4.1 cannot be satisfied (although Condition 1 of Theorems 3.1 and 3.2 can work well so that these theorems still give us some estimates for the Carathéodory dimension and upper Carathéodory capacity). There are also some general conditions in this case that can be used to obtain the coincidence of the Carathéodory dimension characteristics .

Theorem 4.2. *Assume that there are numbers $\beta_1, \beta_2 \in \mathbb{R}$, $\beta_1 < \beta_2$ such that for μ-almost every $x \in X$*

(1) $\underline{d}_{C,\mu,\alpha}(x) = \overline{d}_{C,\mu,\alpha}(x) \stackrel{\text{def}}{=} d(\alpha)$ *for any $\alpha \in [\beta_1, \beta_2]$ and $d(\alpha) \in C^1$ on $[\beta_1, \beta_2]$*
(2) *the equation $d(\alpha) = \alpha$ has a unique root $\alpha = \beta \in (\beta_1, \beta_2)$ and $\beta \neq 0$; moreover,*

$$0 < d'(\beta) < 1 \ \ if \ \beta > 0, \quad d'(\beta) > 1 \ \ if \ \beta < 0;$$

(3) *there exists $\varepsilon(x) > 0$ such that $\xi(U(x,\varepsilon))\eta(U(x,\varepsilon))^\alpha < 1$ for any $\alpha \in [\beta_1, \beta_2]$ and any $U(x,\varepsilon) \in \mathcal{F}'$ with $\varepsilon \leq \varepsilon(x)$; moreover, the function $\varepsilon(x)$ is measurable;*
(4) *Condition 3 of Theorem 3.1 and Condition 3 of Theorem 3.2 hold.*

Then $\dim_C \mu = \underline{\text{Cap}}_C \mu = \overline{\text{Cap}}_C \mu = \beta$.

Proof. Without loss of generality we can assume that $\beta_1 > 0$ if $\beta > 0$ and $\beta_2 < 0$ if $\beta < 0$. We can also assume that for all $\alpha \in [\beta_1, \beta_2]$, $0 < d'(\alpha) < 1$ if $\beta > 0$ and $d'(\alpha) > 1$ if $\beta < 0$ (otherwise the interval $[\beta_1, \beta_2]$ can be replaced by a smaller subinterval $[\beta_1', \beta_2'] \ni \beta$ for which this assumption holds). Let Λ be the set of points $x \in X$ for which Condition $A5$ and Conditions 1 and 3 of the theorem are satisfied. Denote

$$s = \begin{cases} \max \{d'(\alpha): \ \alpha \in [\beta_1, \beta_2]\} & \text{if } \beta > 0 \\ \min \{d'(\alpha): \ \alpha \in [\beta_1, \beta_2]\} & \text{if } \beta < 0. \end{cases}$$

Note that $s < 1$ if $\beta > 0$ and $s > 1$ if $\beta < 0$. It follows from Condition 1 of the theorem that given $\widetilde{\gamma} > 0$, $\alpha \in [\beta_1, \beta_2]$, and $x \in \Lambda$, there exists $\varepsilon_1(x)$, $0 < \varepsilon_1(x) \le \varepsilon(x)$ such that for any ε, $0 < \varepsilon \le \varepsilon_1(x)$,

$$d(\alpha) - \widetilde{\gamma} \le \frac{\alpha \log \mu(U(x, \varepsilon))}{\log \left(\xi(U(x, \varepsilon)) \eta(U(x, \varepsilon))^\alpha \right)}. \tag{4.1}$$

Let us fix $x \in \Lambda$, $0 < \varepsilon \le \varepsilon_1(x)$, and a number γ satisfying

$$\begin{aligned} 0 < \gamma \le \frac{s}{2} \ \min \{\beta - \beta_1, \beta_2 - \beta\} & \quad \text{if } \beta > 0 \\ 0 < \gamma \le \frac{1}{2} \ \min \{\beta - \beta_1, \beta_2 - \beta\} & \quad \text{if } \beta < 0. \end{aligned} \tag{4.2}$$

Considering $\alpha = \beta - \frac{\gamma}{s} \in [\beta_1, \beta_2]$ in (4.1) and putting $\widetilde{\gamma} = (\frac{1}{s} - 1)\gamma$ if $\beta > 0$ and $\widetilde{\gamma} = (1 - \frac{1}{s})\gamma$ if $\beta < 0$ we obtain

$$d\left(\beta - \frac{\gamma}{s}\right) \le \left(\beta - \frac{\gamma}{s}\right) \frac{\log \mu(U(x, \varepsilon))}{\log \left(\xi(U(x, \varepsilon)) \eta(U(x, \varepsilon))^{\beta - \frac{\gamma}{s}} \right)} + \gamma \left(\frac{1}{s} - 1\right) \ \text{if } \beta > 0$$

$$\tag{4.3}$$

$$d\left(\beta - \frac{\gamma}{s}\right) \ge \left(\beta - \frac{\gamma}{s}\right) \frac{\log \mu(U(x, \varepsilon))}{\log \left(\xi(U(x, \varepsilon)) \eta(U(x, \varepsilon))^{\beta - \frac{\gamma}{s}} \right)} - \gamma \left(1 - \frac{1}{s}\right) \ \text{if } \beta < 0.$$

Consider the function $a(\gamma) = d(\beta - \frac{\gamma}{s})$. We have $a(0) = d(\beta) = \beta$ and $a'(\gamma) = d'(\beta - \frac{\gamma}{s})(-\frac{1}{s})$. Therefore, Condition 2 of the theorem implies that for all γ satisfying (4.2)

$$a'(\gamma) \ge -1 \ \text{ if } \beta > 0 \quad \text{and} \quad a'(\gamma) \le -1 \ \text{ if } \beta < 0.$$

This gives us that

$$d\left(\beta - \frac{\gamma}{s}\right) \ge \beta - \gamma \ \text{ if } \beta > 0,$$

$$d\left(\beta - \frac{\gamma}{s}\right) \le \beta - \gamma \ \text{ if } \beta < 0.$$

Combining this with (4.3) and taking into account that $\beta - \frac{\gamma}{s} > 0$ if $\beta > 0$ and $\beta - \frac{\gamma}{s} < 0$ if $\beta < 0$ we have

$$\log \mu(U(x,\varepsilon)) \Big/ \log \left(\xi(U(x,\varepsilon))\eta(U(x,\varepsilon))^{\beta - \frac{\gamma}{s}} \right) \geq 1.$$

In view of Condition 3 of the theorem it follows that

$$\mu(U(x,\varepsilon)) \leq \xi(U(x,\varepsilon))\eta(U(x,\varepsilon))^{\beta - \frac{\gamma}{s}}. \tag{4.4}$$

Repeating the argument, presented at the end of the proof of Theorem 3.1, and using (4.4) instead of (3.6), we obtain the lower bound $\dim_C \mu \geq \beta$.

We now proceed with the upper bound and prove that $\overline{\mathrm{Cap}}_C \mu \leq \beta$. It follows from Condition 1 of the theorem that given $\tilde{\gamma} > 0$, $\alpha \in [\beta_1, \beta_2]$, and $x \in \Lambda$, there exists $\varepsilon_1(x)$, $0 < \varepsilon_1(x) \leq \varepsilon(x)$ such that for any ε, $0 < \varepsilon \leq \varepsilon_1(x)$,

$$\frac{\alpha \log \mu(U(x,\varepsilon))}{\log \left(\xi(U(x,\varepsilon))\eta(U(x,\varepsilon))^\alpha \right)} \leq d(\alpha) + \tilde{\gamma}.$$

Let us fix $x \in \Lambda$, $0 < \varepsilon \leq \varepsilon_1(x)$ and choose a number γ satisfying (4.2). Set $\alpha = \beta + \frac{\gamma}{s} \in [\beta_1, \beta_2]$. By virtue of (4.1) we have that if $\beta < 0$ then

$$d\left(\beta + \frac{\gamma}{s}\right) \geq \left(\beta + \frac{\gamma}{s}\right) \frac{\log \mu(U(x,\varepsilon))}{\log \left(\xi(U(x,\varepsilon))\eta(U(x,\varepsilon))^{\beta + \frac{\gamma}{s}} \right)} - \left(\frac{1}{s} - 1\right)\gamma \tag{4.5}$$

and if $\beta > 0$ then

$$d\left(\beta + \frac{\gamma}{s}\right) \leq \left(\beta + \frac{\gamma}{s}\right) \frac{\log \mu(U(x,\varepsilon))}{\log \left(\xi(U(x,\varepsilon))\eta(U(x,\varepsilon))^{\beta + \frac{\gamma}{s}} \right)} + \left(1 - \frac{1}{s}\right)\gamma. \tag{4.6}$$

Considering the function $c(\gamma) = d(\beta + \frac{\gamma}{s})$ and repeating the arguments presented above one can show that

$$d\left(\beta + \frac{\gamma}{s}\right) \leq \beta + \gamma \quad \text{if } \beta > 0$$
$$d\left(\beta + \frac{\gamma}{s}\right) \geq \beta + \gamma \quad \text{if } \beta < 0.$$

Using these inequalities, (4.5) and (4.6), and the fact that $\beta + \frac{\gamma}{s} > 0$ if $\beta > 0$ and $\beta + \frac{\gamma}{s} < 0$ if $\beta < 0$ we obtain

$$\log \mu(U(x,\varepsilon)) \Big/ \log \left(\xi(U(x,\varepsilon))\eta(U(x,\varepsilon))^{\beta + \frac{\gamma}{s}} \right) \leq 1. \tag{4.7}$$

It follows from (4.7) and Condition 3 of the theorem that

$$\xi(U(x,\varepsilon))\eta(U(x,\varepsilon))^{\beta + \frac{\gamma}{s}} \leq \mu(U(x,\varepsilon)). \tag{4.8}$$

Now, in order to obtain the inequality $\overline{\mathrm{Cap}}_C \mu \leq \beta$ we need only to repeat the arguments presented at the end of the proof of Theorem 3.2 and to use (4.8) instead of (3.8). ∎

5. Lower and Upper Bounds for Carathéodory
Dimension of Sets; Carathéodory Dimension Spectrum

We shall use results from the previous section to produce sharp lower and upper estimates for the Carathéodory dimension of a set by considering measures supported on the set. We begin with the upper bound.

Let X be a separable topological space, $\mathcal{F}$ a collection of Borel subsets of X. Assume that $\eta, \psi \colon \mathcal{F} \to \mathbb{R}^+$ are set functions satisfying Conditions $A1, A2$, and $A3'$, and $\xi \colon \mathcal{F} \to \mathbb{R}^+$ is a set function. Let μ be a Borel probability measure on X satisfying Condition $A5$.

We fix a subcollection $\mathcal{F}' = \{U(x,\varepsilon) \in \mathcal{F} : x \in X, 0 < \varepsilon \leq \epsilon\}$ as in Section 3 (i.e., $x \in U(x,\varepsilon)$ and $\psi(U(x,\varepsilon)) = \varepsilon$) and consider the lower and upper α-pointwise Carathéodory dimensions of the measure μ specified by the subcollection $\mathcal{F}'$.

Given a Borel set Z and $\delta > 0$, we call a cover $\mathcal{G} \subset \mathcal{F}'$ of Z a **Besicovitch cover** if for any $x \in Z$ there exists $0 < \varepsilon = \varepsilon(x) \leq \delta$ such that the set $U(x,\varepsilon) \in \mathcal{G}$.

Theorem 5.1. *Assume that there exist numbers $\beta, \beta_0 \in \mathbb{R}$, $\beta_0 > 0$ such that for μ-almost every $x \in X$*

(1) *if $\beta \geq 0$ then $\underline{d}_{C,\mu,\alpha}(x) \leq \beta$ and if $\beta < 0$ then $\overline{d}_{C,\mu,\alpha}(x) \geq \beta$;*

(2) *there exists $\varepsilon(x) > 0$ such that $\xi(U(x,\varepsilon))\eta(U(x,\varepsilon))^\alpha < 1$ for any $U(x,\varepsilon) \in \mathcal{F}'$ with $\varepsilon \leq \varepsilon(x)$ and any $\alpha \in [\beta, \beta + \beta_0]$ if $\beta \geq 0$ and $\alpha \in [\beta - \beta_0, \beta]$ if $\beta < 0$; moreover, the function $\varepsilon(x)$ is measurable;*

(3) *there exist $K > 0$ and $\varepsilon_0 > 0$ such that for any measurable set $Z \subset X$ of positive measure and any Besicovitch cover $\mathcal{G} \subset \mathcal{F}'$ of Z with $\psi(\mathcal{G}) \leq \varepsilon_0$ one can find a subcover $\widetilde{\mathcal{G}} \subset \mathcal{G}$ of Z for which*

$$\sum_{U \in \widetilde{\mathcal{G}}} \mu(U) \leq K. \tag{5.1}$$

If Λ is the set of points for which Condition $A5$ and Conditions 1 and 2 hold then $\dim_C \Lambda \leq \beta$.

Proof. Let us choose numbers $\delta > 0$ and $0 < \gamma \leq \beta_0$. Given $x \in \Lambda$, one can find a set $U(x,\varepsilon) \in \mathcal{F}'$ with $\varepsilon = \varepsilon(x) \leq \delta$ such that

$$\frac{\alpha \log \mu(U(x,\varepsilon))}{\log\left(\xi(U(x,\varepsilon))\eta(U(x,\varepsilon))^\alpha\right)} \leq \beta + \gamma$$

if $\beta \geq 0$ and

$$\frac{\alpha \log \mu(U(x,\varepsilon))}{\log\left(\xi(U(x,\varepsilon))\eta(U(x,\varepsilon))^\alpha\right)} \geq \beta - \gamma$$

if $\beta < 0$. Set $\alpha = \beta + \gamma$ if $\beta \geq 0$ and $\alpha = \beta - \gamma$ if $\beta < 0$. Note that $\beta + \gamma > 0$ if $\beta \geq 0$ and $\beta - \gamma < 0$ if $\beta < 0$. Therefore, by the second condition of the theorem, we obtain

$$\xi(U(x,\varepsilon))\eta(U(x,\varepsilon))^{\beta+\gamma} \leq \mu(U(x,\varepsilon)) \tag{5.2}$$

if $\beta \geq 0$ and

$$\xi(U(x,\varepsilon))\eta(U(x,\varepsilon))^{\beta-\gamma} \leq \mu(U(x,\varepsilon)) \tag{5.3}$$

if $\beta < 0$. The sets $\{U(x,\varepsilon) : x \in X,\ \varepsilon = \varepsilon(x)\}$ comprise a cover $\mathcal{G}$ of Λ which is clearly a Besicovitch cover. Therefore, if δ is sufficiently small then by the third condition of the theorem, one can find a subcover $\widetilde{\mathcal{G}} \subset \mathcal{G}$ satisfying (5.1). In the case $\beta \geq 0$ it follows that

$$\sum_{U \in \widetilde{\mathcal{G}}} \xi(U)\eta(U)^{\beta+\gamma} \leq \sum_{U \in \widetilde{\mathcal{G}}} \mu(U) \leq K.$$

Hence $M_C(\Lambda, \beta + \gamma, \varepsilon) \leq K$. Passing to the limit as $\varepsilon \to \infty$ we obtain that $m_C(\Lambda, \beta + \gamma) \leq K$. This implies that $\dim_C \Lambda \leq \beta + \gamma$. Since γ is arbitrary the result follows. The case $\beta < 0$ is considered in a similar fashion. ∎

We shall now obtain a lower bound for the Carathéodory dimension of a set $Z \subset X$ in the special case $\xi(U) = 1$ for any $U \in \mathcal{F}$.

Given a Borel measure μ on X, denote by Λ_μ the set of points for which Condition A5 holds. In this case, $\underline{d}_{C,\mu,\alpha}(x) \geq 0$ for any $x \in \Lambda_\mu$ and it does not depend on α, i.e., $\underline{d}_{C,\mu,\alpha}(x) \stackrel{\text{def}}{=} d_{C,\mu}(x)$. Denote also by $\mathcal{M}(Z)$ the set of Borel measures μ on X with $\mu(Z) = 1$ (implicitly, we assume that Z is measurable with respect to μ). Put

$$\beta = \sup_{\mu \in \mathcal{M}(Z)} \inf_{x \in \Lambda_\mu} d_{C,\mu}(x).$$

Theorem 5.2. $\dim_C Z \geq \beta$.

Proof. It follows from the definition of β that for any $\varepsilon > 0$ one can find a measure μ on X with $\mu(Z) = 1$ such that $\underline{d}_{C,\mu}(x) \geq \beta - \varepsilon$ for μ-almost every $x \in Z$. Theorem 3.3 implies now that $\dim_C Z \geq \dim_C \mu \geq \beta - \varepsilon$ and the desired result follows. ∎

We still assume that $\xi(U) = 1$ for any $U \in \mathcal{F}$. The previous results give rise to the following notion. Given $\alpha \geq 0$, define

$$\mathcal{D}_\alpha = \{x \in \Lambda_\mu : d_{C,\mu}(x) = \alpha\}.$$

The function $f_\mu(\alpha) = \dim_C \mathcal{D}_\alpha$ is called the **Carathéodory dimension spectrum** specified by the measure μ. Some particular examples will be given in Chapters 6 and 7. Let us notice that if $\mu(\mathcal{D}_\alpha) > 0$ for some $\alpha \geq 0$ then $f_\mu(\alpha) = \alpha$.

Let $\mathcal{M} \subset \mathcal{M}(Z)$ be a subset. For any $\mu \in \mathcal{M}(Z)$ we have

$$\dim_C \mu \leq \dim_C Z.$$

Hence,

$$\sup_{\mu \in \mathcal{M}} \dim_C \mu \leq \dim_C Z.$$

We say that the Carathéodory dimension of Z admits the **variational principle** (with respect to $\mathcal{M}$) if

$$\sup_{\mu \in \mathcal{M}} \dim_C \mu = \dim_C Z.$$

We say that a measure $\nu \in \mathcal{M}$ is the **measure of full Carathéodory dimension** (specified by $\mathcal{M}$) if

$$\dim_C \nu = \dim_C Z. \tag{5.4}$$

In the following chapters of the book we will present explicit versions of the variational principle for Carathéodory dimension when the parameters of the general construction $\mathcal{F}, \xi, \eta$, and ψ are fixed. Examples will include the *classical variational principles for topological pressure* and *topological entropy* in statistical physics (see Theorem A2.1 in Appendix II). In this case the class of measures $\mathcal{M} \subset \mathcal{M}(Z)$ consists of measures invariant under the underlying dynamical system. We will see that any *equilibrium measure* is the measure of full Carathéodory dimension. Therefore, the above approach enables us to obtain the "dimension" interpretation of the thermodynamic formalism of statistical physics.

Another example is the *variational principle for the Hausdorff dimension* of a set invariant under a dynamical system (with $\mathcal{M}$ to be the collection of all invariant measures). We will establish this principle and the existence of an invariant measure of full (Carathéodory) dimension for the *limit sets of some geometric constructions* (see Theorem 13.1) and *conformal repellers* (see Theorem 20.1).

Chapter 2

C-Structures Associated with Metrics: Hausdorff Dimension and Box Dimension

We begin now to exploit the general construction described in Chapter 1 in order to produce different Carathéodory dimension characteristics. This means that we should make a choice of the parameters of the construction, i.e., the collection of subsets $\mathcal{F}$ and the set functions ξ, η, ψ which satisfy Conditions $A1, A2, A3$ (or $A3'$).

In this chapter we consider the C-structure induced by the standard metric (or an equivalent one) in the Euclidean space $\mathbb{R}^m$ (in Appendix I we consider C-structures which are induced by metrics in *general* metric spaces). This structure generates Carathéodory dimension characteristics that are well-known in dimension theory, i.e., the Hausdorff dimension and lower and upper box dimensions. We describe some of their properties and present some methods for their calculation. Our aim is to demonstrate how they can be used to characterize sets and measures invariant under dynamical systems. There are two aspects of this problem. One is to use some geometric constructions in order to estimate the Hausdorff dimension and box dimension of invariant sets and measures (see Chapters 7 and 8). The second is, conversely, to use some ideas in the theory of dynamical systems in order to compute the Hausdorff dimension and box dimension of "geometrically constructed" sets (see Chapters 5 and 6).

One of the challenging problems in dimension theory is the problem of coincidence of the Hausdorff dimension and lower and upper box dimensions of sets. It is now accepted by most experts in the field that the coincidence is a relatively rare phenomenon and can occur only in some "rigid" situations. One well-known class of sets for which the coincidence usually takes place is the class of limit sets for some geometric constructions (see Chapter 5).

For subsets which are invariant under a dynamical system one can pose another problem of the coincidence of the Hausdorff dimension and box dimension of invariant measures. In order to explain this let us consider a map $f: U \to \mathbb{R}^m$, where $U \subset \mathbb{R}^m$ is an open domain. Assume that f preserves an ergodic Borel probability measure μ with compact support $Z \subset U$ (i.e., Z is a compact invariant set and $\mu(Z) = 1$). The stochastic properties of the map $f|Z$ are closely related to the topological structure of the set Z that, in many "physically" interesting situations, resembles a Cantor-like set. The relevant quantitative characteristics, which can be used to describe complexity of the topological structure of Z, are the Hausdorff dimension and box dimension of the measure μ. If they coincide the common value can be used to characterize the "fractal" structure of the invariant set, more precisely, the part of the set where the measure is concentrated; this also gives rise to the problem of multifractality which we consider in Chapter 6.

The notion of Hausdorff dimension was introduced by Hausdorff [H] in 1919 to characterize sets with "pathologically complicated" topological structures (like the famous middle-third Cantor set) but some very basic ideas behind this notion were formulated earlier by Carathéodory [C] in 1914. The properties of the Hausdorff dimension have been intensively studied for a long time, within the framework of function theory, by Besicovitch and his students and collaborators, who have made significant progress and obtained several fundamental results. (Different aspects of their study are described in [Fe, Ro]; the contemporary exposition and further references can be found in [F5]). Apparently, they knew about box dimension but did not find it sufficiently useful. In 1932, Pontryagin and Shnirel'man [PS] proved a fundamental result involving the lower box dimension (which they called the *metric order*). Namely, the infimum of lower box dimensions of a compact subset Z in a compact metric space X, taken over all possible metrics on X, coincides with the topological dimension of Z. Il'yashenko [Il] considered the lower box dimension (under the name *entropy dimension*) in connection with the study of dimension of maximal attractors for some partial differential equations. Bakhtin [B] examined some properties of the lower box dimension. Furstenberg [Fu] obtained some relations between box dimension and topological entropy for symbolic dynamical systems.

The general definition of lower and upper box dimensions was given by Young [Y1,Y2], who also studied relations between the Hausdorff dimension and lower and upper box dimensions. The significance of lower and upper box dimensions is acknowledged now by experts in the field (see, for example, [F5]; the lower and upper box dimensions have several other names: lower and upper box-counting dimensions, capacities, Minkowski dimensions; there is no special reason to prefer any of them, but the name "box dimension" seems most appropriate). One of the most successful methods in dimension theory for estimating the Hausdorff dimension of sets was suggested by Frostman [Fr] and is known as the mass distribution principle. A similar (but, in a way, more general) method for estimating the Hausdorff dimension of measures was formulated by Young [Y2]. Theorem 4.2 is an extension of her result to other characteristics of dimension type.

6. Hausdorff Dimension and Box Dimension of Sets

We introduce a C-structure on the Euclidean space $\mathbb{R}^m$ endowed with a metric ρ which is equivalent to the standard metric. Let $\mathcal{F}$ be the collection of all *open* subsets of $\mathbb{R}^m$. For $U \in \mathcal{F}$ define

$$\xi(U) = 1, \quad \eta(U) = \psi(U) = \operatorname{diam} U. \tag{6.1}$$

A direct verification shows that the collection of subset $\mathcal{F}$ and set functions η and ψ satisfy Conditions $A1, A2, A3$, and $A3'$. Hence they determine a C-structure $\tau = (\mathcal{F}, \xi, \eta, \psi)$ on $\mathbb{R}^m$. The corresponding Carathéodory set function $m_C(\cdot, \alpha)$ (see (1.1) and (1.2)) is an outer measure for any $\alpha \geq 0$. It is known as the

α-**Hausdorff outer measure** of Z and is denoted by $m_H(Z, \alpha)$. We have for any $Z \subset \mathbb{R}^m$ and $\alpha \geq 0$,

$$m_H(Z, \alpha) = \lim_{\varepsilon \to 0} \inf_{\mathcal{G}} \left\{ \sum_{U \in \mathcal{G}} (\operatorname{diam} U)^\alpha \right\}, \qquad (6.2)$$

where the infimum is taken over all finite or countable covers $\mathcal{G}$ of Z by open sets with $\operatorname{diam} \mathcal{G} \leq \varepsilon$.

The set function $m_H(\cdot, \alpha)$ is an outer measure on $\mathbb{R}^m$ and hence it induces a σ-additive measure on $\mathbb{R}^m$ called the α-**Hausdorff measure**. This measure can be shown to be Borel (i.e., all Borel sets in $\mathbb{R}^m$ are measurable). Moreover, it is also a regular measure (see Appendix V).

Further, for a subset Z, the above C-structure generates the Carathéodory dimension of Z called the **Hausdorff dimension of the set** Z. We denote it by $\dim_H Z$. According to (1.3), we have

$$\dim_H Z = \inf\{\alpha : m_H(Z, \alpha) = 0\} = \sup\{\alpha : m_H(Z, \alpha) = \infty\}. \qquad (6.3)$$

Moreover, the above C-structure generates also the lower and upper Carathéodory capacities of Z called the **lower** and **upper box dimensions of the set** Z. We denote them by $\underline{\dim}_B Z$ and $\overline{\dim}_B Z$ respectively. According to (2.1) we have

$$\underline{\dim}_B Z = \inf\{\alpha : \underline{r}_B(Z, \alpha) = 0\} = \sup\{\alpha : \underline{r}_B(Z, \alpha) = \infty\},$$
$$\overline{\dim}_B Z = \inf\{\alpha : \overline{r}_B(Z, \alpha) = 0\} = \sup\{\alpha : \overline{r}_B(Z, \alpha) = \infty\}, \qquad (6.4)$$

where

$$\underline{r}_B(Z, \alpha) = \underline{\lim_{\varepsilon \to 0}} \inf_{\mathcal{G}} \left\{ \sum_{U \in \mathcal{G}} \varepsilon^\alpha \right\}, \quad \overline{r}_B(Z, \alpha) = \overline{\lim_{\varepsilon \to 0}} \inf_{\mathcal{G}} \left\{ \sum_{U \in \mathcal{G}} \varepsilon^\alpha \right\}$$

and the infimum is taken over all finite or countable covers $\mathcal{G}$ of Z by open sets of diameter ε.

The set function $\overline{r}_B(Z, \cdot)$ can be shown to be a finite sub-additive outer measure on $\mathbb{R}^m$ while the set function $\underline{r}_B(Z, \cdot)$ may not have this property (see Remark in Section 2; Example 6.2 below provides a counterexample).

In the above definition of the C-structure τ one can choose $\mathcal{F}$ to be the collection of all *closed* subsets of $\mathbb{R}^m$ or even *all* subsets of $\mathbb{R}^m$ to obtain the same value of the α-Hausdorff measure. If, instead, one chooses $\mathcal{F}$ to be the collection of all open or closed *balls* in $\mathbb{R}^m$ then the value of the corresponding α-Hausdorff measure can change but the Hausdorff dimension and lower and upper box dimensions of Z remain the same.

Obviously, the set functions η and ψ satisfy Condition A4. Therefore, Theorem 2.2 produces another equivalent definition of the lower and upper box dimensions of a set $Z \subset \mathbb{R}^m$; namely,

$$\underline{\dim}_B Z = \underline{\lim_{\varepsilon \to 0}} \frac{\log N(Z, \varepsilon)}{\log(1/\varepsilon)}, \quad \overline{\dim}_B Z = \overline{\lim_{\varepsilon \to 0}} \frac{\log N(Z, \varepsilon)}{\log(1/\varepsilon)}, \qquad (6.5)$$

where, in accordance with (2.3), $N(Z, \varepsilon)$ is the least number of balls of radius ε needed to cover Z.

We formulate the main properties of the Hausdorff dimension and lower and upper box dimensions of sets. They are immediate corollaries of the definitions and Theorems 1.1, 2.1, and 2.4.

Theorem 6.1.

(1) $\dim_H \varnothing = 0$; $\dim_H Z \geq 0$ for any $Z \subset \mathbb{R}^m$.

(2) $\dim_H Z_1 \leq \dim_H Z_2$ if $Z_1 \subset Z_2$.

(3) $\dim_H(\bigcup_{i \geq 1} Z_i) = \sup_{i \geq 1} \dim_H Z_i$, $i = 1, 2, \ldots$.

(4) If Z is a finite or countable set then $\dim_H Z = 0$.

Theorem 6.2.

(1) $\underline{\dim}_B \varnothing = \overline{\dim}_B \varnothing = 0$; $\underline{\dim}_B Z \geq 0$ and $\overline{\dim}_B Z \geq 0$ for any $Z \subset \mathbb{R}^m$.

(2) $\dim_H Z \leq \underline{\dim}_B Z \leq \overline{\dim}_B Z$.

(3) $\underline{\dim}_B Z_1 \leq \underline{\dim}_B Z_2$ and $\overline{\dim}_B Z_1 \leq \overline{\dim}_B Z_2$ if $Z_1 \subset Z_2$.

(4) $\underline{\dim}_B(\bigcup_{i \geq 1} Z_i) \geq \sup_{i \geq 1} \underline{\dim}_B Z_i$ and $\overline{\dim}_B(\bigcup_{i \geq 1} Z_i) \geq \sup_{i \geq 1} \overline{\dim}_B Z_i$, $i = 1, 2, \ldots$.

(5)
$$\overline{\dim}_B \left(\bigcup_{i=1}^{n} Z_i \right) = \max_{1 \leq i \leq n} \overline{\dim}_B Z_i$$

and if $\underline{\dim}_B Z_i = \overline{\dim}_B Z_i$ then

$$\underline{\dim}_B \left(\bigcup_{i=1}^{n} Z_i \right) = \max_{1 \leq i \leq n} \underline{\dim}_B Z_i.$$

(6) If the set Z is finite then $\underline{\dim}_B Z = \overline{\dim}_B Z = 0$.

(7) $\underline{\dim}_B Z = \underline{\dim}_B \overline{Z}$, $\overline{\dim}_B Z = \overline{\dim}_B \overline{Z}$, where $\overline{Z}$ is the closure of the set Z.

We remark that any Lipschitz continuous homeomorphism of $\mathbb{R}^m$ with a Lipschitz continuous inverse preserves the C-structure τ. This fact and Theorems 1.3 and 2.5 imply the following statement.

Theorem 6.3. *The Hausdorff dimension and lower and upper box dimensions are invariant with respect to a Lipschitz continuous homeomorphism with a Lipschitz continuous inverse.*

We now illustrate that Statements 2, 4, 5, and 6 of Theorem 6.2 cannot be improved. The set of rational points on $[0, 1]$ is countable and hence it has Hausdorff dimension 0. By Statement 7 of Theorem 6.2 its lower and upper box dimensions equal 1. This shows that Statement 6 may not be true for a countable set of points and strict inequalities may occur in Statement 4.

We present now an example which demonstrates that Statement 2 of Theorem 6.2 may fail for a set of a very simple topological structure. In this example the Hausdorff dimension of the set is zero while the lower and upper box dimensions are positive and distinct. A more sophisticated example where all three characteristics are positive and distinct is constructed in Section 16.2. In fact, one can expect the non-coincidence of the Hausdorff dimension and lower and upper box dimensions to be a typical phenomenon in a sense.

Example 6.1. *For given $0 < \alpha \leq \beta < 1$ there is a closed countable set $Z \subset [0, 1]$ such that $\underline{\dim}_B Z = \alpha$, $\overline{\dim}_B Z = \beta$ while $\dim_H Z = 0$.*

Proof. First we need the following lemma.

Lemma. *There exist sequences of positive numbers $\{a_n\}$ and $\{b_n\}$, $n \geq 1$ such that*

    (1) *a_n decrease monotonically and $a_n \to 0$; b_n increase monotonically and $b_n \to \infty$;*

    (2) *$\sum\limits_{n=1}^{\infty} a_n b_n \leq 1$;*

    (3) *$\alpha = \varliminf\limits_{n \to \infty} \log S_n / \log \frac{1}{a_n}$, $\beta = \varlimsup\limits_{n \to \infty} \log S_n / \log \frac{1}{a_n}$, where $S_n = \sum\limits_{k=1}^{n} b_k$;*

    (4) *there exists $C > 0$ such that for any $n > 0$*

$$\frac{1}{a_n S_n} \left(\sum_{k=n}^{\infty} a_k b_k \right) \leq C.$$

Proof of the lemma. Set $a_n = a^n$, where $0 < a \leq \frac{1}{3}$ and choose a number M such that

$$a^{-\beta} < M < a^{-1}. \tag{6.6}$$

Let n_k be a growing sequence of integers. Define

$$b_n = \begin{cases} a^{-\alpha n} & \text{if } n_{4k} \leq n < n_{4k+1} \\ M b_{n-1} & \text{if } n_{4k+1} \leq n < n_{4k+2} \\ a^{-\beta n} & \text{if } n_{4k+2} \leq n < n_{4k+3} \\ b_{n_{4k+3}-1} & \text{if } n_{4k+3} \leq n < n_{4k+4} \end{cases}$$

We now specify the sequence n_k. Set $n_0 = 0$. Assuming that n_{4k} is given we choose n_{4k+1} so large that

$$\sum_{\ell=n_{4k}}^{n_{4k+1}} b_\ell \Big/ \sum_{\ell=1}^{n_{4k+1}} b_\ell \geq \frac{1}{2}. \tag{6.7}$$

Let n_{4k+2} be the smallest integer such that

$$a^{-\alpha n_{4k+1}} M^{n_{4k+2}-n_{4k+1}} \geq a^{-\beta n_{4k+2}}. \tag{6.8}$$

This inequality is equivalent to

$$(a^\alpha M)^{n_{4k+1}} \leq (a^\beta M)^{n_{4k+2}}$$

and can be solved by an appropriate choice of n_{4k+2} since by (6.6),

$$a^\alpha M > a^\beta M > 1.$$

Now we choose n_{4k+3} so large that

$$\sum_{\ell=n_{4k+2}}^{n_{4k+3}} b_\ell \Big/ \sum_{\ell=1}^{n_{4k+3}} b_\ell \geq \frac{1}{2}. \tag{6.9}$$

Let n_{4k+4} be the smallest integer such that

$$a^{-\beta n_{4k+3}} \le a^{-\alpha n_{4k+4}}. \tag{6.10}$$

This inequality can be solved since $1 > a^\alpha > a^\beta$.

The definition of the numbers b_n can be better understood if one imagines the motion of the point (n, b_n) when n grows starting from $n = n_{4k}$. This point first moves along the graph of the function $y = a^{-\alpha x}$ until $n = n_{4k+1}$; then it jumps up $(n_{4k+2} - n_{4k+1})$ times with the coefficient M until it reaches the graph of the function $y = e^{-\beta x}$ (by virtue of (6.8) this happens when $n = n_{4k+2}$); then it moves along this graph until $n = n_{4k+3}$; then it stays on the line $y = b_{n_{4k+3}}$ until this line crosses the graph of the function $y = a^{-\alpha x}$ (by virtue of (6.10) it happens when $n = n_{4k+4}$); then the process is repeated.

Clearly, the sequences a_n and b_n satisfy Conditions 1–2. Condition 3 follows from (6.7) and (6.9). We have according to (6.6) that

$$\frac{1}{a_n S_n}\left(\sum_{k=n}^\infty a_k b_k\right) \le \frac{C}{a^n b_n}\left(\sum_{k=n}^\infty a^k b_k\right) \le C\sum_{k=n}^\infty \frac{b_k}{b_n} a^{k-n} \le C\sum_{k=n}^\infty (Ma)^{k-n} < \infty,$$

where $C > 0$ is a constant. This implies Condition 4 and completes the proof of the lemma. ∎

In order to construct the required set Z we define by induction the sets Z_n. Namely, set $Z_0 = \{0\}$ and then

$$Z_n = Z_{n-1} \cup \{P_{n-1} + a_n, \ldots, P_{n-1} + [b_n] a_n\},$$

where $P_0 = 0$, $P_n = \sum_{k=1}^n a_k [b_k]$ and $[b_k]$ denotes the entire part of b_k. Let

$$Z = \bigcup_{n \ge 0} Z_n \cup \{P_\infty\},$$

where $P_\infty = \sum_{k=1}^\infty a_k [b_k]$. We wish to show that $\underline{\dim}_B Z = \alpha$ and $\overline{\dim}_B Z = \beta$. Given $\varepsilon > 0$, consider the sets

$$T_1 = \bigcup_{n=0}^{m(\varepsilon)} Z_n, \quad T_2 = Z \setminus T_1,$$

where $m(\varepsilon)$ is the smallest integer for which $a_{m(\varepsilon)} \le \varepsilon$. These sets are disjoint and $|z_1 - z_2| \ge \varepsilon$ for any distinct $z_1, z_2 \in T_1$. It is easy to see that $N(T_1, \varepsilon/2) = S_{m(\varepsilon)}$ and

$$N(T_2, \varepsilon/2) \le \frac{C_1}{\varepsilon}\left(\sum_{k=m(\varepsilon)}^\infty a_k b_k\right),$$

where $C_1 > 0$ is a constant. Condition 4 of the lemma implies that

$$N(T_2, \varepsilon/2) \le C_2 S_{m(\varepsilon)},$$

where $C_2 > 0$ is a constant. Now the desired result follows from Condition 3 of the lemma. ∎

We now show that Statement 5 of Theorem 6.2 cannot be improved.

Example 6.2. *There are two closed countable disjoint sets $Z_1, Z_2 \subset \mathbb{R}$ such that*

$$\underline{\dim}_B(Z_1 \cup Z_2) > \max\{\underline{\dim}_B Z_1, \ \underline{\dim}_B Z_2\}.$$

Proof. By analyzing the construction of Example 6.1 one can show that there are countable closed sets $Z_1 \subset [0,1]$ and $Z_2 \subset [2,3]$ satisfying the following conditions:

(a) $\underline{\dim}_B Z_1 = \underline{\dim}_B Z_2 = \alpha$, where $\alpha \in (0,1)$ is a given number;
(b) given $i = 1$ or 2 and a sequence $\varepsilon_n \to 0$ for which the limit

$$\lim_{n \to \infty} \frac{\log N(Z_i, \varepsilon_n)}{\log(1/\varepsilon_n)} = \alpha$$

exists, we have

$$\varliminf_{n \to \infty} \frac{\log N(Z_j, \varepsilon_n)}{\log(1/\varepsilon_n)} > \alpha,$$

where $j = 2$ if $i = 1$ and $j = 1$ if $i = 2$.

Indeed, defining the sequence b_n in the proof of the lemma, we can make the point (n, b_n) start moving along the graph of the function $y = a^{-\alpha x}$ while constructing the set Z_1 and start moving along the graph of the function $y = a^{-\beta x}$ while constructing the set Z_2. Obviously, Z_1 and Z_2 have the required properties. ∎

We formulate a simple but useful criterion that guarantees the coincidence of the lower and upper box dimensions of a set $Z \subset \mathbb{R}^m$. It is a simple corollary of Theorem 2.3.

Theorem 6.4. *Assume that ε_n is a monotonically decreasing sequence of numbers, $\varepsilon_n \to 0$, and $\varepsilon_{n+1} \geq C\varepsilon_n$ for some $C > 0$ which is independent of n. Assume also that there exists the limit*

$$\lim_{n \to 0} \frac{\log N(Z, \varepsilon_n)}{\log(1/\varepsilon_n)} \overset{\text{def}}{=} d.$$

Then, $\underline{\dim}_B Z = \overline{\dim}_B Z = d$.

Let $E \subset U \subset \mathbb{R}^n$ and $F \subset V \subset \mathbb{R}^m$ be two Borel sets (U and V are bounded open subsets). It was a long-standing problem in dimension theory whether the Hausdorff dimension of the Cartesian product of E and F is equal to the sum of their Hausdorff dimensions. In general, this is false, and we state some positive results in this directions (see [F5] for detailed discussion and proofs).

Theorem 6.5.

(1) $\dim_H(E \times F) \geq \dim_H E + \dim_H F$.
(2) $\dim_H(E \times F) \leq \dim_H E + \overline{\dim}_B F$.
(3) $\overline{\dim}_B(E \times F) \leq \overline{\dim}_B E + \overline{\dim}_B F$.
(4) *If $\dim_H E = \overline{\dim}_B E$ then $\dim_H(E \times F) = \dim_H E + \dim_H F$ for any Borel set F.*

The C-structure defined by (6.1) (which induces the Hausdorff measure on $\mathbb{R}^m$) has a trivial weight function ξ. We now introduce another C-structure with non-trivial weight function ξ in order to illustrate some new phenomena which can occur in this case.

Example 6.3.

Let $\mathcal{F}$ be the collection of all open sets of $\mathbb{R}^m$. For $U \in \mathcal{F}$ set

$$\xi(U) = h(\operatorname{diam} U), \quad \eta(U) = \psi(U) = \operatorname{diam} U,$$

where h is a positive continuous function on $\mathbb{R}$. It is easy to verify that Conditions $A1, A2, A3$, and $A3'$ hold. Assuming that $h(x) = \exp(-\frac{1}{x})$, one can show that $m_C(\mathbb{R}^m, \alpha) = 0$ for any $\alpha \in \mathbb{R}$ and hence $\dim_H \mathbb{R}^m = -\infty$. On the other hand, if $h(x) = \exp(\frac{1}{x})$, then $m_C(\mathbb{R}^m, \alpha) = \infty$ for any $\alpha \in \mathbb{R}$ and hence $\dim_H \mathbb{R}^m = \infty$.

7. Hausdorff Dimension and Box Dimension of Measures; Pointwise Dimension; Mass Distribution Principle

Let μ be a Borel finite measure on $\mathbb{R}^m$. The C-structure $\tau = (\mathcal{F}, \xi, \eta, \psi)$ given by (6.1) produces, in accordance with (3.1) and (3.2), the Carathéodory dimension of μ and lower and upper Carathéodory capacities of μ. They are called respectively the **Hausdorff dimension of the measure** and **lower** and **upper box dimensions of the measure** and are denoted by $\dim_H \mu$, $\underline{\dim}_B \mu$, and $\overline{\dim}_B \mu$ respectively. Thus, we have

$$\dim_H \mu = \inf\{\dim_H Z : \ \mu(Z) = 1\},$$
$$\underline{\dim}_B \mu = \lim_{\delta \to 0} \inf\{\underline{\dim}_B Z : \ \mu(Z) \geq 1 - \delta\}, \tag{7.1}$$
$$\overline{\dim}_B \mu = \lim_{\delta \to 0} \inf\{\overline{\dim}_B Z : \ \mu(Z) \geq 1 - \delta\}.$$

One can also define the Hausdorff dimension of the measure as follows (see (3.3))

$$\dim_H \mu = \lim_{\delta \to 0} \inf\{\dim_H Z : \mu(Z) \geq 1 - \delta\}.$$

On the other hand, in general,

$$\underline{\dim}_B \mu \leq \inf\{\underline{\dim}_B Z : \mu(Z) = 1\}, \quad \overline{\dim}_B \mu \leq \inf\{\overline{\dim}_B Z : \mu(Z) = 1\}.$$

As the following example shows the strict inequalities can occur. It also demonstrates that the Hausdorff dimension and box dimension of the measure can be strictly less than the Hausdorff dimension and box dimension of the support of the measure (i.e., the smallest closed subset of full measure).

Example 7.1. *There is a Borel finite measure μ on $[0,1]$ for which $\underline{\dim}_B \mu = \overline{\dim}_B \mu$ and*

$$\underline{\dim}_B \mu < \inf\{\underline{\dim}_B Z : \mu(Z) = 1, \ Z \subset [0,1]\},$$
$$\overline{\dim}_B \mu < \inf\{\overline{\dim}_B Z : \mu(Z) = 1, \ Z \subset [0,1]\}.$$

Proof. Let Q be the set of rational points in $[0,1]$. We number points in Q such that $Q = \{r_1, r_2, \dots\}$. Define an atomic measure μ on $[0,1]$ by setting $\mu(r_n) = 1/2^{n+1}$. Clearly, μ is a Borel finite measure on $[0,1]$. Moreover, Q is the only set of full measure. Since $\underline{\dim}_B Q = \underline{\dim}_B \overline{Q} = 1$ and also $\overline{\dim}_B Q = \overline{\dim}_B \overline{Q} = 1$ we have

$$\inf\{\underline{\dim}_B Z : \mu(Z) = 1\} = \inf\{\overline{\dim}_B Z : \mu(Z) = 1\} = 1.$$

On the other hand, for every $\delta > 0$ there is a finite set $Z_\delta \subset Q$ for which $\mu(Z_\delta) \geq 1 - \delta$. We have $\underline{\dim}_B Z_\delta = \overline{\dim}_B Z_\delta = 0$. Thus, $\underline{\dim}_B \mu = \overline{\dim}_B \mu = 0$. ■

It follows from definitions (7.1) that

$$\dim_H \mu \le \underline{\dim}_B \mu \le \overline{\dim}_B \mu.$$

Below we will construct examples where strict inequalities occur. On the other hand, using Theorem 4.1 we will obtain a powerful criterion that guarantees the coincidence of the Hausdorff dimension and lower and upper box dimensions of measures.

Consider the collection of balls

$$\mathcal{F}' = \{B(x,r) : x \in \mathbb{R}^m, r > 0\}.$$

In accordance with Section 3 (see (3.5)) we define the **lower** and **upper pointwise dimensions** of μ at x (specified by $\mathcal{F}'$) by

$$\underline{d}_\mu(x) = \varliminf_{r \to 0} \frac{\log \mu(B(x,r))}{\log r}, \quad \overline{d}_\mu(x) = \varlimsup_{r \to 0} \frac{\log \mu(B(x,r))}{\log r}. \tag{7.2}$$

Clearly, $\underline{d}_\mu(x) \le \overline{d}_\mu(x)$. The following statements are consequences of Theorems 3.3 and 4.1. They were first established by Young in [Y2].

Theorem 7.1. *Let μ be a Borel finite measure on $\mathbb{R}^m$. Then the following statements hold:*

(1) *if $\underline{d}_\mu(x) \ge d$ for μ-almost every x then $\dim_H \mu \ge d$;*

(2) *if $\overline{d}_\mu(x) \le d$ for μ-almost every x then $\overline{\dim}_B \mu \le d$;*

(3) *if $\underline{d}_\mu(x) = \overline{d}_\mu(x) = d$ for μ-almost every x then $\dim_H \mu = \underline{\dim}_B \mu = \overline{\dim}_B \mu = d$.*

Proof. Assume first that μ is a non-atomic measure. For any set Z of positive measure and any $\varepsilon > 0$ there exists a finite multiplicity cover of Z by balls of radius ε (see Appendix V). This implies (3.7). The validity of the second conditions of Theorems 3.1 and 3.2 is obvious. The desired result follows now from Theorems 3.3 and 4.1. If μ is an atomic measure then, obviously, $\underline{d}_\mu(x) = \overline{d}_\mu(x) = 0$ and $\dim_H \mu = \underline{\dim}_B \mu = \overline{\dim}_B \mu = 0$. ∎

The direct calculation of the Hausdorff dimension of a set $Z \subset \mathbb{R}^m$ based on its definition is usually very difficult. In order to obtain an upper estimate of the Hausdorff dimension one must present a specific "good" cover $U = \{U_i\}$ of Z with elements of small diameter for which $\sum(\operatorname{diam} U_i)^d$ is finite (so that d provides an upper bound for the Hausdorff dimension). In many cases a keen-minded person can guess how to choose a "good" cover with an appropriate value d (which often turns out to be the actual value for the Hausdorff dimension). Another method for obtaining an upper estimate for the Hausdorff dimension of Z is based on Theorem 5.1.

Theorem 7.2. *Let μ be a Borel finite measure on $\mathbb{R}^m$. Assume that there exists a number $d > 0$ such that $\underline{d}_\mu(x) \le d$ for **every** $x \in Z$. Then $\dim_H Z \le d$.*

Proof. One can easily check that the first two conditions of Theorem 5.1 hold while the third condition follows directly from the Besicovitch Covering Lemma (see Appendix V). The result follows from Theorem 5.1. ∎

Theorem 7.2 is an effective tool which allows one to establish an efficient upper estimate of the Hausdorff dimension and it is useful in applications (see, for example, Theorem 18.3). Moreover, in view of Theorems 7.1 and 7.2, one can actually compute the Hausdorff dimension of a set Z if one can find a measure μ and a number d such that $\mu(Z) > 0$, $\underline{d}_\mu(x) \geq d$ for μ-*almost every* $x \in Z$, and $\underline{d}_\mu(x) \leq d$ for *every* $x \in Z$ (we will use this fact in the proof of Lemma 2 in Theorem 21.1 and Lemma 1 in Theorem 24.1).

There is a stronger version of Theorem 7.2 which is often used in applications. It claims the following:

Assume that there are numbers $d > 0$, $C > 0$, and a Borel finite measure μ on Z such that for **every** *$x \in Z$ and $r > 0$*

$$\mu(B(x,r)) \geq Cr^d;$$

then $\dim_H Z \leq d$.

In order to obtain a lower estimate of the Hausdorff dimension one must work with all open covers of Z with elements of small diameter. There is an intelligent way of producing lower bounds which is based on the first statement of Theorem 7.1. Namely, assume that one can construct a Borel finite measure μ concentrated on Z such that $\underline{d}_\mu(x) \geq d$ for μ-almost every $x \in Z$ and some $d \geq 0$ then

$$\dim_H Z \geq \dim_H \mu \geq d.$$

This simple corollary of Theorem 7.1 can be reformulated as the **non-uniform mass distribution principle**:

Assume that there are a number $d > 0$ and a Borel finite measure μ on Z such that for any $\varepsilon > 0$ and μ-almost every $x \in Z$ one can find a constant $C(x, \varepsilon) > 0$ satisfying for any $r > 0$

$$\mu(B(x,r)) \leq C(x,\varepsilon)r^{d-\varepsilon}; \tag{7.3}$$

then $\dim_H Z \geq d$.

It is surprising that in many interesting cases one can construct a measure that satisfies a stronger version of (7.3) known as the **(uniform) mass distribution principle** (see [Fr]):

Assume that there are numbers $d > 0$, $C > 0$, and a Borel finite measure μ on Z such that for μ-almost every $x \in Z$ and $r > 0$

$$\mu(B(x,r)) \leq Cr^d; \tag{7.4}$$

then $\dim_H Z \geq d$.

It is easier to establish the non-uniform mass distribution principle than the uniform one. On the other hand, as soon as the uniform mass distribution principle is established for a given set Z, it is usually more effective and allows one to obtain more information about the Hausdorff dimension of Z. For example, (7.4) immediately produces the positivity of the d-Hausdorff measure of Z, namely

$$m_H(Z, d) \geq 1/C.$$

One can construct an example of a set Z of Hausdorff dimension d for which the non-uniform mass distribution principle (7.3) holds but the d-Hausdorff measure of Z is zero (see [PW1]).

In [Fr], Frostman developed another method of estimating the Hausdorff dimension of a subset Z of $\mathbb{R}^m$ using measures supported on Z. It is known as the **potential theoretic method**. Let μ be a Borel finite measure on $\mathbb{R}^m$. For $s \geq 0$ the s-potential at a point $x \in \mathbb{R}^m$, due to the measure μ, is defined as

$$\varphi_s(x) = \int_{\mathbb{R}^m} |x - y|^{-s} d\mu(y).$$

The integral

$$I_s(\mu) = \int_{\mathbb{R}^m} \varphi_s(x) \, d\mu(x) = \int_{\mathbb{R}^m \times \mathbb{R}^m} |x - y|^{-s} d\mu(x) \times d\mu(y) \qquad (7.5)$$

is known as the s-energy of the measure μ. The **potential principle** claims (see [F5]) that *for any $Z \subset \mathbb{R}^m$*

   (a) *if there is a measure μ on Z with $I_s(\mu) < \infty$ then $m_H(Z, s) = \infty$;*
   (b) *if $m_H(Z, s) > 0$ then there exists a measure μ on Z with $I_t(\mu) < \infty$ for all $t < s$.*

This principle can be restated as follows. Define the s-capacity of a set Z by

$$C_s(Z) = \sup_\mu \left\{ \frac{1}{I_s(\mu)} : \mu(Z) = 1 \right\}.$$

Then

$$\dim_H Z = \inf \{s : C_s(Z) = 0\} = \sup \{s : C_s(Z) > 0\}. \qquad (7.6)$$

We now proceed with Statement 3 of Theorem 7.1. A measure μ is said to be **exact dimensional** if it satisfies (7.3) for μ-almost every x. This notion was introduced by Cutler (see [Cu]). Let us emphasize that exact dimensionality includes two conditions:

   (1) the limit

$$\lim_{r \to 0} \frac{\log \mu(B(x, r))}{\log r} \stackrel{\text{def}}{=} d_\mu(x) \qquad (7.7)$$

   exists almost everywhere;
   (2) $d_\mu(x)$ is constant almost everywhere.

We shall discuss the relationships between these two conditions. Let us notice that for any $x \in \mathbb{R}^m$ the function $\psi(x, r) = \log \mu(B(x, r)) / \log r$ is right-continuous in r for every x. Therefore, the functions $\underline{d}_\mu(x)$ and $\overline{d}_\mu(x)$ are measurable.

We consider the case when μ is an invariant measure for a diffeomorphism f of a smooth Riemannian manifold M.

Theorem 7.3. *The functions $\underline{d}_\mu(x)$ and $\overline{d}_\mu(x)$ are invariant under f, i.e., $\underline{d}_\mu(f(x)) = \underline{d}_\mu(x), \overline{d}_\mu(x) = \overline{d}_\mu(f(x))$.*

Proof. The statement follows immediately from the obvious implications: for any $x \in M$ and $r > 0$,

$$B(f(x), C_2 r) \subset f(B(x, r)) \subset B(f(x), C_1 r),$$

where $C_1 > 0$ and $C_2 > 0$ are constants independent of x and r. $\blacksquare$

Thus, if μ is ergodic we have $\underline{d}_\mu(x) = \text{const} \stackrel{\text{def}}{=} \underline{d}(\mu)$ and $\overline{d}_\mu(x) = \text{const} \stackrel{\text{def}}{=} \overline{d}(\mu)$ almost everywhere. As Example 25.3 illustrates one can construct a smooth map of the unit interval with $\underline{d}_\mu(x) < \overline{d}_\mu(x)$ for almost all x (indeed, these functions are constant almost everywhere).

In Section 26 we will show that an ergodic Borel measure with non-zero Lyapunov exponents invariant under a $C^{1+\alpha}$-diffeomorphism of a smooth Riemannian manifold is exact dimensional. The assumption, that the map is smooth, is crucial: as Example 25.2 demonstrates there exists a Hölder continuous homeomorphism (whose Hölder exponent can be made arbitrarily close to one) with positive topological entropy and unique measure of maximal entropy whose lower and upper pointwise dimensions are different at almost every point.

Theorem 7.3 is not, in general, true for continuous maps on compact metric spaces. As Example 25.1 shows there exists a Hölder continuous map f preserving an ergodic Borel probability measure μ such that the limit (7.7) exists almost everywhere but is not essentially constant.

Remarks.

(1)   As we saw above the (lower and upper) pointwise dimension allows one to estimate (and sometimes to determine precisely) the value of the Hausdorff dimension and box dimension of sets. There is an alternative approach based on the notion of density of a measure at a point which simulates the notion of the Lebesgue density (see [Fe]).

Let μ be a finite Borel measure on $\mathbb{R}^m$. Given a point $x \in \mathbb{R}^m$, we call the quantities

$$\underline{D}_\mu(\alpha, x) = \varliminf_{r \to 0} \frac{\mu(B(x,r))}{r^\alpha},$$

$$\overline{D}_\mu(\alpha, x) = \varlimsup_{r \to 0} \frac{\mu(B(x,r))}{r^\alpha}$$

the **lower** and **upper** α-**densities** of the measure μ at the point x. It is straightforward to check that if $\underline{D}_\mu(\alpha, x) \geq c$ for every point x of a set $Z \subset \mathbb{R}^m$ of positive measure ($c > 0$ is a constant) then $\overline{d}_\mu(x) \leq \alpha$. Similarly, if $\overline{D}_\mu(\alpha, x) \leq c$ for every point $x \in Z$ ($c > 0$ is a constant) then $\underline{d}_\mu(x) \geq \alpha$. Thus, in view of Theorems 7.1 and 7.2, estimating the lower and upper densities provides upper and lower bounds for the Hausdorff dimension and box dimension of Z. In fact, it gives more and allows one to obtain lower and upper bounds for the α-Hausdorff measure of Z.

Theorem 7.4. *Let Z be a Borel subset of positive measure.*

(1) *If $\overline{D}_\mu(\alpha, x) \leq c$ for μ-almost every $x \in Z$ then $m_H(Z, \alpha) \geq \mu(Z)/c > 0$.*
(2) *If $\overline{D}_\mu(\alpha, x) \geq c$ for **every** $x \in Z$ then $m_H(Z, \alpha) \leq 8^\alpha c^{-1} \mu(\mathbb{R}^m) < \infty$.*

Proof. Let $\tilde{Z}$ denote the set of points $x \in Z$ for which $\overline{D}_\mu(\alpha, x) \leq c$. Given $\delta > 0$, consider the set Z_δ which consists of points $x \in \tilde{Z}$ satisfying

$$\mu(B(x,r)) \leq (c + \varepsilon)r^\alpha$$

for all $0 < r \leq \delta$ and some $\varepsilon > 0$. It follows from the definition of the upper density that

$$\tilde{Z} = \bigcup_{\delta > 0} Z_\delta.$$

Let $\{U_i\}$ be a cover of Z by open sets of diameter $\leq \delta$. For each U_i containing a point $x \in Z_\delta$, consider the ball $B(x, |U_i|)$. Since $U_i \subset B(x, |U_i|)$, by the definition of the set Z_δ, we obtain that

$$\mu(U_i) \leq \mu(B(x, |U_i|)) \leq (c + \varepsilon)|U_i|^\alpha.$$

It follows that

$$\mu(Z_\delta) \leq \sum_i \{\mu(U_i) : U_i \cap Z_\delta \neq \varnothing\} \leq (c + \varepsilon) \sum_i |U_i|^\alpha.$$

This implies that $\mu(Z_\delta) \leq (c + \varepsilon)m_H(Z, \alpha)$. Since this inequality holds for all δ and ε the first statement of the theorem follows.

We proceed with the second statement of the theorem. Note that it is sufficient to assume that the set Z is bounded. Given $\delta > 0$ and $\varepsilon > 0$, let $\mathcal{A}$ be the collection of balls $B(x, r)$ centered at points $x \in Z$ of radius $0 < r \leq \delta$ for which

$$\mu(B(x, r)) \geq (c - \varepsilon)r^\alpha.$$

It follows from the definition of the upper density that

$$Z = \bigcup_{B(x,r) \in \mathcal{A}} B(x, r).$$

Applying the Vitali Covering Lemma (see Appendix V) we can find a subcollection $\mathcal{B} = \{B_i\} \subset \mathcal{A}$ consisting of disjoint balls for which

$$\bigcup_{B(x,r) \in \mathcal{A}} B(x, r) \subset \bigcup_i \tilde{B}_i,$$

where $\tilde{B}_i$ denotes the closed ball concentric with B_i whose radius is four times the radius of B_i. We have that

$$\sum_i |\tilde{B}_i|^\alpha \leq 8^\alpha(c - \varepsilon)^{-1} \sum_i \mu(B_i) \leq 8^\alpha(c - \varepsilon)^{-1}\mu(\mathbb{R}^m).$$

This implies that $m_H(Z, \alpha) \leq 8^\alpha(c - \varepsilon)^{-1}\mu(\mathbb{R}^m)$. Since ε is chosen arbitrarily the second statement of the theorem follows. ∎

(2)   A subset $Z \subset \mathbb{R}^m$ is called an α-**set** $(0 \le \alpha \le m)$ if its α-Hausdorff measure is positive and finite. One can now be interested in estimating the densities of the measure at points in Z. We describe a result in this direction referring the reader to [F5].

Theorem 7.5. *Let Z be a Borel α-set. Then*

(1) $\overline{D}_\mu(\alpha, x) = 0$ *for $m_H(\cdot, \alpha)$-almost all x outside Z;*
(2) $1 \le \overline{D}_\mu(\alpha, x) \le 2^\alpha$ *for $m_H(\cdot, \alpha)$-almost all $x \in Z$.*

Let Z be an α-set. A point $x \in Z$ is called **regular** if $\underline{D}_\mu(\alpha, x) = \overline{D}_\mu(\alpha, x) = 1$. Otherwise x is called **irregular**. An α-set Z is called **regular** if $m_H(\cdot, \alpha)$-almost every point $x \in Z$ is regular; otherwise Z is called **irregular**. Characterizing α-sets is one of the main directions of study in the dimension theory. One of the main results claims that an *α-set cannot be regular unless α is an integer* (see [F5]).

(3)   Let μ be the measure that is constructed in Example 7.1. We have that

$$0 = \dim_H \mu = \underline{\dim}_B \mu = \overline{\dim}_B \mu <$$
$$1 = \dim_H S(\mu) = \underline{\dim}_B S(\mu) = \overline{\dim}_B S(\mu),$$

where $S(\mu)$ is the support of μ.

Another example of a measure (which is invariant under a dynamical system), whose Hausdorff dimension is strictly less than the Hausdorff dimension of its support, is given in Section 23 (see baker's transformations).

Chapter 3

C-Structures Associated with Metrics and Measures: Dimension Spectra

In this Chapter we introduce and study C-structures on the Euclidean space $\mathbb{R}^m$ induced by measures and metrics on $\mathbb{R}^m$ (in Appendix I we consider C-structures which are generated by measures and metrics on *general* metric spaces). Namely, given a measure μ, we define one-parameter families of Carathéodory dimension characteristics that we call q-dimension and lower and upper q-box dimensions of the sets or measures respectively.

In Chapter 6 we will demonstrate an important role that these characteristics play in the theory of dynamical systems. In particular, we will show in Section 18 that the lower and upper q-box dimensions of the support of μ are intimately related to the well-known dimension spectra specified by the measure μ (i.e., the Hentschel–Procaccia spectrum for dimensions and the Rényi spectrum for dimensions). In the case when μ is an invariant measure for a dynamical system f acting on $\mathbb{R}^m$, the q-dimension and lower and upper q-box dimensions are invariants of f which are completely specified by μ. Among them the most valuable are the correlation dimensions of order q (see Section 17).

8. q-Dimension and q-Box Dimension of Sets

Let μ be a Borel finite measure on the Euclidean space $\mathbb{R}^m$. We assume that $\mathbb{R}^m$ is endowed with a metric ρ which is equivalent to the standard metric. Given numbers $q \geq 0$ and $\gamma > 0$, we introduce a C-structure on $\mathbb{R}^m$ generated by ρ and μ. Namely, let $\mathcal{F}$ be the collection of open balls in $\mathbb{R}^m$. For any ball $B(x, \varepsilon) \in \mathcal{F}$, we define

$$\xi(B(x,\varepsilon)) = \begin{cases} \mu(B(x, \gamma\varepsilon))^q & \text{if } q > 0 \\ 1 & \text{if } q = 0 \end{cases}, \quad \eta(B(x,\varepsilon)) = \psi(B(x,\varepsilon)) = \varepsilon. \quad (8.1)$$

It is easy to verify that the collection $\mathcal{F}$ and the set functions ξ, η, and ψ satisfy conditions $A1, A2, A3$, and $A3'$. Hence they define a C-structure in $\mathbb{R}^m$, $\tau_{q,\gamma} = (\mathcal{F}, \xi, \eta, \psi)$. The corresponding Carathéodory set function $m_C(Z, \alpha)$, (where $Z \subset \mathbb{R}^m$ and $\alpha \in \mathbb{R}$) is called the (q, γ)-set function. We denote it by $m_{q,\gamma}(Z, \alpha)$. By virtue of (1.1) and (1.2) this function is given as follows:

$$m_{q,\gamma}(Z, \alpha) = \lim_{\varepsilon \to 0} \inf_{\mathcal{G}} \left\{ \sum_{B(x_i,\varepsilon_i) \in \mathcal{G}} \mu(B(x_i, \gamma\varepsilon_i))^q \varepsilon_i^\alpha \right\}, \quad (8.2)$$

where the infimum is taken over all finite or countable covers $\mathcal{G} \subset \mathcal{F}$ of Z by balls $B(x_i, \varepsilon_i)$ with $\varepsilon_i \leq \varepsilon$.

If $m_{q,\gamma}(\varnothing, \alpha) = 0$ (this holds true for $\alpha > 0$ but can also happen for some negative α) the set function $m_{q,\gamma}(\cdot, \alpha)$ becomes an outer measure on $\mathbb{R}^m$ (see Appendix V). Hence it induces a σ-additive measure on $\mathbb{R}^m$ that we call the (q, γ)-**measure**. This measure can be shown to be Borel (i.e., all Borel sets are measurable; see Appendix V).

Further, the above C-structure produces the Carathéodory dimension of Z. We call it the (q, γ)-**dimension of the set** Z and denote it by $\dim_{q,\gamma} Z$. According to (1.3), we have

$$\dim_{q,\gamma} Z = \inf\{\alpha : m_{q,\gamma}(Z, \alpha) = 0\} = \sup\{\alpha : m_{q,\gamma}(Z, \alpha) = \infty\}. \qquad (8.3)$$

Moreover, the above C-structure generates also the lower and upper Carathéodory capacities of Z. We call them the **lower** and **upper** (q, γ)-**box dimensions of the set** Z and denote them by $\underline{\dim}_{q,\gamma} Z$ and $\overline{\dim}_{q,\gamma} Z$ respectively. According to (2.1) we have

$$\begin{aligned}
\underline{\dim}_{q,\gamma} Z &= \inf\{\alpha : \underline{r}_{q,\gamma}(Z, \alpha) = 0\} = \sup\{\alpha : \underline{r}_{q,\gamma}(Z, \alpha) = \infty\}, \\
\overline{\dim}_{q,\gamma} Z &= \inf\{\alpha : \overline{r}_{q,\gamma}(Z, \alpha) = 0\} = \sup\{\alpha : \overline{r}_{q,\gamma}(Z, \alpha) = \infty\},
\end{aligned} \qquad (8.4)$$

where

$$\underline{r}_{q,\gamma}(Z, \alpha) = \varliminf_{\varepsilon \to 0} \inf_{\mathcal{G}} \left\{ \sum_{U \in \mathcal{G}} \varepsilon^\alpha \right\}, \quad \overline{r}_{q,\gamma}(Z, \alpha) = \varlimsup_{\varepsilon \to 0} \inf_{\mathcal{G}} \left\{ \sum_{U \in \mathcal{G}} \varepsilon^\alpha \right\}$$

and the infimum is taken over all finite or countable covers $\mathcal{G}$ of Z by balls of radius ε.

It is obvious that the set functions η and ψ satisfy Condition A4. Therefore, Theorem 2.2 gives us another equivalent definition of the lower and upper (q, γ)-box dimensions of sets:

$$\underline{\dim}_{q,\gamma} Z = \varliminf_{\varepsilon \to 0} \frac{\log \Lambda_{q,\gamma}(Z, \varepsilon)}{\log(1/\varepsilon)}, \quad \overline{\dim}_{q,\gamma} Z = \varlimsup_{\varepsilon \to 0} \frac{\log \Lambda_{q,\gamma}(Z, \varepsilon)}{\log(1/\varepsilon)}, \qquad (8.5)$$

where in accordance with (2.3),

$$\Lambda_{q,\gamma}(Z, \varepsilon) = \inf_{\mathcal{G}} \left\{ \sum_{B(x_i,\varepsilon) \in \mathcal{G}} \mu(B(x_i, \gamma \varepsilon))^q \right\} \qquad (8.6)$$

(here the infimum is taken over all finite or countable covers $\mathcal{G}$ of Z by balls of radius ε).

The main properties of the (q, γ)-dimension and lower and upper (q, γ)-box dimensions of sets are listed below. They follow immediately from the definitions and Theorems 1.1, 2.1, and 2.4.

Theorem 8.1.

(1) $\dim_{q,\gamma} Z_1 \le \dim_{q,\gamma} Z_2$ if $Z_1 \subset Z_2$.

(2)

$$\dim_{q,\gamma} \left(\bigcup_{i\geq 1} Z_i \right) = \sup_{i\geq 1} \dim_{q,\gamma} Z_i.$$

(3) If Z is a finite or countable set then $\dim_{q,\gamma} Z \le 0$; if x is an atom of μ then $\dim_{q,\gamma}\{x\} = 0$.

Theorem 8.2.

(1) $\dim_{q,\gamma} Z \le \underline{\dim}_{q,\gamma} Z \le \overline{\dim}_{q,\gamma} Z$ for any $Z \subset \mathbb{R}^m$.

(2) $\underline{\dim}_{q,\gamma} Z_1 \le \underline{\dim}_{q,\gamma} Z_2$ and $\overline{\dim}_{q,\gamma} Z_1 \le \overline{\dim}_{q,\gamma} Z_2$ if $Z_1 \subset Z_2$.

(3)

$$\underline{\dim}_{q,\gamma} \left(\bigcup_{i\geq 1} Z_i \right) \ge \sup_{i\geq 1} \underline{\dim}_{q,\gamma} Z_i, \quad \overline{\dim}_{q,\gamma} \left(\bigcup_{i\geq 1} Z_i \right) \ge \sup_{i\geq 1} \overline{\dim}_{q,\gamma} Z_i.$$

(4)

$$\overline{\dim}_{q,\gamma} \left(\bigcup_{i=1}^{n} Z_i \right) = \max_{1\leq i\leq n} \overline{\dim}_{q,\gamma} Z_i$$

and if $\underline{\dim}_{q,\gamma} Z_i = \overline{\dim}_{q,\gamma} Z_i$ then

$$\underline{\dim}_{q,\gamma} \left(\bigcup_{i=1}^{n} Z_i \right) = \max_{1\leq i\leq n} \underline{\dim}_{q,\gamma} Z_i.$$

(5) If Z is a finite set then $\underline{\dim}_{q,\gamma} Z \le \overline{\dim}_{q,\gamma} Z \le 0$; if x is an atom of μ then $\underline{\dim}_{q,\gamma}\{x\} = \overline{\dim}_{q,\gamma}\{x\} = 0$.

(6) $\underline{\dim}_{q,\gamma} Z = \underline{\dim}_{q,\gamma} \overline{Z}$, $\overline{\dim}_{q,\gamma} Z = \overline{\dim}_{q,\gamma} \overline{Z}$, where $\overline{Z}$ is the closure of Z.

We remark that any Lipschitz continuous homeomorphism of $\mathbb{R}^m$ with a Lipschitz continuous inverse that moves the measure μ into an equivalent measure is an isomorphism of the C-structure $\tau_{q,\gamma}$. This fact and Theorems 1.3 and 2.5 imply the following statement.

Theorem 8.3. The (q,γ)-dimension and lower and upper (q,γ)-box dimensions are invariant with respect to any Lipschitz continuous homeomorphism with a Lipschitz continuous inverse that moves the measure μ into an equivalent measure.

We describe the behavior of the functions $\dim_{q,\gamma} Z$, $\underline{\dim}_{q,\gamma} Z$, and $\overline{\dim}_{q,\gamma} Z$ over $\gamma > 0$ for a fixed set $Z \subset \mathbb{R}^m$ and a number $q \ge 1$. First let us notice that these functions are non-decreasing, i.e., for any $0 < \gamma_1 \le \gamma_2$

$$\dim_{q,\gamma_1} Z \le \dim_{q,\gamma_2} Z, \quad \underline{\dim}_{q,\gamma_1} Z \le \underline{\dim}_{q,\gamma_2} Z, \quad \overline{\dim}_{q,\gamma_1} Z \le \overline{\dim}_{q,\gamma_2} Z.$$

We now obtain a formula that allows one to compute the lower and upper (q, γ)-box dimensions in the case $q \geq 1$. Note that the function $x \mapsto \mu(B(x, \varepsilon))^{q-1}$ is measurable (this can easily be seen by decomposing μ into discrete and continuous parts). Since it is bounded, it is integrable. For any measurable set $Z \subset \mathbb{R}^m$, let us set

$$\varphi_q(Z, \varepsilon) = \int_Z \mu(B(x, \varepsilon))^{q-1} d\mu(x). \tag{8.7}$$

Theorem 8.4. *The following statements hold.*

(1) *For any $q \geq 1$, $\gamma > 1$, and any measurable set $Z \subset \mathbb{R}^m$*

$$\underline{\dim}_{q,\gamma} Z \geq \varliminf_{\varepsilon \to 0} \frac{\log \varphi_q(Z, \varepsilon)}{\log(1/\varepsilon)}, \quad \overline{\dim}_{q,\gamma} Z \geq \varlimsup_{\varepsilon \to 0} \frac{\log \varphi_q(Z, \varepsilon)}{\log(1/\varepsilon)}.$$

(2) *For any $q \geq 1$, $\gamma \geq 1$, and any measurable set $Z \subset \mathbb{R}^m$,*

$$\underline{\dim}_{q,\gamma} Z \leq \varliminf_{\varepsilon \to 0} \frac{\log \varphi_q((Z)_\varepsilon, \varepsilon)}{\log(1/\varepsilon)}, \quad \overline{\dim}_{q,\gamma} Z \leq \varlimsup_{\varepsilon \to 0} \frac{\log \varphi_q((Z)_\varepsilon, \varepsilon)}{\log(1/\varepsilon)},$$

where $(Z)_\varepsilon = \bigcup_{x \in Z} B(x, \varepsilon)$ is the ε-neighborhood of Z.

Proof. For any $\delta > 0$ and $\varepsilon > 0$ one can find a cover $\mathcal{G}$ of Z by balls $B(x_i, \varepsilon)$ such that

$$\sum_{B(x_i, \varepsilon) \in \mathcal{G}} \mu(B(x_i, \gamma\varepsilon))^q \leq \Lambda_{q,\gamma}(Z, \varepsilon) + \delta.$$

We have that

$$\sum_{B(x_i, \varepsilon) \in \mathcal{G}} \mu(B(x_i, \gamma\varepsilon))^q = \sum_{B(x_i, \varepsilon) \in \mathcal{G}} \int_{B(x_i, \gamma\varepsilon)} \mu(B(x_i, \gamma\varepsilon))^{q-1} d\mu(x)$$

$$\geq \sum_{B(x_i, \varepsilon) \in \mathcal{G}} \int_{B(x_i, \varepsilon)} \mu(B(x, (\gamma-1)\varepsilon))^{q-1} d\mu(x) \geq \varphi_q(Z, (\gamma-1)\varepsilon).$$

We use here the fact that $B(x_i, \gamma\varepsilon) \supset B(x, (\gamma-1)\varepsilon)$ for any $x \in B(x_i, \varepsilon)$. Since δ can be chosen arbitrarily small it follows that

$$\Lambda_{q,\gamma}(Z, \varepsilon) \geq \varphi_q(Z, (\gamma-1)\varepsilon).$$

This implies the first statement.

We now prove the second statement. Given $\varepsilon > 0$, one can choose a cover $\mathcal{G} = \{B(x_i, \varepsilon)\}$ of Z by balls of the same radius ε which has finite multiplicity independent of ε, q, and γ (see Appendix V). Since the metric in $\mathbb{R}^m$ is equivalent to the standard metric the cover of Z by the balls $\{B(x_i, \gamma\varepsilon)\}$ has finite multiplicity, which is independent of ε, q, and γ. We denote this multiplicity by K. We have

$$\Lambda_{q,\gamma}(Z,\varepsilon) \leq \sum_{B(x_i,\varepsilon)\in\mathcal{G}} \mu(B(x_i,\gamma\varepsilon))^q$$

$$= \sum_{B(x_i,\varepsilon)\in\mathcal{G}} \int_{B(x_i,\gamma\varepsilon)} \mu(B(x_i,\gamma\varepsilon))^{q-1} d\mu(x)$$

$$\leq \sum_{B(x_i,\varepsilon)\in\mathcal{G}} \int_{B(x_i,\gamma\varepsilon)} \mu(B(x,2\gamma\varepsilon))^{q-1} d\mu(x)$$

(we use here the fact that $B(x_i,\gamma\varepsilon) \subset B(x,2\gamma\varepsilon)$ for any point $x \in B(x_i,\gamma\varepsilon)$)

$$\leq K \int_{\bigcup_i B(x_i,\gamma\varepsilon)} \mu(B(x,2\gamma\varepsilon))^{q-1} d\mu(x)$$

$$\leq K \int_{(Z)_{\gamma\varepsilon}} \mu(B(x,2\gamma\varepsilon))^{q-1} d\mu(x) \leq K\varphi_q((Z)_{2\gamma\varepsilon}, 2\gamma\varepsilon).$$

The desired result now follows.   ∎

As an immediate consequence of Theorem 8.4 we obtain that for any set Z of full measure, any $q \geq 1$, and $\gamma > 1$,

$$\underline{\dim}_{q,\gamma} Z = \varliminf_{\varepsilon \to 0} \frac{\log \varphi_q(Z,\varepsilon)}{\log(1/\varepsilon)}, \quad \overline{\dim}_{q,\gamma} Z = \varlimsup_{\varepsilon \to 0} \frac{\log \varphi_q(Z,\varepsilon)}{\log(1/\varepsilon)} \tag{8.8}$$

(since one can replace integral over the ε-neighborhood of Z by the integral over the set Z). This gives rise to the notion of q-**dimension of the set** Z and **lower** and **upper** q-**box dimensions of the set** Z. Namely, for $q \geq 0$ and any set $Z \subset \mathbb{R}^m$, we define

$$\dim_q Z = \inf_{\gamma>1} \dim_{q,\gamma} Z = \lim_{\gamma\downarrow 1} \dim_{q,\gamma} Z,$$

$$\underline{\dim}_q Z = \inf_{\gamma>1} \underline{\dim}_{q,\gamma} Z = \lim_{\gamma\downarrow 1} \underline{\dim}_{q,\gamma} Z, \tag{8.9}$$

$$\overline{\dim}_q Z = \inf_{\gamma>1} \overline{\dim}_{q,\gamma} Z = \lim_{\gamma\downarrow 1} \overline{\dim}_{q,\gamma} Z.$$

These quantities have the properties established in Theorems 8.1, 8.2, and 8.3. Moreover, if one considers these quantities as functions over $q \geq 0$ then

(1) *they are non-increasing:* $\dim_{q_1} Z \geq \dim_{q_2} Z$, $\underline{\dim}_{q_1} Z \geq \underline{\dim}_{q_2} Z$, *and* $\overline{\dim}_{q_1} Z \geq \overline{\dim}_{q_2} Z$ *for any* $0 \leq q_1 \leq q_2$;

(2) $\dim_0 Z = \dim_H Z \geq 0$, *and* $\underline{\dim}_0 Z = \underline{\dim}_B Z \geq 0$, $\overline{\dim}_0 Z = \overline{\dim}_B Z \geq 0$;

(3) *if* $\mu(Z) = 0$ *then* $\dim_1 Z \leq \underline{\dim}_1 Z \leq \overline{\dim}_1 Z \leq 0$ *and hence,* $\dim_q Z \leq \underline{\dim}_q Z \leq \overline{\dim}_q Z \leq 0$ *for any* $q \geq 1$;

(4) *if* $\mu(Z) > 0$ *then* $\dim_1 Z = \underline{\dim}_1 Z = \overline{\dim}_1 Z = 0$ *and hence,*

$$0 \leq \dim_q Z \leq \underline{\dim}_q Z \leq \overline{\dim}_q Z \quad \text{if } 0 \leq q \leq 1$$

and

$$\dim_q Z \leq \underline{\dim}_q Z \leq \overline{\dim}_q Z \leq 0 \quad \text{if } q \geq 1.$$

Properties (1) and (2) are obvious. By (8.6) we obtain for $q = 1$ and every $\gamma > 1$ that

$$\mu(Z) \leq \Lambda_{1,\gamma}(Z,\varepsilon) \leq \mu((Z)_{\gamma\varepsilon}) \leq 1.$$

This implies Properties (3) and (4).

One can obtain formulae for computing the lower and upper q-box dimensions of sets using equalities (8.5): for $q \geq 0$ and $Z \subset \mathbb{R}^m$,

$$\underline{\dim}_q Z = \inf_{\gamma > 1} \varliminf_{\varepsilon \to 0} \log \frac{\Lambda_{q,\gamma}(Z,\varepsilon)}{\log(1/\varepsilon)},$$

$$\overline{\dim}_q Z = \inf_{\gamma > 1} \varlimsup_{\varepsilon \to 0} \log \frac{\Lambda_{q,\gamma}(Z,\varepsilon)}{\log(1/\varepsilon)},$$

where $\Lambda_{q,\gamma}(Z,\varepsilon)$ is given by (8.6).

Given a set Z and $\alpha \in \mathbb{R}$, we set $m_q(Z,\alpha) = \inf_{\gamma > 1} m_{q,\gamma}(Z,\alpha)$, where $m_{q,\gamma}(Z,\alpha)$ is defined by (8.2). The set function $m_q(\cdot,\alpha)$ (α is fixed) has the properties described by Proposition 1.1 and the function $m_q(Z,\cdot)$ (Z is fixed) has the properties described by Proposition 1.2 with the critical value $\alpha_C = \dim_q Z$. The values $\underline{\dim}_q Z$, and $\overline{\dim}_q Z$ can be obtained in a similar fashion.

Another description of the lower and upper q-box dimensions of sets for $q \geq 1$ is based on Theorem 8.4 (see also (8.8)). Namely, for any set Z of full measure

$$\underline{\dim}_q Z = \varliminf_{\varepsilon \to 0} \frac{\log \int_{\mathbb{R}^m} \mu(B(x,\varepsilon))^{q-1} d\mu(x)}{\log(1/\varepsilon)},$$

$$\overline{\dim}_q Z = \varlimsup_{\varepsilon \to 0} \frac{\log \int_{\mathbb{R}^m} \mu(B(x,\varepsilon))^{q-1} d\mu(x)}{\log(1/\varepsilon)}. \tag{8.10}$$

One can easily see that $\underline{\dim}_q Z \leq \overline{\dim}_q Z$ for every compact set Z and every $q \geq 1$. We construct an example of a measure μ for which the strict inequality occurs on an arbitrary interval in q.

Example 8.1. *For any $Q > 0$ there exists a finite Borel measure μ on the interval $I = [0,p]$, for some $p > 0$, such that*

(1) *μ is equivalent to the Lebesgue measure;*
(2) *$\underline{\dim}_q I < \overline{\dim}_q I$ for any $1 < q \leq Q$.*

Proof. We first choose any three numbers α, β, γ such that $0 < \alpha < \beta < \gamma < 1$. Let n_k be an increasing sequence of integers. Define $a_n = \alpha^n$ and

$$b_n = \begin{cases} \gamma^{-n} & \text{if } n_{4k} \leq n < n_{4k+1} \\ Mb_{n-1} & \text{if } n_{4k+1} \leq n < n_{4k+2} \\ \beta^{-n} & \text{if } n_{4k+2} \leq n < n_{4k+3} \\ b_{n_{4k+3}-1} & \text{if } n_{4k+3} \leq n < n_{4k+4} \end{cases}$$

where $M > 0$ is a number satisfying $M > \beta^{-1}$. One can choose a sequence n_k such that for any $Q > q > 1$,

$$\varlimsup_{n \to \infty} \frac{\log A_n}{n} = \log(\alpha\beta^{-q}), \quad \varliminf_{n \to \infty} \frac{\log A_n}{n} = \log(\alpha\gamma^{-q}), \tag{8.11}$$

where $A_n = \sum_{i=1}^{n} a_i b_i^q$. Since $\alpha \gamma^{-1} < \alpha \beta^{-1} < 1$ we have that $p \stackrel{\text{def}}{=} \sum_{n=1}^{\infty} a_n b_n < \infty$.

Let μ be the measure on the interval I that is absolutely continuous with respect to the Lebesgue measure with the density function $f(x)$ given by

$$f(x) = b_n \text{ if } r_n \leq x < r_{n-1},$$

where $r_n = \sum_{i=1}^{n} a_i = \frac{\alpha^n}{1-\alpha}$. We first compute the value $\varphi_q(I, r_n)$. Consider an interval $I_i = [r_i, r_{i-1}]$ of length $|I_i| = a_i$, where $1 \leq i \leq n$. We decompose it into three subintervals

$$I_i^{(1)} = [r_i, r_i + r_n], \ I_i^{(2)} = [r_i + r_n, \ r_{i-1} - r_n], \ I_i^{(3)} = [r_{i-1} - r_n, \ r_{i-1}].$$

It is easy to see that

$$\mu(B(x, r_n)) = \begin{cases} r_n b_i + (x - r_i) b_i + (r_i - x + r_n) b_{i+1} & \text{if } x \in I_i^{(1)} \\ 2 r_n b_i & \text{if } x \in I_i^{(2)} \\ r_n b_i + (r_{i-1} - x) b_i + (x + r_n - r_{i-1}) b_{i-1} & \text{if } x \in I_i^{(3)}. \end{cases}$$

Integrating $\mu(B(x, r_n))^{q-1}$ over $x \in I_i$ with respect to μ for $q > 1$ we have

$$\int_{I_i} \mu(B(x, r_n))^{q-1} d\mu(x) = 2^{q-1} r_n^{q-1} b_i^q (a_i - 2 r_n) + r_n^q (B_i - B_{i-1}),$$

where

$$B_i = b_i (2^q b_i^q - (b_i + b_{i+1})^q) q^{-1} (b_i - b_{i+1})^{-1}$$

if $b_i \neq b_{i+1}$ and

$$B_i = 2^q b_i^q$$

if $b_i = b_{i+1}$. Taking the sum over $i = 1, \ldots, n$ we obtain

$$\int_{r_n}^{p} \mu(B(x, r_n))^{q-1} d\mu(x) = 2^{q-1} r_n^{q-1} A_n - 2^q r_n^q \sum_{i=1}^{n} b_i^q + r_n^q (B_n - B_1). \quad (8.12)$$

One can easily verify that

$$\left(\sum_{k=n}^{\infty} a_k b_k \right)^q \leq \int_0^{r_n} \mu(B(x, r_n))^{q-1} d\mu(x)$$

$$\leq \left(\sum_{k=n-1}^{\infty} a_k b_k \right)^{q-1} \left(\sum_{k=n}^{\infty} a_k b_k \right). \quad (8.13)$$

We also have that for all sufficiently large n

$$r_n \sum_{i=1}^{n} b_i^q \leq \frac{1}{2} A_n, \quad r_n B_n \leq \frac{1}{2} A_n. \quad (8.14)$$

It follows from (8.12), (8.13), and (8.14) that

$$C_2 r_n^{q-1} A_n \leq \int_0^p \mu(B(x,r_n))^{q-1} d\mu(x) \leq C_1 r_n^{q-1} A_n,$$

where $C_1 > 0$ and $C_2 > 0$ are constants. This implies (see also (8.11)) that

$$\varliminf_{n \to \infty} \frac{\log \varphi_q(I, r_n)}{\log(1/r_n)} = \left(1 - q + \varliminf_{n \to \infty} \frac{\log A_n}{n \log \alpha^{-1}}\right) = q \left(\frac{\log \gamma}{\log \alpha} - 1\right),$$

$$\varlimsup_{n \to \infty} \frac{\log \varphi_q(I, r_n)}{\log(1/r_n)} = \left(1 - q + \varlimsup_{n \to \infty} \frac{\log A_n}{n \log \alpha^{-1}}\right) = q \left(\frac{\log \beta}{\log \alpha} - 1\right).$$

Let ρ_k be any decreasing sequence of numbers, $\rho_k \to 0$. Given $k > 0$ one can find $n = n(k)$ such that $r_{n(k)} \leq \rho_k \leq r_{n(k)-1}$. Since the function $\varphi(I, r)$ is non-decreasing and $r_{n(k)-1} \leq \text{const} \times r_{n(k)}$ it follows that

$$\varliminf_{k \to \infty} \frac{\log \varphi_q(I, \rho_k)}{\log(1/\rho_k)} = \varliminf_{n \to \infty} \frac{\log \varphi_q(I, r_n)}{\log(1/r_n)},$$

$$\varlimsup_{k \to \infty} \frac{\log \varphi_q(I, \rho_k)}{\log(1/\rho_k)} = \varlimsup_{n \to \infty} \frac{\log \varphi_q(I, r_n)}{\log(1/r_n)}.$$

This implies the desired result. ∎

We discuss now the problem of coincidence of the lower and upper $(q, 1)$-box dimensions with the lower and upper q-box dimensions respectively. As we know $\underline{\dim}_{q,1} Z \leq \underline{\dim}_q Z$ and $\overline{\dim}_{q,1} Z \leq \overline{\dim}_q Z$ for any $q \geq 0$ and any set $Z \subset \mathbb{R}^m$. Whether they coincide or not depends on rather delicate relations between metrics and measures on $\mathbb{R}^m$.

Following [Fe] we call a Borel finite measure μ **diametrically regular** if it satisfies the following condition:

   there exist constants $\gamma_0 > 1$ and $C_0 > 0$ such that for any point x and any $r > 0$

$$\mu(B(x, \gamma_0 r)) \leq C_0 \mu(B(x, r)). \tag{8.15}$$

Such a measure is sometimes called a **Federer measure**; in harmonic analysis it is also known as a **doubling measure**. Given $\gamma > 1$, choose the least positive integer $n = n(\gamma)$ such that $\gamma \gamma_0^{-n} < 1$. It follows from (8.15) that

$$\mu(B(x, \gamma r)) \leq C_0^n \mu(B(x, r))$$

for any point x and $r > 0$. This immediately implies the following statement.

Theorem 8.5. *If μ is a diametrically regular measure on $\mathbb{R}^m$, then*

$$\underline{\dim}_{q,\gamma} Z = \underline{\dim}_q Z = \underline{\dim}_{q,1} Z \quad and \quad \overline{\dim}_{q,\gamma} Z = \overline{\dim}_q Z = \overline{\dim}_{q,1} Z \tag{8.16}$$

for any $Z \subset \mathbb{R}^m$, $q \geq 0$, and $\gamma > 1$.

Although the assumption (8.15) is sufficiently strong one can show that it holds in many interesting cases: for example, for equilibrium measures (corresponding to Hölder continuous functions) for conformal expanding maps, two-dimensional (or more general, conformal) Axiom A diffeomorphisms, and symbolic dynamical systems (see Propositions 19.1, 21.4, and 24.1). If Assumption (8.15) is violated the $(q, 1)$-box dimension of a set Z may be strictly less than the q-box dimension of Z as the following example shows.

Example 8.2. *Given a number $q_0 > 1$ there exists a finite Borel measure μ on $[0, 1]$ and a set $Z \subset [0, 1]$ such that*

(1) *μ is absolutely continuous with respect to the Lebesgue measure on $[0, 1]$ but is not diametrically regular;*

(2) *for any $q > q_0$*

$$\underline{\dim}_{q,1} Z = \overline{\dim}_{q,1} Z < \underline{\dim}_q Z = \overline{\dim}_q Z.$$

In other words, the measure μ satisfies

$$\lim_{\varepsilon \to 0} \frac{\log \inf_{\mathcal{G}} \left\{ \sum_{B(x_i, \varepsilon) \in \mathcal{G}} \mu(B(x_i, \varepsilon))^q \right\}}{\log(1/\varepsilon)} < \inf_{\gamma > 1} \lim_{\varepsilon \to 0} \frac{\log \inf_{\mathcal{G}} \left\{ \sum_{B(x_i, \varepsilon) \in \mathcal{G}} \mu(B(x_i, \gamma\varepsilon))^q \right\}}{\log(1/\varepsilon)},$$

where the infimum is taken over all finite or countable covers $\mathcal{G}$ of Z by balls of radius ε.

Proof. Let us fix $q_0 > 1$. Given $0 < \alpha < 1$, let μ be the measure on $[0, 1]$ which is absolutely continuous with respect to the Lebesgue measure with the density function $h(x)$ given by

$$h(x) = \left(\frac{1}{2} - x \right)^{-\alpha} \text{ if } 0 \le x < \frac{1}{2} \text{ and } h(x) = 1 \text{ if } \frac{1}{2} \le x \le 1.$$

Set $Z = [\frac{1}{2}, 1]$. For any $q \ge 0$, we have

$$\underline{\dim}_{q,1} Z = \overline{\dim}_{q,1} Z = 1 - q.$$

On the other hand for any $\gamma > 1$,

$$\underline{\dim}_{q,\gamma} Z = \overline{\dim}_{q,\gamma} Z = \max\{-(1 - \alpha)q, \, 1 - q\}.$$

Thus, the desired result follows if we choose $\alpha = \frac{1}{q_0}$. ■

Note that in the above example the set Z is of positive but *not* full measure. Guysinsky and Yaskolko [GY] constructed another example of a measure in $\mathbb{R}^m$ which is *not* diametrically regular and for which $(q, 1)$-box dimension of Z is strictly less than the q-box dimension of Z for a set Z of *full* measure.

Example 8.3. *There exists a finite Borel measure μ on $\mathbb{R}^2$ with the support inside of the unit square $S = [0, 1] \times [0, 1]$ such that for every $q \ge 1$*

$$\underline{\dim}_{q,1} S = \overline{\dim}_{q,1} S < \underline{\dim}_q S = \overline{\dim}_q S.$$

In other words, the measure μ satisfies

$$\lim_{\varepsilon \to 0} \frac{\log \inf_{\mathcal{G}} \left\{ \sum_{B(x_i, \varepsilon) \in \mathcal{G}} \mu(B(x_i, \varepsilon))^q \right\}}{\log(1/\varepsilon)} < \lim_{\varepsilon \to 0} \frac{\log \int_S \mu(B(x, \varepsilon))^{q-1} d\mu(x)}{\log(1/\varepsilon)},$$

where the infimum is taken over all finite or countable covers $\mathcal{G}$ of S by balls of radius ε.

Proof. Let I be a horizontally spaced interval which lies strictly inside S and μ a measure on S which coincides with the standard Lebesgue measure on I. Observe that

$$\underline{\dim}_q S = \overline{\dim}_q S = \lim_{r \to 0} \frac{\log \int_S (2r)^{q-1} \, d\mu}{\log(1/r)} = 1 - q.$$

In order to compute the $(q,1)$-box dimension of I let us fix $\varepsilon > 0$, $\delta > 0$, and $n > 0$ and let I_n be the interval which is placed horizontally strictly above (or below) I on the distance $a_n = \sqrt{r^2 - \frac{1}{4n^2}}$. Consider the cover of I by balls $B(x_i, \varepsilon)$ whose centers lie on I_n spaced equally on the distance $\frac{1}{n} - \delta$ from each other starting from the left endpoint. A straightforward calculation shows that

$$\sum_{i=1}^{\infty} \mu(B(x_i, \varepsilon))^q \le \frac{n}{1 - \delta n} \frac{1}{n^q}. \tag{8.17}$$

By adding other balls of radius ε, which do not intersect I, we can obtain a cover $\mathcal{G}$ of S. It follows from (8.17) that for every $n > 0$,

$$0 \le \Delta_{q,1}(S, \varepsilon) \le \frac{n}{1 - \delta n} \frac{1}{n^q}$$

Therefore, $\Delta_{q,1}(S, \varepsilon) = 0$ and hence $\underline{\dim}_{q,1} S = \overline{\dim}_{q,1} S = -\infty$. ∎

9. q-Dimension and q-Box Dimension of Measures

Let μ be a Borel finite measure on $\mathbb{R}^m$. For any $q \ge 0$, $\gamma > 0$, and any Borel finite measure ν on $\mathbb{R}^m$ the C-structure $\tau_{q,\gamma}$ defined by (8.1) (and specified by the measure μ) yields, in accordance with (3.1) and (3.2), the Carathéodory dimension of ν and lower and upper Carathéodory capacities of ν. We call them the (q,γ)-**dimension of the measure** ν and the **lower** and **upper** (q,γ)-**box dimensions of the measure** ν and denote them by $\dim_{q,\gamma} \nu$, $\underline{\dim}_{q,\gamma} \nu$, and $\overline{\dim}_{q,\gamma} \nu$ respectively. Thus, we have

$$\begin{aligned}
\dim_{q,\gamma} \nu &= \inf\{\dim_{q,\gamma} Z : \nu(Z) = 1\}, \\
\underline{\dim}_{q,\gamma} \nu &= \lim_{\delta \to 0} \inf\{\underline{\dim}_{q,\gamma} Z : \nu(Z) \ge 1 - \delta\}, \\
\overline{\dim}_{q,\gamma} \nu &= \lim_{\delta \to 0} \inf\{\overline{\dim}_{q,\gamma} Z : \nu(Z) \ge 1 - \delta\}.
\end{aligned} \tag{9.1}$$

It follows from the definitions that

$$\dim_{q,\gamma} \nu \le \underline{\dim}_{q,\gamma} \nu \le \overline{\dim}_{q,\gamma} \nu. \tag{9.2}$$

In the case $\nu = \mu$, we present, based on Theorem 4.2, a powerful criterion that guarantees the coincidence of (q,γ)-dimension and lower and upper (q,γ)-box dimensions of ν. Consider the collection of balls $\mathcal{F}' = \{B(x,r) : x \in \mathbb{R}^m, r > 0\}$. In accordance with Section 3 (see (3.5)) we define the **lower** and **upper** (q,γ)-**pointwise dimension** of ν at x by

$$\underline{d}_{\nu,\alpha,q,\gamma}(x) = \varliminf_{r \to 0} \frac{\alpha \log \nu(B(x,r))}{q \log \mu(B(x,\gamma r)) + \alpha \log r},$$

$$\overline{d}_{\nu,\alpha,q,\gamma}(x) = \varlimsup_{r \to 0} \frac{\alpha \log \nu(B(x,r))}{q \log \mu(B(x,\gamma r)) + \alpha \log r}.$$

Theorem 9.1. *Assume that the measure μ is exact dimensional, i.e.,*

$$\underline{d}_\mu(x) = \overline{d}_\mu(x) = d > 0 \tag{9.3}$$

for μ-almost every x. Choose numbers $q \geq 0$, $q \neq 1$, $\gamma > 0$, and $\varepsilon > 0$ such that $d - \varepsilon > 0$ and the interval $I = [d(1-q) - \varepsilon, \, d(1-q) + \varepsilon]$ does not contain 0. Then for μ-almost every x and any $\alpha \in I$,

(1) $\underline{d}_{\mu,\alpha,q,\gamma}(x) = \overline{d}_{\mu,\alpha,q,\gamma}(x) = d(\alpha) = \alpha d(dq + \alpha)^{-1}$ *(note that $dq + \alpha \geq d - \varepsilon > 0$ although α and hence $d(\alpha)$ may be negative);*

(2) $\dim_{q,\gamma} \mu = \underline{\dim}_{q,\gamma} \mu = \overline{\dim}_{q,\gamma} \mu = d(1-q)$.

Proof. The first statement is a straightforward calculation. For any set Z of positive measure and any $\varepsilon > 0$, there exists a finite multiplicity cover of Z by balls of radius ε (see Appendix V). This implies (3.7). The validity of the second conditions of Theorems 3.1 and 3.2 is obvious. The second statement now follows from the first one and Theorem 4.2. ∎

The functions $\dim_{q,\gamma}\nu$, $\underline{\dim}_{q,\gamma}\nu$, and $\overline{\dim}_{q,\gamma}\nu$ are non-decreasing in γ, i.e., for any $0 < \gamma_1 \leq \gamma_2$,

$$\dim_{q,\gamma_1}\nu \leq \dim_{q,\gamma_2}\nu, \quad \underline{\dim}_{q,\gamma_1}\nu \leq \underline{\dim}_{q,\gamma_2}\nu, \quad \overline{\dim}_{q,\gamma_1}\nu \leq \overline{\dim}_{q,\gamma_2}\nu.$$

We define now the q-**dimension of the measure** ν by setting

$$\dim_q\nu = \inf_{\gamma > 1}\dim_{q,\gamma}\nu = \lim_{\gamma \downarrow 1}\dim_{q,\gamma}\nu$$

and the **lower** and **upper** q-**box dimension of the measure** ν by setting

$$\underline{\dim}_q\nu = \inf_{\gamma > 1}\underline{\dim}_{q,\gamma}\nu = \lim_{\gamma \downarrow 1}\underline{\dim}_{q,\gamma}\nu,$$

$$\overline{\dim}_q\nu = \inf_{\gamma > 1}\overline{\dim}_{q,\gamma}\nu = \lim_{\gamma \downarrow 1}\overline{\dim}_{q,\gamma}\nu.$$

In general, $\dim_q\nu \leq \underline{\dim}_q\nu \leq \overline{\dim}_q\nu$. It follows from Theorem 9.1 that: *if the measure μ satisfies (9.3) for almost every x then for any $q \geq 0$,*

$$\dim_q\mu = \underline{\dim}_q\mu = \overline{\dim}_q\mu = d(1-q).$$

In the general case (when (9.3) does not hold) we establish a refined version of Theorem 9.1 which demonstrates a close connection between q-box dimension of μ and the pointwise dimension of μ.

Theorem 9.2. *Let μ be a Borel finite measure on $\mathbb{R}^m$. Then for any $q \geq 1$,*

$$(1 - q) \operatorname*{ess\,inf}_{x \in \mathbb{R}^m} \overline{d}_\mu(x) \leq \underline{\dim}_q \mu \leq \overline{\dim}_q \mu \leq (1 - q) \operatorname*{ess\,inf}_{x \in \mathbb{R}^m} \underline{d}_\mu(x).$$

Proof. Given $\alpha > 0$ and $\beta > 0$ we define the set

$$X_{\alpha,\beta} = \left\{ x \in \mathbb{R}^m : \underline{d}_\mu(x) - \alpha \leq \frac{\log \mu(B(x,r))}{\log r} \leq \overline{d}_\mu(x) + \alpha \text{ if } 0 < r \leq \beta \right\}.$$

It is easy to see that for any $x \in X_{\alpha,\beta}$ and $0 < r \leq \beta$,

$$r^{\overline{d}_\mu(x)+\alpha} \leq \mu(B(x,r)) \leq r^{\underline{d}_\mu(x)-\alpha}.$$

For any $\delta > 0$ there exist $\alpha = \alpha(\delta)$ and $\beta = \beta(\delta)$ such that $\mu(X_{\alpha,\beta}) \geq 1 - \delta$. We denote

$$\underline{d} = \operatorname*{ess\,inf}_{x \in \mathbb{R}^m} \underline{d}_\mu(x), \quad \overline{d} = \operatorname*{ess\,inf}_{x \in \mathbb{R}^m} \overline{d}_\mu(x).$$

For μ-almost every x we have $\underline{d}_\mu(x) \geq \underline{d} - \alpha$. Furthermore, there exists a set X_0 of positive measure such that $\overline{d}_\mu(x) \leq \overline{d} + \alpha$ for any $x \in X_0$. Let $Z \subset X_{\alpha,\beta}$ be a set of positive measure and $\mathcal{G}$ a cover of Z by balls $B(x_i, r)$ with $r \leq \beta$ of finite multiplicity K. For any $\gamma > 1$ we have that

$$\sum_{B(x_i,\varepsilon)\in\mathcal{G}} \mu(B(x_i, \gamma r))^q = \sum_{B(x_i,r)\in\mathcal{G}} \mu(B(x_i, \gamma r))^{q-1} \mu(B(x_i, \gamma r))$$

$$\leq r^{(\underline{d}-2\alpha)(q-1)} \sum_{B(x_i,r)\in\mathcal{G}} \mu(B(x_i, \gamma r))$$

$$\leq K\, r^{(\underline{d}-2\alpha)(q-1)} \mu((Z)_{\gamma r}) \leq K_1\, r^{(\underline{d}-2\alpha)(q-1)},$$

where $K_1 > 0$ is a constant (we recall that $(Z)_\varepsilon$ is the ε-neighborhood of the set Z). In view of (8.5) this implies that $\overline{\dim}_q Z \leq (\underline{d} - 2\alpha)(1 - q)$ and hence $\overline{\dim}_q \mu \leq (\underline{d} - 2\alpha)(1 - q)$. Since α can be chosen arbitrarily small it follows that $\overline{\dim}_q \mu \leq \underline{d}(1 - q)$. We can choose a set Z with $\mu(Z) \geq 1 - \delta$ for which $\underline{\dim}_q \mu \geq \underline{\dim}_q Z - \alpha$. If δ is sufficiently small the set

$$Y = Z \cap X_{\alpha,\beta} \cap Z_0$$

has positive measure. Let $\mathcal{G}$ be a cover of Y by balls $B(x_i, r)$ with $r \leq \beta$. For any $\gamma > 1$ we have

$$\sum_{B(x_i,r)\in\mathcal{G}} \mu(B(x_i, \gamma r))^q = \sum_{B(x_i,r)\in\mathcal{G}} \mu(B(x_i, \gamma r))^{q-1} \mu(B(x_i, \gamma r))$$

$$\geq r^{(\overline{d}+2\alpha)(q-1)} \mu(Y).$$

This implies that

$$\underline{\dim}_q \mu \geq \underline{\dim}_q Z - \alpha \geq \underline{\dim}_q Y - \alpha \geq (\overline{d} + 2\alpha)(1 - q) - \alpha.$$

Since α can be taken arbitrarily small the last inequalities yield $\underline{\dim}_q \mu \geq \overline{d}(1-q)$ and the desired result follows. ∎

As an immediate consequence of Theorem 9.2 we have: *if the measure μ satisfies*

$$\underline{d}_\mu(x) = \overline{d}_\mu(x) = d_\mu(x)$$

μ-almost everywhere then

$$\underline{\dim}_q\mu = \overline{\dim}_q\mu = (1 - q)\operatorname*{ess\,inf}_{x\in\mathbb{R}^m} d_\mu(x). \tag{9.4}$$

Appendix I

Hausdorff (Box) Dimension and q-(Box) Dimension of Sets and Measures in General Metric Spaces

Hausdorff Dimension and Box Dimension of Sets and Measures

Let X be a complete separable metric space endowed with a metric ρ. Consider the collection of all *open* subsets of X and set functions ξ, η, and ψ defined by (6.1). They satisfy Conditions $A1$, $A2$, $A3$, and $A3'$ and hence determine a C-structure $\tau = (\mathcal{F}, \xi, \eta, \psi)$ on X. For any $\alpha \geq 0$, the corresponding Carathéodory set function $m_C(\cdot, \alpha)$ (see (1.1) and (1.2)) is an outer measure on X and is given by (6.2). It is known as the α-**Hausdorff outer measure** on X. This outer measure induces a σ-additive measure on X called the α-**Hausdorff measure**. The latter can be shown to be a Borel regular measure (see Appendix V).

Further, for a subset $Z \subset \mathbb{R}^m$, the C-structure τ generates the Carathéodory dimension of Z called the **Hausdorff dimension of the set** Z as well as the lower and upper Carathéodory capacities of Z called the **lower** and **upper box dimensions of the set** Z. We denote them by $\dim_H Z$, $\underline{\dim}_B Z$, and $\overline{\dim}_B Z$ respectively. They obey (6.3) and (6.4). Obviously, the set functions η and ψ satisfy Condition $A4$ and thus, the lower and upper box dimensions of Z satisfy (6.5). The main properties of the Hausdorff dimension and lower and upper box dimensions of sets are described in Theorems 6.1, 6.2, and 6.3.

Let μ be a Borel finite measure on X. The C-structure τ produces, in accordance with (3.1) and (3.2), the Carathéodory dimension of μ and lower and upper Carathéodory capacities of μ. They are called respectively the **Hausdorff dimension of the measure** μ and **lower** and **upper box dimensions of the measure** μ and are denoted by $\dim_H \mu$, $\underline{\dim}_B \mu$, and $\overline{\dim}_B \mu$. They satisfy (7.1).

Consider the collection of balls $\mathcal{F}' = \{B(x, r) : x \in X, r > 0\}$ and define in accordance with (3.5) the **lower** and **upper pointwise dimensions** of μ at x by (7.2).

We say that X is **a metric space of finite multiplicity** if the following condition holds:

H1. there exist $K > 0$ and $\varepsilon_0 > 0$ such that for any ε, $0 < \varepsilon \leq \varepsilon_0$ one can find a cover of X by balls of radius ε of multiplicity K.

The analysis of the proof of Theorem 7.1 shows that it holds for any complete separable metric space X of finite multiplicity and any Borel finite measure μ

61

on X. In particular, *if $\underline{d}_\mu(x) = \overline{d}_\mu(x) = d$ for μ-almost every x then $\dim_H \mu =$* $\underline{\dim}_B \mu = \overline{\dim}_B \mu = d$. Moreover, the uniform and non-uniform mass distribution principles can be used to obtain lower bounds for the Hausdorff dimension of sets.

Further, we say that a complete separable metric space X is a **Besicovitch metric space** if the following condition holds:

H2. there exist $K > 0$ and $\varepsilon_0 > 0$ such that for any subset $Z \subset X$ and any cover $\{B(x, \varepsilon(x)) : x \in Z, 0 < \varepsilon(x) \le \varepsilon_0\}$ one can find a subcover of Z of multiplicity K (in other words, the Besicovitch Covering Lemma holds true with respect to the metric on X; see Appendix V).

One can show that Theorem 7.2 holds for any Besicovitch metric space and any Borel finite measure on it.

q-Dimension and q-Box Dimension of Sets and Measures

Let μ be a Borel finite measure on a complete separable metric space X. Given numbers $q \ge 0$ and $\gamma > 0$, consider the collection $\mathcal{F}$ of open balls in X and define set functions ξ, η, and ψ by (8.1). They satisfy conditions $A1, A2, A3$, and $A3'$ and hence define the C-structure $\tau_{q,\gamma}$ in X. The corresponding Carathéodory set function $m_C(Z, \alpha)$ (where $Z \subset X$ and $\alpha \in \mathbb{R}$) is called the (q, γ)-set function. It is denoted by $m_{q,\gamma}(Z, \alpha)$ and is given by (8.2).

Further, the C-structure $\tau_{q,\gamma}$ produces the Carathéodory dimension of Z as well as the lower and upper Carathéodory capacities of Z. We call them the (q, γ)-**dimension of the set** Z and **lower** and **upper** (q, γ)-**box dimensions of the set** Z respectively and denote them by $\dim_{q,\gamma} Z$, $\underline{\dim}_{q,\gamma} Z$, and $\overline{\dim}_{q,\gamma} Z$. They obey (8.3) and (8.4). Moreover, since the set functions η and ψ satisfy Condition $A4$ the lower and upper (q, γ)-box dimensions of sets can be computed using (8.5) and (8.6).

The main properties of the (q, γ)-dimension and lower and upper (q, γ)-box dimensions of sets are stated in Theorems 8.1, 8.2, and 8.3.

Finally, for any $q \ge 0$ and any set $Z \subset X$, we define the q-**dimension of the set** Z and **lower** and **upper** q-**box dimensions of the set** Z by formulae (8.9).

We say that a complete separable metric space X is **isotropic** if the following condition holds:

H3. for every $A > 0$ there exists $B > 0$ such that for any set $Z \subset X$ and any cover $\mathcal{G}$ of Z of finite multiplicity K by balls of radius ε, the cover of Z by concentric balls of radius $A\varepsilon$ is of finite multiplicity BK.

Repeating arguments in the proof of Theorem 8.4 one can show that *if X is a complete separable isotropic metric space of finite multiplicity* (see Conditions (H1) and (H3)) *then for any set Z of full measure the q-dimension of Z and lower and upper q-box dimensions of Z can be computed by formulae (8.10).*

A Borel finite measure μ on X is called **diametrically regular** if it satisfies Condition (8.15). It is easy to see that *if μ is a diametrically regular measure on X, then (8.16) holds.* In [GY], Guysinsky and Yaskolko proved that if a complete separable metric space X admits a Borel finite diametrically regular measure ν

then X is isotropic and of finite multiplicity. In this case for *any* Borel finite measure μ and any set Z of full measure equalities (8.8) and (8.10) hold for any $\gamma > 1$ and $q \geq 1$.

If ν is a Borel finite measure on X then the C-structure $\tau_{q,\gamma}$ defined above (and specified by the measure μ) yields, in accordance with (9.1), the (q,γ)-**dimension of the measure** ν and the **lower** and **upper** (q,γ)-**box dimensions of the measure** ν. We denote them by $\dim_{q,\gamma}\nu$, $\underline{\dim}_{q,\gamma}\nu$, and $\overline{\dim}_{q,\gamma}\nu$ respectively. One can show that *if X is a complete separable metric space of finite multiplicity and μ is a Borel finite measure on X satisfying Condition (9.3) then for every $q \geq 0$, $\dim_{q,\gamma}\mu = \underline{\dim}_{q,\gamma}\mu = \overline{\dim}_{q,\gamma}\mu = d(1 - q)$. Moreover, for every $q \geq 1$, the conclusion of Theorem 9.2 holds.*

Chapter 4

C-Structures Associated with Dynamical Systems: Thermodynamic Formalism

Let f be a continuous map acting on a compact metric space (X, ρ) with metric ρ and φ a continuous function on X. In this chapter we discuss the notion of topological pressure of φ on a subset $Z \subset X$ (specified by f). This notion was brought to the theory of dynamical systems by Ruelle [R1], who was inspired by the theory of Gibbs states in statistical mechanics. Ruelle considered only the case when the set Z is compact and f-invariant (he also assumed that f is a homeomorphism which separates points; the case of general continuous maps was later studied by Walters [W]). The fact that the topological pressure is a characteristic of dimension type was first noticed (implicitly) by Bowen [Bo1]. Pesin and Pitskel' [PP] further developed his approach and extended the notion of topological pressure to arbitrary subsets of X which are not necessarily invariant or compact. In this chapter we systematically use the "dimensional" approach to the notion of topological pressure which is based on a modification of the general Carathéodory construction (we describe the modified version in Section 10).

The topological pressure is a key notion in the thermodynamic formalism (see Appendix II) which is the main tool in studying dimension of invariant sets and measures for dynamical systems and dimension of Cantor-like sets in dimension theory. The "dimension" approach that is faithful to the general Carathéodory construction gives us a new insight on the thermodynamic formalism and allows us to extend the classical notion of topological pressure to non-compact or non-invariant sets. This is an important advantage which we will use in studying dimension. Furthermore, we will use the "dimension" approach to obtain a more general non-additive version of the topological pressure. It was introduced by Barreira in [Bar2]. The associated non-additive thermodynamic formalism is a powerful tool to study dimension of Cantor-like sets with extremely complicated geometric structure where other methods of study failed to work.

Let us outline the "dimension" approach. Given a finite open cover $\mathcal{U}$ of X, we introduce a C-structure τ on X which is specified by the map f, metric ρ, continuous function φ, and cover $\mathcal{U}$. According to Section 10, for any subset $Z \subset X$, this C-structure generates the Carathéodory dimension and lower and upper Carathéodory capacities of Z. We denote them respectively by $P_Z(\varphi, \mathcal{U})$, $\underline{CP}_Z(\varphi, \mathcal{U})$, and $\overline{CP}_Z(\varphi, \mathcal{U})$. We then show that these quantities have limits as $\operatorname{diam}\mathcal{U}$ tends to zero which we call the topological pressure and lower and upper capacity topological pressures of φ on Z and denote them by $P_Z(\varphi)$, $\underline{CP}_Z(\varphi)$, and $\overline{CP}_Z(\varphi)$ respectively. We show that, if Z is f-invariant, then $\underline{CP}_Z(\varphi) = \overline{CP}_Z(\varphi)$ and if, in addition, Z is compact, then

$P_Z(\varphi) = \underline{CP}_Z(\varphi) = \overline{CP}_Z(\varphi)$. In the latter case, the common value coincides with the "classical" topological pressure introduced by Bowen, Ruelle, and Walters.

We stress that for an arbitrary subset Z, one has three, in general distinct, quantities $P_Z(\varphi)$, $\underline{CP}_Z(\varphi)$, and $\overline{CP}_Z(\varphi)$ to be used as a generalization of the classical notion of topological pressure. In view of the variational principle (see Appendix II below) a crucial role is played by the quantity $P_Z(\varphi)$ while lower and upper capacity topological pressures are also often used in computing dimension of invariant sets for dynamical systems (see Chapters 5 and 7).

There is an important particular case when $\varphi = 0$. We call the quantities $P_Z(0)$, $\underline{CP}_Z(0)$, and $\overline{CP}_Z(0)$ the topological entropy and lower and upper capacity topological entropies and we use the generally accepted notations $h_f(Z)$, $\underline{Ch}_f(Z)$, and $\overline{Ch}_f(Z)$. The topological entropy is a well-known invariant of dynamical systems and plays a key role in topological dynamics. For example, for subshifts of finite type (and hence for Axiom A diffeomorphisms) the topological entropy is the exponential growth rate of the number of periodic points.

The first definition of the topological entropy for compact invariant sets was given by Adler, Konheim, and McAndrew [AKM]. In [Bo1], Bowen extended it to non-compact invariant sets and pointed out the "dimensional" nature of this notion. We emphasize that the straightforward generalization of the Adler–Konheim–McAndrew definition of the topological entropy for non-compact sets leads to the quantities $\underline{Ch}_f(Z)$ and $\overline{Ch}_f(Z)$. On the other hand, we show that Bowen's topological entropy coincides with $h_f(Z)$.

Let μ be a Borel probability measure on X. In Section 10 we will show that the C-structure τ generates the Carathéodory dimension and lower and upper Carathéodory capacities of μ. We denote them by $P_\mu(\varphi, \mathcal{U})$, $\underline{CP}_\mu(\varphi, \mathcal{U})$, and $\overline{CP}_\mu(\varphi, \mathcal{U})$ respectively. We show that these quantities have limits as diam$\,\mathcal{U}$ tends to zero and these limits coincide and are equal to $h_\mu(f) + \int_X \varphi \, d\mu$, where $h_\mu(f)$ is the measure-theoretic entropy of f. The expression $h_\mu(f) + \int_X \varphi \, d\mu$ is the potential function in the variational principle (see below Appendix II). In the case $\varphi = 0$ our approach produces, in particular, a "dimension" definition of the measure-theoretic entropy.

10. A Modification of the General Carathéodory Construction

We describe a modification of a general Carathéodory construction. Let X and S be arbitrary sets and $\mathcal{F} = \{U_s : s \in S\}$ a collection of subsets in X. We assume that there exist two functions $\eta, \psi \colon S \to \mathbb{R}^+$ satisfying the following conditions:

A1. there exists $s_0 \in S$ such that $U_{s_0} = \varnothing$; if $U_s = \varnothing$ then $\eta(s) = 0$ and $\psi(s) = 0$; if $U_s \neq \varnothing$ then $\eta(s) > 0$ and $\psi(s) > 0$;

A2. for any $\delta > 0$ one can find $\varepsilon > 0$ such that $\eta(s) \leq \delta$ for any $s \in S$ with $\psi(s) \leq \varepsilon$;

A3. for any $\varepsilon > 0$ there exists a finite or countable subcollection $\mathcal{G} \subset \mathcal{S}$ which covers X (i.e., $\underset{s \in \mathcal{G}}{\cup} U_s \supset X$) and $\psi(\mathcal{G}) \overset{\text{def}}{=} \sup\{\psi(s) : s \in \mathcal{S}\} \leq \varepsilon$.

Let $\xi \colon \mathcal{S} \to \mathbb{R}^+$ be a function. We say that the set $\mathcal{S}$, collection of subsets $\mathcal{F}$, and the functions ξ, η, ψ, satisfying Conditions $A1, A2$, and $A3$, introduce the **Carathéodory dimension structure** or **C-structure** τ on X and write $\tau = (\mathcal{S}, \mathcal{F}, \xi, \eta, \psi)$.

If the map $s \mapsto U_s$ is one-to-one then the functions ξ, η, and ψ can be considered as being defined on the set $\mathcal{F}$ and thus, the above C-structure coincides with the C-structure introduced in Section 1. In the general case one can still follow the approach, described in Chapter 1, to define the Carathéodory dimension and lower and upper Carathéodory capacities generated by the C-structure. We shall briefly outline this approach.

Given a set $Z \subset X$ and numbers $\alpha \in \mathbb{R}$, $\varepsilon > 0$, we define

$$M_C(Z, \alpha, \varepsilon) = \inf_{\mathcal{G}} \left\{ \sum_{s \in \mathcal{G}} \xi(s) \eta(s)^\alpha \right\},$$

where the infimum is taken over all finite or countable subcollections $\mathcal{G} \subset \mathcal{S}$ covering Z with $\psi(\mathcal{G}) \leq \varepsilon$. By Condition $A3$ the function $M_C(Z, \alpha, \varepsilon)$ is correctly defined. It is non-decreasing as ε decreases. Therefore, the following limit exists:

$$m_C(Z, \alpha) = \lim_{\varepsilon \to 0} M_C(Z, \alpha, \varepsilon).$$

One can show that the function $m_C(Z, \alpha)$ satisfies Propositions 1.1 and 1.2. We define the **Carathéodory dimension of the set** Z by (1.3). It has properties stated in Theorem 1.1.

Let X and X' be sets endowed with C-structures $\tau = (\mathcal{S}, \mathcal{F}, \xi, \eta, \psi)$ and $\tau' = (\mathcal{S}, \mathcal{F}', \xi', \eta', \psi')$ respectively. One can show that the Carathéodory dimension of sets is invariant with respect to a bijective map $\chi \colon X \to X'$ which preserves the C-structures τ and τ' (compare to Theorem 1.3).

We shall now assume that the following condition holds:

A3'. there exists $\epsilon > 0$ such that for any $0 < \varepsilon \leq \epsilon$ there exists a finite or countable subcollection $\mathcal{G} \subset \mathcal{S}$ covering X such that $\psi(s) = \varepsilon$ for any $s \in \mathcal{G}$.

Given $\alpha \in \mathbb{R}$ and $\varepsilon > 0$, let us consider a set $Z \subset X$ and define

$$R_C(Z, \alpha, \varepsilon) = \inf_{\mathcal{G}} \left\{ \sum_{s \in \mathcal{G}} \xi(s) \eta(s)^\alpha \right\},$$

where the infimum is taken over all finite or countable subcollections $\mathcal{G} \subset \mathcal{S}$ covering Z such that $\psi(s) = \varepsilon$ for any $s \in \mathcal{G}$. According to $A3'$, $R_C(Z, \alpha, \varepsilon)$ is correctly defined. We set

$$\underline{r}_C(Z, \alpha) = \varliminf_{\varepsilon \to 0} R_C(Z, \alpha, \varepsilon), \quad \overline{r}_C(Z, \alpha) = \varlimsup_{\varepsilon \to 0} R_C(Z, \alpha, \varepsilon).$$

One can show that the functions $\underline{r}_C(Z,\alpha)$ and $\bar{r}_C(Z,\alpha)$ satisfy Proposition 2.1. We define the **lower** and **upper Carathéodory capacities of the set** Z, $\overline{\mathrm{Cap}}_C Z$ and $\overline{\mathrm{Cap}}_C Z$, by (2.1). They have properties stated in Theorem 2.1.

For any $\varepsilon > 0$ and any set $Z \subset X$, let us put

$$\Lambda(Z, \varepsilon) = \inf_{\mathcal{G}} \left\{ \sum_{s \in \mathcal{G}} \xi(s) \right\},$$

where the infimum is taken over all finite or countable subcollections $\mathcal{G} \subset \mathcal{S}$ covering Z for which $\psi(s) = \varepsilon$ for all $s \in \mathcal{G}$.

Let us assume that the function η satisfies the following condition:

A4. $\eta(s_1) = \eta(s_2)$ for any $s_1, s_2 \in \mathcal{S}$ for which $\psi(s_1) = \psi(s_2)$.

One can now correctly define the function $\eta(\varepsilon)$ of a real variable ε by setting $\eta(\varepsilon) = \eta(s)$ if $\psi(s) = \varepsilon$. One can prove that, provided Condition $A4$ holds, the lower and upper Carathéodory capacities of sets satisfy Theorem 2.2, 2.3, and 2.4.

Let X and X' be sets endowed with C-structures $\tau = (\mathcal{S}, \mathcal{F}, \xi, \eta, \psi)$ and $\tau' = (\mathcal{S}, \mathcal{F}', \xi', \eta', \psi')$ respectively. One can show that the lower and upper Carathéodory capacities of sets are invariants with respect to a bijective map $\chi \colon X \to X'$ which preserves C-structures τ and τ' (compare to Theorem 2.5).

Let (X, μ) be a Lebesgue space with a probability measure μ endowed with a C-structure $\tau = (\mathcal{F}, \xi, \eta, \psi)$. Assume that any set $U_s \in \mathcal{F}$ is measurable. We define the **Carathéodory dimension of the measure** μ, $\dim_H \mu$, and **lower and upper Carathéodory capacities of the measure** μ, $\underline{\mathrm{Cap}}_C \mu$ and $\overline{\mathrm{Cap}}_C \mu$, by (3.1) and (3.2) respectively. We have that

$$\dim_C \mu \leq \underline{\mathrm{Cap}}_C \mu \leq \overline{\mathrm{Cap}}_C \mu.$$

We shall now assume that the following condition holds:

A5. for μ-almost every $x \in X$ and any $\varepsilon > 0$, if $s \in \mathcal{S}$ and $U_s \ni x$ is a set with $\psi(s) \leq \varepsilon$ then $\mu(U_s) > 0$ and $\xi(s) > 0$.

For each point $x \in X$ and a number ε, $0 < \varepsilon \leq \epsilon$, we choose $s = s(x, \varepsilon) \in \mathcal{S}$ such that $x \in U_s$ and $\psi(s) = \varepsilon$ (this is possible in view of Condition $A3'$).

Once this choice is made we obtain the subcollection

$$\mathcal{S}' = \{ s(x, \varepsilon) \in \mathcal{S} : x \in X,\, 0 < \varepsilon \leq \epsilon \}.$$

Given $\alpha \in \mathbb{R}$ and $x \in X$, we define now the **lower** and **upper** α-**Carathéodory pointwise dimensions** of μ at x by

$$\underline{d}_{C,\mu,\alpha}(x) = \varliminf_{\varepsilon \to 0} \frac{\alpha \log \mu(U_{s(x,\varepsilon)})}{\log \left(\xi(s(x,\varepsilon)) \eta(s(x,\varepsilon))^\alpha \right)},$$

$$\overline{d}_{C,\mu,\alpha}(x) = \varlimsup_{\varepsilon \to 0} \frac{\alpha \log \mu(U_{s(x,\varepsilon)})}{\log \left(\xi(s(x,\varepsilon)) \eta(s(x,\varepsilon))^\alpha \right)}$$

(see (3.5)). We have that $\underline{d}_{C,\mu,\alpha}(x) \leq \overline{d}_{C,\mu,\alpha}(x)$ for any $x \in X$. It is a simple exercise to prove that the conclusion of Theorems 3.1, 3.2, 3.3, 4.1, 4.2, 5.1, and 5.2 hold (with obvious modifications in the formulations).

We also define

$$
\begin{aligned}
\underline{\mathcal{D}}_{C,\mu,\alpha}(x) &= \varliminf_{\varepsilon \to 0} \inf_{\substack{x \in U_s \\ \psi(s) = \varepsilon}} \frac{\alpha \log \mu(U_s)}{\log(\xi(s)\eta(s)^\alpha)}, \\
\overline{\mathcal{D}}_{C,\mu,\alpha}(x) &= \varlimsup_{\varepsilon \to 0} \sup_{\substack{x \in U_s \\ \psi(s) = \varepsilon}} \frac{\alpha \log \mu(U_s)}{\log(\xi(s)\eta(s)^\alpha)}.
\end{aligned} \tag{10.1}
$$

It is easy to check that conclusions of Theorem 3.5 and 3.6. hold.

11. Dimensional Definition of Topological Pressure; Topological and Measure-Theoretic Entropies

Topological Pressure

Let (X, ρ) be a compact metric space with metric ρ, $f : X \to X$ a continuous map, and $\varphi : X \to \mathbb{R}$ a continuous function. Consider a finite open cover $\mathcal{U}$ of X and denote by $\mathcal{S}_m(\mathcal{U})$ the set of all **strings** $\boldsymbol{U} = \{U_{i_0} \ldots U_{i_{m-1}} : U_{i_j} \in \mathcal{U}\}$ of **length** $m = m(\boldsymbol{U})$. We put $\mathcal{S} = \mathcal{S}(\mathcal{U}) = \cup_{m \geq 0} \mathcal{S}_m(\mathcal{U})$.

To a given string $\boldsymbol{U} = \{U_{i_0} \ldots U_{i_{m-1}}\} \in \mathcal{S}(\mathcal{U})$ we associate the set

$$
X(\boldsymbol{U}) = \{x \in X : f^j(x) \in U_{i_j} \text{ for } j = 0, \ldots, m(\boldsymbol{U}) - 1\}. \tag{11.1}
$$

Define the collection of subsets

$$
\mathcal{F} = \mathcal{F}(\mathcal{U}) = \{X(\boldsymbol{U}) : \boldsymbol{U} \in \mathcal{S}(\mathcal{U})\} \tag{11.2}
$$

and three functions $\xi, \eta, \psi : \mathcal{S}(\mathcal{U}) \to \mathbb{R}$ as follows

$$
\begin{aligned}
\xi(\boldsymbol{U}) &= \exp\left(\sup_{x \in X(\boldsymbol{U})} \sum_{k=0}^{m(\boldsymbol{U})-1} \varphi(f^k(x)) \right), \\
\eta(\boldsymbol{U}) &= \exp(-m(\boldsymbol{U})), \quad \psi(\boldsymbol{U}) = m(\boldsymbol{U})^{-1}.
\end{aligned} \tag{11.3}
$$

It is straightforward to verify that the set $\mathcal{S}$, the collection of subsets $\mathcal{F}$, and the functions η, ξ, and ψ satisfy Conditions $A1, A2, A3$, and $A3'$ in Section 10 and hence they determine a C-structure $\tau = \tau(\mathcal{U}) = (\mathcal{S}, \mathcal{F}, \xi, \eta, \psi)$ on X. The corresponding Carathéodory function $m_C(Z, \alpha)$ (where $Z \subset X$ and $\alpha \in \mathbb{R}$; see Section 10) depends on the cover $\mathcal{U}$ (and the function φ) and is given by

$$
m_C(Z, \alpha) = \lim_{N \to \infty} M(Z, \alpha, \varphi, \mathcal{U}, N),
$$

where

$$
M(Z, \alpha, \varphi, \mathcal{U}, N) =
$$

$$
\inf_{\mathcal{G}} \left\{ \sum_{U \in \mathcal{G}} \exp\left(-\alpha\, m(\boldsymbol{U}) + \sup_{x \in X(\boldsymbol{U})} \sum_{k=0}^{m(\boldsymbol{U})-1} \varphi(f^k(x)) \right) \right\} \tag{11.4}
$$

and the infimum is taken over all finite or countable collections of strings $\mathcal{G} \subset \mathcal{S}(\mathcal{U})$ such that $m(\boldsymbol{U}) \geq N$ for all $\boldsymbol{U} \in \mathcal{G}$ and $\mathcal{G}$ covers Z (i.e., the collection of sets $\{X(\boldsymbol{U}) : \boldsymbol{U} \in \mathcal{G}\}$ covers Z).

Furthermore, the Carathéodory functions $\underline{r}_C(Z, \alpha)$ and $\overline{r}_C(Z, \alpha)$ (where $Z \subset X$ and $\alpha \in \mathbb{R}$; see Section 10) depend on the cover $\mathcal{U}$ and are given by

$$\underline{r}_C(Z, \alpha) = \lim_{N \to \infty} R(Z, \alpha, \varphi, \mathcal{U}, N), \quad \overline{r}_C(Z, \alpha) = \overline{\lim_{N \to \infty}} R(Z, \alpha, \varphi, \mathcal{U}, N),$$

where

$$R(Z, \alpha, \varphi, \mathcal{U}, N) = \inf_{\mathcal{G}} \left\{ \sum_{\boldsymbol{U} \in \mathcal{G}} \exp \left(-\alpha N + \sup_{x \in X(\boldsymbol{U})} \sum_{k=0}^{N-1} \varphi(f^k(x)) \right) \right\} \quad (11.4')$$

and the infimum is taken over all finite or countable collections of strings $\mathcal{G} \subset \mathcal{S}(\mathcal{U})$ such that $m(\boldsymbol{U}) = N$ for all $\boldsymbol{U} \in \mathcal{G}$ and $\mathcal{G}$ covers Z.

According to Section 10, given a set $Z \subset X$, the *C*-structure τ generates the Carathéodory dimension of Z and lower and upper Carathéodory capacities of Z specified by the cover $\mathcal{U}$ and the map f. We denote them by $P_Z(\varphi, \mathcal{U})$, $\underline{CP}_Z(\varphi, \mathcal{U})$, and $\overline{CP}_Z(\varphi, \mathcal{U})$ respectively. We have that (compare to (1.3) and (2.1))

$$P_Z(\varphi, \mathcal{U}) = \inf\{\alpha : m_C(Z, \alpha) = 0\} = \sup\{\alpha : m_C(Z, \alpha) = \infty\},$$

$$\underline{CP}_Z(\varphi, \mathcal{U}) = \inf\{\alpha : \underline{r}_C(Z, \alpha) = 0\} = \sup\{\alpha : \underline{r}_C(Z, \alpha) = \infty\},$$

$$\overline{CP}_Z(\varphi, \mathcal{U}) = \inf\{\alpha : \overline{r}_C(Z, \alpha) = 0\} = \sup\{\alpha : \overline{r}_C(Z, \alpha) = \infty\}.$$

Let $|\mathcal{U}| = \max\{\operatorname{diam} U_i : U_i \subset \mathcal{U}\}$ be the diameter of the cover $\mathcal{U}$.

Theorem 11.1. *For any set $Z \subset X$ the following limits exist:*

$$P_Z(\varphi) \overset{\text{def}}{=} \lim_{|\mathcal{U}| \to 0} P_Z(\varphi, \mathcal{U}),$$

$$\underline{CP}_Z(\varphi) \overset{\text{def}}{=} \lim_{|\mathcal{U}| \to 0} \underline{CP}_Z(\varphi, \mathcal{U}),$$

$$\overline{CP}_Z(\varphi) \overset{\text{def}}{=} \lim_{|\mathcal{U}| \to 0} \overline{CP}_Z(\varphi, \mathcal{U}).$$

Proof. Let $\mathcal{V}$ be a finite open cover of X with diameter smaller than the Lebesgue number of $\mathcal{U}$. One can see that each element $V \in \mathcal{V}$ is contained in some element $U(V) \in \mathcal{U}$. To any string $\boldsymbol{V} = \{V_{i_0} \dots V_{i_m}\} \in \mathcal{S}(\mathcal{V})$ we associate the string $\boldsymbol{U}(\boldsymbol{V}) = \{U(V_{i_0}) \dots U(V_{i_m})\} \in \mathcal{S}(\mathcal{U})$. If $\mathcal{G} \subset \mathcal{S}(\mathcal{V})$ covers a set $Z \subset X$ then $\boldsymbol{U}(\mathcal{G}) = \{\boldsymbol{U}(\boldsymbol{V}) : \boldsymbol{V} \in \mathcal{G}\} \subset \mathcal{S}(\mathcal{U})$ also covers Z. Let

$$\gamma = \gamma(\mathcal{U}) = \sup\{|\varphi(x) - \varphi(y)| : x, y \in U \text{ for some } U \in \mathcal{U}\}.$$

One can verify using (11.4) that for every $\alpha \in \mathbb{R}$ and $N > 0$

$$M(Z, \alpha, \varphi, \mathcal{U}, N) \leq M(Z, \alpha - \gamma, \varphi, \mathcal{V}, N). \tag{11.5}$$

This implies that
$$P_Z(\varphi, \mathcal{U}) - \gamma \leq P_Z(\varphi, \mathcal{V}).$$
Since X is compact it has finite open covers of arbitrarily small diameter. Therefore,
$$P_Z(\varphi, \mathcal{U}) - \gamma \leq \varliminf_{|\mathcal{V}| \to 0} P_Z(\varphi, \mathcal{V}).$$
If $|\mathcal{U}| \to 0$ then $\gamma(\mathcal{U}) \to 0$ and hence
$$\varliminf_{|\mathcal{U}| \to 0} P_Z(\varphi, \mathcal{U}) \leq \varliminf_{|\mathcal{V}| \to 0} P_Z(\varphi, \mathcal{V}).$$

This implies the existence of the first limit. The existence of two other limits can be proved in a similar fashion by using the inequality which is an analog of (11.5)
$$R(Z, \alpha, \varphi, \mathcal{U}, N) \leq R(Z, \alpha - \gamma, \varphi, \mathcal{V}, N). \tag{11.6}$$
This completes the proof of the theorem.   ∎

We call the quantities $P_Z(\varphi)$, $\underline{CP}_Z(\varphi)$, and $\overline{CP}_Z(\varphi)$, respectively the **topological pressure** and **lower** and **upper capacity topological pressures** of the function φ on the set Z (with respect to f). Sometimes more explicit notations $P_{Z,f}(\varphi)$, $\underline{CP}_{Z,f}(\varphi)$, and $\overline{CP}_{Z,f}(\varphi)$ will be used to emphasize the dependence on the map f.

We emphasize that the set Z can be arbitrary and need not be compact or invariant under the map f. If f is a homeomorphism then for any set $Z \subset X$ its topological pressure coincides with topological pressure on the invariant hull of Z (i.e., the set $\bigcup_{n \in \mathbb{Z}} f^n(Z)$; this follows from Theorem 11.2 below). However, this may not be true for lower and upper capacity topological pressures (see Example 11.2 below).

We formulate the basic properties of topological pressure and lower and upper capacity topological pressures. They are immediate corollaries of the definitions and Theorems 1.1 and 2.1.

Theorem 11.2.

(1) $P_\varnothing(\varphi) \leq 0$.
(2) $P_{Z_1}(\varphi) \leq P_{Z_2}(\varphi)$ if $Z_1 \subset Z_2 \subset X$.
(3) $P_Z(\varphi) = \sup_{i \geq 1} P_{Z_i}(\varphi)$, where $Z = \bigcup_{i \geq 1} Z_i$ and $Z_i \subset X$, $i = 1, 2, \ldots$.
(4) If f is a homeomorphism then $P_Z(\varphi) = P_{f(Z)}(\varphi)$.

Theorem 11.3.

(1) $\underline{CP}_\varnothing(\varphi) \leq 0$, $\overline{CP}_\varnothing(\varphi) \leq 0$.
(2) $\underline{CP}_{Z_1}(\varphi) \leq \underline{CP}_{Z_2}(\varphi)$ and $\overline{CP}_{Z_1}(\varphi) \leq \overline{CP}_{Z_2}(\varphi)$ if $Z_1 \subset Z_2 \subset X$.
(3) $\underline{CP}_Z(\varphi) \geq \sup_{i \geq 1} \underline{CP}_{Z_i}(\varphi)$ and $\overline{CP}_Z(\varphi) \geq \sup_{i \geq 1} \overline{CP}_{Z_i}(\varphi)$, where $Z = \bigcup_{i \geq 1} Z_i$ and $Z_i \subset X$, $i = 1, 2, \ldots$.
(4) If $h : X \to X$ is a homeomorphism which commutes with f (i.e., $f \circ h = h \circ f$) then
$$P_Z(\varphi) = P_{h(Z)}(\varphi \circ h^{-1}),$$
$$\underline{CP}_Z(\varphi) = \underline{CP}_{h(Z)}(\varphi \circ h^{-1}), \quad \overline{CP}_Z(\varphi) = \overline{CP}_{h(Z)}(\varphi \circ h^{-1})$$

Obviously, the functions η and ψ satisfy Condition $A4$ in Section 10. Therefore, by Theorems 2.2 and 11.1, we have for any $Z \subset X$ that

$$
\begin{aligned}
\underline{CP}_Z(\varphi) &= \lim_{|\mathcal{U}| \to 0} \varliminf_{N \to \infty} \frac{1}{N} \log \Lambda(Z, \varphi, \mathcal{U}, N), \\
\overline{CP}_Z(\varphi) &= \lim_{|\mathcal{U}| \to 0} \varlimsup_{N \to \infty} \frac{1}{N} \log \Lambda(Z, \varphi, \mathcal{U}, N),
\end{aligned}
\tag{11.7}
$$

where in accordance with (2.3), (11.1), and (11.3)

$$
\Lambda(Z, \varphi, \mathcal{U}, N) = \inf_{\mathcal{G}} \left\{ \sum_{U \in \mathcal{G}} \exp \left(\sup_{x \in X(U)} \sum_{k=0}^{N-1} \varphi(f^k(x)) \right) \right\}
\tag{11.8}
$$

and the infimum is taken over all finite or countable collections of strings $\mathcal{G} \subset \mathcal{S}(\mathcal{U})$ such that $m(U) = N$ for all $U \in \mathcal{G}$ and $\mathcal{G}$ covers Z.

We also point out the *continuity property* of the topological pressure and lower and upper capacity topological pressures.

Theorem 11.4. *For any two continuous functions φ and ψ on X*

$$
\begin{aligned}
|P_Z(\varphi) - P_Z(\psi)| &\leq \|\varphi - \psi\|, \\
|\underline{CP}_Z(\varphi) - \underline{CP}_Z(\psi)| &\leq \|\varphi - \psi\|, \\
|\overline{CP}_Z(\varphi) - \overline{CP}_Z(\psi)| &\leq \|\varphi - \psi\|
\end{aligned}
$$

where $\|\cdot\|$ denotes the supremum norm in the space of continuous functions on X.

Proof. Given $N > 0$, we have that

$$
\frac{1}{N} \sup_{x \in X} \sum_{k=0}^{N-1} |\varphi(f^k(x)) - \psi(f^k(x))| \leq \|\varphi - \psi\|.
$$

It follows that

$$
M(Z, \alpha + \|\varphi - \psi\|, \psi, \mathcal{U}, N) \leq M(Z, \alpha, \varphi, \mathcal{U}, N) \leq M(Z, \alpha - \|\varphi - \psi\|, \psi, \mathcal{U}, N).
$$

This implies that

$$
P_Z(\psi, \mathcal{U}) - \|\varphi - \psi\| \leq P_Z(\varphi, \mathcal{U}) \leq P_Z(\psi, \mathcal{U}) + \|\varphi - \psi\|
$$

and concludes the proof of the first inequality. The proof of the other two inequalities is similar. ∎

One can easily see that

$$P_Z(\varphi) \leq \underline{CP}_Z(\varphi) \leq \overline{CP}_Z(\varphi). \tag{11.9}$$

Below we will give an example where the strict inequalities occur (see Examples 11.1 and 11.2). The situation for invariant and compact sets is different.

Theorem 11.5.

(1) *For any f-invariant set $Z \subset X$ we have $\underline{CP}_Z(\varphi) = \overline{CP}_Z(\varphi)$; moreover, for any open cover $\mathcal{U}$ of X, we have $\underline{CP}_Z(\varphi, \mathcal{U}) = \overline{CP}_Z(\varphi, \mathcal{U})$.*

(2) *For any compact invariant set $Z \subset X$ we have $P_Z(\varphi) = \underline{CP}_Z(\varphi) = \overline{CP}_Z(\varphi)$; moreover, for any open cover $\mathcal{U}$ of X, we have $P_Z(\varphi, \mathcal{U}) = \underline{CP}_Z(\varphi, \mathcal{U}) = \overline{CP}_Z(\varphi, \mathcal{U})$.*

Proof. Let $Z \subset X$ be an f-invariant set. Choose two collections of strings $\mathcal{G}_m \subset \mathcal{S}_m(\mathcal{U})$ and $\mathcal{G}_n \subset \mathcal{S}_n(\mathcal{U})$ which cover Z and consider

$$\mathcal{G}_{m,n} \stackrel{\text{def}}{=} \{UV : U \in \mathcal{G}_m, V \in \mathcal{G}_n\} \subset \mathcal{S}_{m+n}(\mathcal{U}).$$

Since Z is f-invariant the collection of strings $\mathcal{G}_{m,n}$ also covers Z. We wish to estimate $\Lambda(Z, \varphi, \mathcal{U}, m+n)$ using (11.8). We have

$$\Lambda(Z, \varphi, \mathcal{U}, m+n) \leq \sum_{UV \in \mathcal{G}_{m+n}} \exp \left(\sup_{x \in X(UV)} \sum_{k=0}^{m+n-1} \varphi(f^k(x)) \right)$$

$$\leq \sum_{U \in \mathcal{G}_m} \exp \left(\sup_{x \in X(U)} \sum_{k=0}^{m-1} \varphi(f^k(x)) \right) \times \sum_{V \in \mathcal{G}_n} \exp \left(\sup_{x \in X(V)} \sum_{k=0}^{n-1} \varphi(f^k(x)) \right).$$

This implies that

$$\Lambda(Z, \varphi, \mathcal{U}, m+n) \leq \Lambda(Z, \varphi, \mathcal{U}, m) \times \Lambda(Z, \varphi, \mathcal{U}, n).$$

Let $a_m = \log \Lambda(Z, \varphi, \mathcal{U}, m)$. Note that $\Lambda(Z, \varphi, \mathcal{U}, m) \geq e^{-m\|\varphi\|}$. Therefore, $\inf_{m \geq 1} \frac{a_m}{m} \geq -\|\varphi\| > -\infty$. The desired result is now a direct consequence of (11.7) and the following lemma (we leave its proof to the reader; see Lemma 1.18 in [Bo2]).

Lemma. *Let a_m, $m = 1, 2, \ldots$ be a sequence of numbers satisfying $\inf_{m \geq 1} \frac{a_m}{m} > -\infty$ and $a_{m+n} \leq a_m + a_n$ for all $m, n \geq 1$. Then the limit $\lim_{m \to \infty} \frac{a_m}{m}$ exists and coincides with $\inf_{m \geq 1} \frac{a_m}{m}$.*

Choose any $\alpha > P_Z(\varphi, \mathcal{U})$. There exist $N > 0$ and $\mathcal{G} \subset \mathcal{S}(\mathcal{U})$ such that $\mathcal{G}$ covers Z and

$$Q(Z, \alpha, \mathcal{G}) \stackrel{\text{def}}{=} \sum_{U \in \mathcal{G}} \exp \left(-\alpha m(U) + \sup_{x \in X(U)} \sum_{k=0}^{m(U)-1} \varphi(f^k(x)) \right) < 1.$$

Since Z is compact we can choose $\mathcal{G}$ to be finite and hence

$$\mathcal{G} \subset \bigcup_{m=1}^{M} \mathcal{S}_m(\mathcal{U})$$

for some $M \geq 1$. Put

$$\mathcal{G}^n = \{\boldsymbol{U}_1 \ldots \boldsymbol{U}_n : \boldsymbol{U}_i \in \mathcal{G}\} \text{ and } \Gamma = \bigcup_{n=1}^{\infty} \mathcal{G}^n.$$

Since Z is invariant Γ covers Z. It is a simple exercise (which we leave to the reader) to check that for every $n > 0$,

$$Q(Z, \alpha, \mathcal{G}^n) \leq Q(Z, \alpha, \mathcal{G})^n.$$

Hence,

$$Q(Z, \alpha, \Gamma) = \sum_{n=1}^{\infty} Q(Z, \alpha, \mathcal{G}^n) < \infty.$$

Let us fix some $N > 0$ and consider a point $x \in Z$. Since Γ covers Z there exists a string $\boldsymbol{U} \in \Gamma$ such that $x \in X(\boldsymbol{U})$ and $N \leq m(\boldsymbol{U}) < N + M$. Denote by $\boldsymbol{U}^*$ the substring that consists of the first N symbols of the string $\boldsymbol{U}$. We have that

$$\sup_{y \in X(\boldsymbol{U}^*)} \sum_{k=0}^{N-1} \varphi(f^k(y)) \leq \sup_{y \in X(\boldsymbol{U})} \sum_{k=0}^{m(\boldsymbol{U})-1} \varphi(f^k(y)) + M\|\varphi\|.$$

If Γ_N denotes the collection of all substrings $\boldsymbol{U}^*$ constructed above then

$$e^{-\alpha N} \sum_{\boldsymbol{U}^* \in \Gamma_N} \exp \sup_{y \in X(\boldsymbol{U}^*)} \sum_{k=0}^{N-1} \varphi(f^k(y)) \leq \max\{1, e^{-\alpha M}\} e^{M\|\varphi\|} Q(Z, \alpha, \Gamma) < \infty.$$

By (11.7) we obtain that $\alpha > \overline{CP}_Z(\varphi)$, and hence the desired result follows from (11.9). ∎

Theorem 11.5 shows that for a compact invariant set Z the topological pressure and lower and upper capacity topological pressures coincide and the common value yields the classical topological pressure (see, for example, [Bo2]). It is worth pointing out that this common value is a topological invariant (i.e., $P_X(\varphi) = P_X(\varphi \circ h)$, where h is a homeomorphism which commutes with f). This means that the pressure does not depend on the metric on X.

If a set Z is neither invariant nor compact one has three, in general distinct, quantities: the topological pressure, $P_Z(\varphi)$, and lower and upper capacity topological pressures, $\underline{CP}_Z(\varphi)$ and $\overline{CP}_Z(\varphi)$. The latter coincide if the set Z is invariant and may not otherwise (see Example 11.1 below). Furthermore, they are defined by formulae (11.7) and (11.8) which are a straightforward generalization of the classical definition of the topological pressure. In view of the variational principle, the topological pressure $P_Z(\varphi)$ seems more adapted to the case of non-compact sets and plays a crucial role in the thermodynamic formalism (see Appendix II).

Remarks.

(1) We describe another approach to the definition of topological pressure. Let (X, ρ) be a compact metric space with metric ρ, $f: X \to X$ a continuous map, and $\varphi: X \to \mathbb{R}$ a continuous function. Fix a number $\delta > 0$. Given $n > 0$ and a point $x \in X$, define the (n, δ)-ball at x by

$$B_n(x, \delta) = \{y \in X : \rho(f^i(x), f^i(y)) \le \delta, \text{ for } 0 \le i \le n\}. \tag{11.10}$$

Put $\mathcal{S} = X \times \mathbb{N}$. We define the collection of subsets

$$\mathcal{F} = \{B_n(x, \delta) : x \in X, n \in \mathbb{N}\}$$

and three functions $\xi, \eta, \psi: \mathcal{S} \to \mathbb{R}$ as follows

$$\xi(x, n) = \exp\left(\sup_{y \in B_n(x, \delta)} \sum_{k=0}^{n-1} \varphi(f^k(y))\right),$$

$$\eta(x, n) = \exp(-n), \quad \psi(x, n) = n^{-1}.$$

One can directly verify that the set $\mathcal{S}$, the collection of subsets $\mathcal{F}$, and functions η, ξ, and ψ satisfy Conditions $A1, A2, A3$, and $A3'$ in Section 10 and hence determine a C-structure $\tau = (\mathcal{S}, \mathcal{F}, \xi, \eta, \psi)$ on X. According to Section 10, given a set $Z \subset X$, this C-structure generates the Carathéodory dimension of Z and lower and upper Carathéodory capacities of Z which depend on δ. We denote them by $P_Z(\varphi, \delta)$, $\underline{CP}_Z(\varphi, \delta)$, and $\overline{CP}_Z(\varphi, \delta)$ respectively.

Let $\mathcal{U}$ be a finite open cover of X and $\delta(\mathcal{U})$ its Lebesgue number. It is easily seen that for every $x \in X$, if $x \in X(\boldsymbol{U})$ for some $\boldsymbol{U} \in \mathcal{S}(\mathcal{U})$ then

$$B_{m(\boldsymbol{U})}(x, \frac{1}{2}\delta(\mathcal{U})) \subset X(\boldsymbol{U}) \subset B_{m(\boldsymbol{U})}(x, 2|\mathcal{U}|). \tag{11.11}$$

It follows now from Theorem 11.1 that

$$P_Z(\varphi) = \lim_{\delta \to 0} P_Z(\varphi, \delta),$$

$$\underline{CP}_Z(\varphi) = \lim_{\delta \to 0} \underline{CP}_Z(\varphi, \delta), \quad \overline{CP}_Z(\varphi) = \lim_{\delta \to 0} \overline{CP}_Z(\varphi, \delta).$$

(2) If the map $f: X \to X$ is a homeomorphism we can consider the topological pressure and lower and upper capacity topological pressures for the map f as well as for the inverse map f^{-1}. If Z is an invariant subset of X then for any continuous function $\varphi: X \to \mathbb{R}$,

$$\underline{CP}_{Z,f}(\varphi) = \underline{CP}_{Z,f^{-1}}(\varphi), \quad \overline{CP}_{Z,f}(\varphi) = \overline{CP}_{Z,f^{-1}}(\varphi).$$

These equalities hold no matter whether Z is compact or not but may fail to be true if Z is not invariant (see Example 11.3 below). If Z is invariant and compact then in addition we have that

$$P_{Z,f}(\varphi) = P_{Z,f^{-1}}(\varphi)$$

and this may fail if Z is not compact (although still invariant; see Example 11.3 below).

Topological Entropy

We consider the special case $\varphi = 0$. Given a set $Z \subset X$, we call the quantities

$$h_Z(f) \overset{\text{def}}{=} P_Z(0), \quad \underline{Ch}_Z(f) \overset{\text{def}}{=} \underline{CP}_Z(0), \quad \overline{Ch}_Z(f) \overset{\text{def}}{=} \overline{CP}_Z(0)$$

respectively, the **topological entropy** and **lower** and **upper capacity topological entropies** of the map f on Z. We stress again that the set Z can be arbitrary and need not be compact or invariant under f. It follows from (11.9) that

$$h_Z(f) \leq \underline{Ch}_Z(f) \leq \overline{Ch}_Z(f). \tag{11.12}$$

If the set Z is f-invariant, we have

$$\underline{Ch}_Z(f) = \overline{Ch}_Z(f) \overset{\text{def}}{=} Ch_Z(f).$$

By (11.7) we obtain for an invariant set Z that

$$\begin{aligned}
Ch_Z(f) &= \lim_{|\mathcal{U}| \to 0} \varliminf_{N \to \infty} \frac{1}{N} \log \Lambda(Z, 0, \mathcal{U}, N) \\
&= \lim_{|\mathcal{U}| \to 0} \varlimsup_{N \to \infty} \frac{1}{N} \log \Lambda(Z, 0, \mathcal{U}, N),
\end{aligned} \tag{11.13}$$

where, in accordance with (11.8), $\Lambda(Z, 0, \mathcal{U}, N)$ is the smallest number of strings U of length N, for which the sets $X(U)$ cover Z. Formula (11.13) reveals the meaning of the quantity $Ch_Z(f)$: it is the exponential rate of growth in N of the smallest number of strings U of length N, for which the sets $X(U)$ cover Z. For a compact invariant set Z we have by Theorem 11.5, that $h_Z(f) = \underline{Ch}_Z(f) = \overline{Ch}_Z(f)$.

The topological entropy and lower and upper capacity topological entropies have properties stated in Theorems 11.2 and 11.3 (applied to $\varphi = 0$). In particular, they are invariant under a homeomorphism of X which commutes with f.

We now proceed with the inequalities (11.12). In examples below we consider the symbolic dynamical system (Σ_p, σ), where Σ_p is the space of two-sided infinite sequences on p symbols and σ is the (two-sided) shift. We recall that the cylinder set $C_{i_m \ldots i_l}$ consists of all sequences $\omega = (j_k)$ for which $j_m = i_m, \ldots, j_l = i_l$ (see more detailed description in Appendix II below).

Example 11.1. *There exists a compact non-invariant set $Z \subset \Sigma_3$ for which* $\underline{Ch}_Z(\sigma) < \overline{Ch}_Z(\sigma)$.

Proof. Let n_k be a strictly increasing sequence of integers. Define the set

$$Z = \big\{\omega = (\omega_n) \in \Sigma_3 : \omega_n = 1 \text{ or } 2 \quad \text{if } n_{2\ell} \leq n < n_{2\ell+1} \text{ and}$$
$$\omega_n = 1, 2, \text{ or } 3 \quad \text{if } n_{2\ell+1} \leq n < n_{2\ell+2} \text{ for some } \ell\big\}.$$

Obviously, the set Z is compact. Consider the sequence a_n defined as follows: $a_n = 2$ if $n_{2\ell} \leq n < n_{2\ell+1}$ and $a_n = 3$ if $n_{2\ell+1} \leq n < n_{2\ell+2}$. Set $S_n = \prod_{k=1}^n a_k$. We choose the sequence n_k growing so fast that

$$C_1^{-1} 3^{n_{2\ell}} \leq S_{n_{2\ell}} \leq C_1 3^{n_{2\ell}}, \quad C_1^{-1} 2^{n_{2\ell+1}} \leq S_{n_{2\ell+1}} \leq C_1 2^{n_{2\ell+1}},$$

where $C_1 > 0$ is a constant independent of ℓ.

Given $m \geq 0$, consider the cover $\mathcal{U}_m$ of Σ_3 by cylinder sets $C_{i_{-m}\cdots i_m}$. Notice that for every string U the set $X(U)$ is a cylinder. Therefore, in accordance with (11.8), $\Lambda(Z, 0, \mathcal{U}_m, N)$ is the smallest number of cylinders of length $m + N$ needed to cover the set Z. It follows that if $n_k \leq N < n_{k+1}$, then

$$C_2^{-1}(m) S_{n_k} \leq \Lambda(Z, 0, \mathcal{U}_m, N) \leq C_2(m) S_{n_{k+1}},$$

where $C_2(m) > 0$ is a constant independent of n. Applying (11.13) with $\mathcal{U} = \mathcal{U}_m$ (and $|\mathcal{U}| \to 0$ as $m \to 0$) yields

$$\underline{Ch}_Z(\sigma) = \log 2, \quad \overline{Ch}_Z(\sigma) = \log 3.$$

The desired result follows.   ∎

Example 11.2. *There is an invariant set $Z \subset \Sigma_2$ for which* $h_Z(\sigma) < Ch_Z(\sigma)$.

Proof. Define the sets

$$Z_k = \big\{\omega = (\omega_n) \in \Sigma_2 : \omega_n = 1 \text{ for all } |n| \geq k\big\}, \quad Z = \bigcup_{k \in \mathbb{Z}} Z_k.$$

It is easy to see that the set Z is invariant and everywhere dense in Σ_2. Therefore, $Ch_Z(\sigma) = Ch_{\Sigma_2}(\sigma) = \log 2$.

Given $m \geq 0$, consider the cover $\mathcal{U}_m$ of Σ_2 by cylinder sets $C_{i_{-m}\cdots i_m}$. It is easy to see that $\Lambda(Z_k, 0, \mathcal{U}_m, N) \leq C$, where $C > 0$ is a constant independent of N. Therefore, $\underline{Ch}_{Z_m}(\sigma) = \overline{Ch}_{Z_m}(\sigma) = 0$ and hence $h_{Z_m}(\sigma) = 0$ for all m. This implies that $h_Z(\sigma) = 0$.   ∎

Note that the set Z in this example is the invariant hull of the set Z_0 and $\underline{Ch}_{Z_0}(\sigma) = \overline{Ch}_{Z_0}(\sigma) = 0$ while $Ch_Z(\sigma) = \log 2$.

Example 11.3.

(1) There is an invariant (non-compact) set $Z \subset \Sigma_2$ for which $h_Z(\sigma) = \log 2$ while $h_Z(\sigma^{-1}) = 0$.

(2) There is a compact (non-invariant) set $Z \subset \Sigma_2$ for which

$$h_Z(\sigma) = \underline{Ch}_Z(\sigma) = \overline{Ch}_Z(\sigma) = \log 2$$

while

$$h_Z(\sigma^{-1}) = \underline{Ch}_Z(\sigma^{-1}) = \overline{Ch}_Z(\sigma^{-1}) = 0.$$

Proof. Define the sets

$$Z_k = \big\{\omega = (\omega_n) \in \Sigma_2 : \omega_n = 1 \text{ for all } n \leq k\big\}, \quad Z = \bigcup_{k \in \mathbb{Z}} Z_k.$$

Obviously, the set Z is invariant (but not compact) and the set Z_0 is compact (but not invariant). We leave it as an exercise to the reader to show that Z fulfills requirements in Statement 1 and so does Z_0 in Statement 2. ∎

Remark.

Let $\mathcal{U}$ be a finite open cover of X. Given a set $Z \subset X$, the quantities

$$h_Z(f,\mathcal{U}) \overset{\text{def}}{=} P_Z(0,\mathcal{U}), \quad \underline{Ch}_Z(f,\mathcal{U}) \overset{\text{def}}{=} \underline{CP}_Z(0,\mathcal{U}), \quad \overline{Ch}_Z(f,\mathcal{U}) \overset{\text{def}}{=} \overline{CP}_Z(0,\mathcal{U})$$

are called the **topological entropy** and **lower** and **upper capacity topological entropies** of f on Z with respect to $\mathcal{U}$. By Theorem 11.5, if Z is invariant, then $\underline{Ch}_Z(f,\mathcal{U}) = \overline{Ch}_Z(f,\mathcal{U}) \overset{\text{def}}{=} Ch_Z(f,\mathcal{U})$ and if, in addition, Z is compact then $h_Z(f,\mathcal{U}) = Ch_Z(f,\mathcal{U})$.

Let $\mathcal{V}$ be a finite open cover of X whose diameter does not exceed the Lebesgue number of $\mathcal{U}$. Applying (11.5) with $\varphi = 0$ we obtain that

$$h_X(f,\mathcal{U}) \leq h_X(f,\mathcal{V}).$$

In [BGH], Blanchard, Glasner, and Host obtained a significantly stronger statement. Namely, *let ξ be a finite Borel partition of X such that each element of ξ is contained in an element of the cover $\mathcal{U}$. Then there exists an f-invariant ergodic measure μ on X for which $h_X(f,\mathcal{U}) \leq h_\mu(f,\xi)$.*

Measure-Theoretic Entropy

Let μ be a Borel probability measure on X (not necessarily invariant under f). Consider a finite open cover $\mathcal{U}$ of X. According to Section 10, the C-structure $\tau = (\mathcal{S}, \mathcal{F}, \xi, \eta, \psi)$ on X, introduced by (11.1), (11.2), and (11.3), generates the Carathéodory dimension of μ and lower and upper Carathéodory capacities of μ specified by the cover $\mathcal{U}$ and the map f. We denote them by $P_\mu(\varphi,\mathcal{U})$, $\underline{CP}_\mu(\varphi,\mathcal{U})$, and $\overline{CP}_\mu(\varphi,\mathcal{U})$ respectively. We have that

$$
\begin{aligned}
P_\mu(\varphi,\mathcal{U}) &= \inf\{P_Z(\varphi,\mathcal{U}) : \mu(Z) = 1\}, \\
\underline{CP}_\mu(\varphi,\mathcal{U}) &= \lim_{\delta\to 0} \inf\{\underline{CP}_Z(\varphi,\mathcal{U}) : \mu(Z) \geq 1 - \delta\}, \\
\overline{CP}_\mu(\varphi,\mathcal{U}) &= \lim_{\delta\to 0} \inf\{\overline{CP}_Z(\varphi,\mathcal{U}) : \mu(Z) \geq 1 - \delta\}.
\end{aligned}
\tag{11.14}
$$

It follows from Theorem 11.1 that there exist the limits

$$
\begin{aligned}
P_\mu(\varphi) &\overset{\text{def}}{=} \lim_{|\mathcal{U}|\to 0} P_\mu(\varphi,\mathcal{U}), \\
\underline{CP}_\mu(\varphi) &\overset{\text{def}}{=} \lim_{|\mathcal{U}|\to 0} \underline{CP}_\mu(\varphi,\mathcal{U}), \\
\overline{CP}_\mu(\varphi) &\overset{\text{def}}{=} \lim_{|\mathcal{U}|\to 0} \overline{CP}_\mu(\varphi,\mathcal{U}).
\end{aligned}
\tag{11.15}
$$

Given a point $x \in X$, we set in accordance with (10.1)

$$
\underline{\mathcal{D}}_{C,\mu,\alpha}(x,\varphi,\mathcal{U}) = \lim_{N\to\infty} \inf_{U} \frac{\alpha \log \mu(X(U))}{-N\alpha + \displaystyle\sup_{y\in X(U)} \sum_{k=0}^{N-1} \varphi(f^k(y))},
$$

$$
\overline{\mathcal{D}}_{C,\mu,\alpha}(x,\varphi,\mathcal{U}) = \varlimsup_{N\to\infty} \sup_{U} \frac{\alpha \log \mu(X(U))}{-N\alpha + \displaystyle\sup_{y\in X(U)} \sum_{k=0}^{N-1} \varphi(f^k(y))},
$$

where the infimum and supremum are taken over all strings U with $x \in X(U)$ and $m(U) = N$.

Proposition 11.1. *If μ is a Borel probability measure on X invariant under the map f and ergodic, then for every $\alpha \in \mathbb{R}$ and μ-almost every $x \in X$*

$$
\lim_{|\mathcal{U}|\to 0} \underline{\mathcal{D}}_{C,\mu,\alpha}(x,\varphi,\mathcal{U}) = \lim_{|\mathcal{U}|\to 0} \overline{\mathcal{D}}_{C,\mu,\alpha}(x,\varphi,\mathcal{U}) = \frac{\alpha h_\mu(f)}{\alpha - \int_X \varphi\, d\mu},
$$

where $h_\mu(f)$ is the measure-theoretic entropy of f.

Proof. We need the following statement known as the Brin–Katok local entropy formula (see [BK]).

Lemma. *For μ-almost every $x \in X$ we have*

$$
h_\mu(f) = \lim_{\delta\to 0} \varliminf_{n\to\infty} \frac{-\log \mu(B_n(x,\delta))}{n} = \lim_{\delta\to 0} \varlimsup_{n\to\infty} \frac{-\log \mu(B_n(x,\delta))}{n},
$$

where $B_n(x,\delta)$ is the (n,δ)-ball at x (see (11.10)).

Proof of the lemma. For the sake of reader's convenience we present a simplified version of the proof in [BK] which exploits the fact that the measure

μ is ergodic. Fix $\delta > 0$ and consider a finite measurable partition ξ with diam $\xi \overset{\text{def}}{=} \max\{\text{diam}\, C_\xi : C_\xi \in \xi\} \leq \delta$. Denote by $C_{\xi_n}(x)$ the element of the partition

$$\xi_n = \xi \vee f^{-1}\xi \vee \cdots \vee f^{-n}\xi$$

containing x. Obviously, $C_{\xi_n}(x) \subset B_n(x, \delta)$.

By the Shannon–McMillan–Breiman theorem the following limit exists for μ-almost every $x \in X$:

$$\lim_{n\to\infty} \frac{-\log\mu(C_{\xi_n}(x))}{n} = h_\mu(f, \xi),$$

where $h_\mu(f, \xi)$ is the measure-theoretic entropy of f with respect to ξ. It follows that

$$\varlimsup_{\delta\to 0}\, \varlimsup_{n\to\infty} \frac{-\log\mu(B_n(x, \delta))}{n} \leq h_\mu(f, \xi) \leq h_\mu(f).$$

We proceed now with the estimate from below. Fix $\varepsilon > 0$. One can show that there exists a finite measurable partition ξ of X satisfying:

(1) $h_\mu(f, \xi) \geq h_\mu(f) - \varepsilon$;

(2) $\mu(\partial\xi) = 0$, where $\partial\xi$ denotes the boundary of the partition ξ.

For $\delta > 0$ let

$$U_\delta(\xi) = \{x \in X : \text{ the ball } B(x, \delta) \text{ is not contained in } C_\xi(x)\}.$$

Since $\bigcap_{\delta > 0} U_\delta(\xi) = \partial\xi$ we obtain that $\mu(U_\delta(\xi)) \to 0$ as $\delta \to 0$. Therefore, one can choose $\delta_0 > 0$ such that $\mu(U_\delta(\xi)) \leq \varepsilon$ for any $0 < \delta \leq \delta_0$. Hence, by the Birkhoff ergodic theorem, for μ-almost every $x \in X$ there exists $N_1(x)$ such that for any $n \geq N_1(x)$,

$$\frac{1}{n}\sum_{i=0}^{n-1}\chi_{U_\delta(\xi)}(f^i(x)) \leq \varepsilon.$$

Let $A_\ell = \{x \in X : N_1(x) \leq \ell\}$. Clearly, the sets A_ℓ are nested and exhaust X up to a set of measure zero. Therefore, there exists $\ell_0 > 1$ such that $\mu(A_\ell) \geq 1 - \varepsilon$ for any $\ell \geq \ell_0$.

Fix $\ell \geq \ell_0$. Given a point $x \in X$, we call the collection

$$(C_\xi(x), C_\xi(f(x)), \ldots, C_\xi(f^{n-1}(x)))$$

the (ξ, n)-name of x. If $y \in B_n(x, \delta)$ then for any $0 \leq i \leq n - 1$ either $f^i(x)$ and $f^i(y)$ belong to the same element of ξ or $f^i(x) \in U_\delta(\xi)$. Hence, if $x \in A_\ell$ and $y \in B_n(x, \delta)$, then the *Hamming distance* between (ξ, n)-names of x and y does not exceed ε (recall that the Hamming distance is defined as follows: $\frac{1}{n}\sum_{i=0}^{n-1}\mu(C_\xi(f^i(x))\Delta C_\xi(f^i(y))))$. Furthermore, for $x \in A_\ell$, Bowen's ball $B_n(x, \delta)$ is contained in the set of points y whose (ξ, n)-names are ε-close to the (ξ, n)-name of x. It can be shown that the total number L_n of such (ξ, n)-names admits the following estimate:

$$L_n \leq \exp(K_1\varepsilon n),$$

where $K_1 > 1$ is a constant independent of x and n. We wish to estimate the measure of those points in A_ℓ whose (ξ, n)-names have an element of the partition ξ_n of measure greater than $\exp\left((-h_\mu(f, \xi) + 2K_1\varepsilon)n\right)$ in their Hamming ε-neighborhood. Obviously, the total number of such elements does not exceed $\exp\left((h_\mu(f, \xi) - 2K_1\varepsilon)n\right)$. Hence, the total number Q_n of elements in their Hamming ε-neighborhood satisfies

$$Q_n \leq \exp\left((h_\mu(f, \xi) + K_1\varepsilon - 2K_1\varepsilon)n\right) \leq \exp\left((h_\mu(f, \xi) - K_1\varepsilon)n\right).$$

By the Shannon–McMillan–Breiman theorem for μ-almost every $x \in X$ there exists $N_2(x)$ such that for any $n \geq N_2(x)$,

$$\frac{-\log\mu(C_{\xi_n}(x))}{n} \geq h_\mu(f, \xi) - \varepsilon.$$

Let $B_k = \{x \in X : N_2(x) \leq k\}$. Clearly, the sets B_k are nested and exhaust X up to a set of measure zero. Therefore, there exists $k > 1$ such that $\mu(B_k) \geq 1 - \varepsilon$ for any $k \geq k_0$. Fix such a number k and consider those of the Q_n elements of ξ_n whose intersection with $A_\ell \cap B_k$ have positive measure. To estimate their total measure S_n we multiply their number by the upper bound of their measure

$$S_n \leq \exp\left((h_\mu(f, \xi) - K_1\varepsilon - h_\mu(f, \xi) + \varepsilon)n\right) \leq \exp\left(-(K_1 - 1)\varepsilon n\right).$$

This implies that for any sufficiently small $\varepsilon > 0$ and $\delta > 0$

$$\varliminf_{n \to \infty} \frac{-\log\mu(B_n(x, \delta))}{n} \geq h_\mu(f, \xi) - K_2\varepsilon \geq h_\mu(f) - \varepsilon - K_2\varepsilon$$

for every point x in a subset $D \subset A_\ell \cap B_k$ of measure $\geq \mu(A_\ell \cap B_k) - K_3\varepsilon$ (here $K_2 > 0$ and $K_3 > 0$ are constants independent of x, δ, and ε). Therefore,

$$\lim_{\delta \to 0}\varliminf_{n \to \infty} \frac{-\log\mu(B_n(x, \delta))}{n} \geq h_\mu(f).$$

This completes the proof of the statement.   ∎

We continue the proof of the proposition. Let $\mathcal{U}$ be a finite open cover of X and $\delta(\mathcal{U})$ its Lebesgue number. Since $\delta(\mathcal{U}) \to 0$ as $|\mathcal{U}| \to 0$ it follows from (11.11) and the lemma that

$$\begin{aligned}
h_\mu(f) &= \lim_{|\mathcal{U}| \to 0}\varliminf_{N \to \infty}\inf_{\boldsymbol{U}} \frac{-\log\mu(X(\boldsymbol{U}))}{N} \\
&= -\lim_{|\mathcal{U}| \to 0}\varlimsup_{N \to \infty}\sup_{\boldsymbol{U}} \frac{-\log\mu(X(\boldsymbol{U}))}{N},
\end{aligned} \tag{11.16}$$

where the infimum and supremum are taken over all strings $\boldsymbol{U}$ for which $x \in X(\boldsymbol{U})$ and $m(\boldsymbol{U}) = N$. Let us fix a number $\varepsilon > 0$. Since φ is continuous on X there exists a number $\delta > 0$ such that $|\varphi(x) - \varphi(y)| \leq \varepsilon$ for any two points

$x, y \in X$ with $\rho(x, y) \leq \delta$. Therefore, if $|\mathcal{U}| \leq \delta$ then by view of Birkhoff ergodic theorem, we obtain for μ-almost every $x \in X$ that

$$\left| \varliminf_{N \to \infty} \inf_{\mathcal{U}} \sup_{y \in X(\mathcal{U})} \frac{1}{N} \sum_{k=0}^{N-1} \varphi(f^k(y)) - \int_X \varphi \, d\mu \right| \leq \varepsilon,$$

$$\left| \varlimsup_{N \to \infty} \sup_{\mathcal{U}} \sup_{y \in X(\mathcal{U})} \frac{1}{N} \sum_{k=0}^{N-1} \varphi(f^k(y)) - \int_X \varphi \, d\mu \right| \leq \varepsilon,$$

where the infimum and supremum are taken over all strings $\mathbf{U}$ for which $x \in X(\mathbf{U})$ and $m(\mathbf{U}) = N$. Since ε is arbitrary this implies that

$$\lim_{|\mathcal{U}| \to 0} \varliminf_{N \to \infty} \inf_{\mathcal{U}} \sup_{y \in X(\mathbf{U})} \frac{1}{N} \sum_{k=0}^{N-1} \varphi(f^k(y))$$
$$= \lim_{|\mathcal{U}| \to 0} \varlimsup_{N \to \infty} \sup_{\mathbf{U}} \sup_{y \in X(\mathbf{U})} \frac{1}{N} \sum_{k=0}^{N-1} \varphi(f^k(y)) = \int_X \varphi \, d\mu. \tag{11.17}$$

The desired result follows immediately from (11.16) and (11.17). ∎

We now use Proposition 11.1 to prove the following result.

Theorem 11.6. *Let f be a homeomorphism of a compact metric space X and μ a non-atomic Borel ergodic measure on X. Then*

$$P_\mu(\varphi) = \underline{CP}_\mu(\varphi) = \overline{CP}_\mu(\varphi) = h_\mu(f) + \int_X \varphi \, d\mu.$$

Proof. Set $h = h_\mu(f) \geq 0$ and $a = \int_X \varphi d\mu$. We first assume that $a > 0$. We wish to use Theorems 3.4 and 3.5 to obtain the proper lower bound for $P_\mu(\varphi)$ and upper bound for $\overline{CP}_\mu(\varphi)$. To do so we need to find estimates of $\underline{\mathcal{D}}_{C,\mu,\alpha}(x, \varphi, \mathcal{U})$ and $\overline{\mathcal{D}}_{C,\mu,\alpha}(x, \varphi, \mathcal{U})$ from below and above respectively which do not depend on α.

Fix ε, $0 < \varepsilon < \frac{a}{2}$. By Proposition 11.1 one can choose $\delta > 0$ such that for μ-almost every $x \in X$,

$$\underline{\mathcal{D}}_{C,\mu,\alpha}(x, \varphi, \mathcal{U}) \geq \frac{\alpha h}{\alpha - a} - \varepsilon.$$

Note that the function $g(\alpha) = \alpha h(\alpha - a)^{-1} - \varepsilon$ is decreasing. Assuming that α varies on the interval $[h + a - \varepsilon, h + a]$, we obtain that for μ-almost every $x \in X$,

$$\underline{\mathcal{D}}_{C,\mu,\alpha}(x, \varphi, \mathcal{U}) \geq h + a - 2\varepsilon.$$

We conclude, using Theorem 3.4, that $P_\mu(\varphi, \mathcal{U}) \geq h + a - 2\varepsilon$ and hence $P_\mu(\varphi, \mathcal{U}) \geq h + a$. Since this holds for every finite open cover $\mathcal{U}$ by (11.15) we obtain that $P_\mu(\varphi) \geq h + a$.

We now show that $\overline{CP}_\mu(\varphi) \leq h + a$. Fix $\varepsilon > 0$. Let $\xi = \{C_1, \ldots, C_p\}$ be a finite measurable partition of X with $|h_\mu(f, \xi) - h| \leq \varepsilon$ and $\mathcal{U} = \{U_1, \ldots, U_p\}$ a finite open cover of X of diameter $\leq \varepsilon$ for which $C_i \subset U_i$, $i = 1, \ldots, p$.

By the Birkhoff ergodic theorem for μ-almost every $x \in X$ there exists a number $N_1(x) > 0$ such that for any $n \geq N_1(x)$,

$$\left| \frac{1}{n} \sum_{k=0}^{n-1} \varphi(f^k(y)) - a \right| \leq \varepsilon. \tag{11.18}$$

By the Shannon–McMillan–Breiman theorem for μ-almost every $x \in X$ there exists a number $N_2(x) > 0$ such that for any $n \geq N_2(x)$,

$$\left| \frac{1}{n} \log \mu(C_{\xi_n}(x)) + h_\mu(f, \xi) \right| \leq \varepsilon. \tag{11.19}$$

Let Δ be the set of points for which (11.18) and (11.19) hold. Given $N > 0$, consider the set $\Delta_N = \{x \in \Delta : N_1(x) \leq N \text{ and } N_2(x) \leq N\}$. We have that $\Delta_N \subset \Delta_{N+1}$ and $\Delta = \cup_{N \geq 0} \Delta_N$. Therefore, given $\delta > 0$, one can find $N_0 > 0$ for which $\mu(\Delta_{n_0}) \geq 1 - \delta$. Fix a number $N \geq N_0$ and a point $x \in \Delta_N$. Let U be a string of length $m(U) = N$ for which $x \in X(U)$. It follows from (11.18) that

$$\left| \frac{1}{N} \sup_{y \in X(U)} \sum_{k=0}^{N-1} \varphi(f^k(y)) - a \right| \leq \varepsilon + \gamma, \tag{11.20}$$

where $\gamma = \gamma(\mathcal{U})$. Furthermore, using (11.19) we obtain that

$$\mu(C_{\xi_N}(x)) \geq \exp(-h - 2\varepsilon)N.$$

This implies that the number of elements of the partition ξ_N that have non-empty intersection with the set Δ_N does not exceed $\exp(h + 2\varepsilon)N$.

To each element C_{ξ_N} of the partition ξ_N we associate a string U of length $m(U) = N$ for which $C_{\xi_N} \subset X(U)$. The collection of such strings consists of at most $\exp(h + 2\varepsilon)N$ elements which comprise a cover $\mathcal{G}$ of Δ_N. By (11.8) and (11.20) we obtain that

$$\Lambda(\Delta_N, \varphi, \mathcal{U}, N) \leq \sum_{U \in \mathcal{G}} \exp\left(\sup_{y \in X(U)} \sum_{k=0}^{N-1} \varphi(f^k(y)) \right)$$
$$\leq \exp(a + h + 3\varepsilon + \gamma)N.$$

In view of (11.7) this means that

$$\overline{CP}_{\Delta_N}(\varphi, \mathcal{U}) \leq a + h + 3\varepsilon + \gamma.$$

This implies that $\overline{CP}_\mu(\varphi, \mathcal{U}) \leq a + h + 3\varepsilon + \gamma$. Passing to the limit as $\operatorname{diam} \mathcal{U} \to 0$ yields that $\overline{CP}_\mu(\varphi) \leq a + h + 3\varepsilon$. It remains to note that ε can be chosen arbitrarily small to conclude that $\overline{CP}_\mu(\varphi) \leq a + h$.

In the case $a \leq 0$, let us consider a function $\psi = \varphi + C$, where C is chosen such that $\int_X \psi d\mu > 0$. Note that $P_\mu(\psi, \mathcal{U}) = P_\mu(\varphi, \mathcal{U}) + C$ and $\overline{CP}_\mu(\psi, \mathcal{U}) = \overline{CP}_\mu(\varphi, \mathcal{U}) + C$, and the desired result follows. ∎

As an immediate consequence of Theorem 11.6 we obtain that

$$h_\mu(f) = P_\mu(0) = \underline{CP}_\mu(0) = \overline{CP}_\mu(0).$$

These relations reveal the "dimension" nature of the notion of measure-theoretic entropy, introduced by Kolmogorov and Sinai within the framework of general measure theory. One can obtain another "dimension" interpretation of measure-theoretic entropy using Proposition 11.1. Namely,

$$h_\mu(f) = \lim_{|\mathcal{U}| \to 0} \underline{d}_{C,\mu,\alpha}(x,0,\mathcal{U}) = \lim_{|\mathcal{U}| \to 0} \overline{d}_{C,\mu,\alpha}(x,0,\mathcal{U}).$$

In conclusion, we point out a remarkable application of relations (11.14) and (11.15) known as the **inverse variational principle for topological pressure**:

$$h_\mu(f) + \int_X \varphi \, d\mu = \inf \{ P_Z(\varphi) : \mu(Z) = 1 \}.$$

In particular, when $\varphi = 0$ this gives the **inverse variational principle for topological entropy**

$$h_\mu(f) = \inf \{ h_Z(f) : \mu(Z) = 1 \}.$$

This result was first established by Bowen [Bo1].

Let us also point out that the requirement in Proposition 11.1 and Theorem 11.6, that μ is ergodic, is crucial; they may not hold true otherwise.

12. Non-additive Thermodynamic Formalism

Let (X, ρ) be a compact metric space with metric ρ, $f : X \to X$ a continuous map, and $\varphi = \{\varphi_n : X \to \mathbb{R}\}$ a sequence of continuous functions. Consider a finite open cover $\mathcal{U}$ of X and define for each $n \geq 1$

$$\gamma_n(\varphi, \mathcal{U}) = \sup \{ |\varphi_n(x) - \varphi_n(y)| : x, y \in X(\mathbf{U}) \text{ for some } \mathbf{U} \in \mathcal{S}_n(\mathcal{U}) \}.$$

We assume that the following property holds:

$$\lim_{|\mathcal{U}| \to 0} \varlimsup_{n \to \infty} \frac{\gamma_n(\varphi, \mathcal{U})}{n} = 0 \tag{12.1}$$

(since $\gamma_n(\varphi, \mathcal{U}) \geq 0$ one can show that the limit exists as $|\mathcal{U}| \to 0$).

We define now the collection of subsets $\mathcal{F}$ by (11.1) and (11.2) and three functions $\xi, \eta, \psi : \mathcal{S}(\mathcal{U}) \to \mathbb{R}$ as follows

$$\xi(\mathbf{U}) = \exp \big(\sup_{x \in X(\mathbf{U})} \varphi_{m(\mathbf{U})}(x) \big), \quad \eta(\mathbf{U}) = \exp(-m(\mathbf{U})),$$

$$\psi(\mathbf{U}) = m(\mathbf{U})^{-1}. \tag{12.2}$$

One can verify using (12.1) that the collection of subsets $\mathcal{F}$ and the functions η, ξ, and ψ satisfy Conditions $A1, A2, A3$, and $A3'$ in Section 10 and hence determine a C-structure $\tau = (\mathcal{S}, \mathcal{F}, \xi, \eta, \psi)$ on X. The corresponding Carathéodory function $m_C(Z, \alpha)$ ($Z \subset X$ and $\alpha \in \mathbb{R}$) depends on the cover $\mathcal{U}$ and is given by

$$m_C(Z, \alpha) = \lim_{N \to \infty} M(Z, \alpha, \varphi, \mathcal{U}, N),$$

where

$$M(Z, \alpha, \varphi, \mathcal{U}, N) = \inf_{\mathcal{G}} \left\{ \sum_{U \in \mathcal{G}} \exp \left(-\alpha\, m(U) + \sup_{x \in X(U)} \varphi_{m(U)}(x) \right) \right\}$$

and the infimum is taken over all finite or countable collections of strings $\mathcal{G} \subset \mathcal{S}(\mathcal{U})$ such that $m(U) \geq N$ for all $U \in \mathcal{G}$ and $\mathcal{G}$ covers Z.

Furthermore, the Carathéodory functions $\underline{r}_C(Z, \alpha)$ and $\overline{r}_C(Z, \alpha)$ (where $Z \subset X$ and $\alpha \in \mathbb{R}$) depend on the cover $\mathcal{U}$ and are given by

$$\underline{r}_C(Z, \alpha) = \varliminf_{N \to \infty} R(Z, \alpha, \varphi, \mathcal{U}, N), \quad \overline{r}_C(Z, \alpha) = \varlimsup_{N \to \infty} R(Z, \alpha, \varphi, \mathcal{U}, N),$$

where

$$R(Z, \alpha, \varphi, \mathcal{U}, N) = \inf_{\mathcal{G}} \left\{ \sum_{U \in \mathcal{G}} \exp \left(-\alpha N + \sup_{x \in X(U)} \varphi_N(x) \right) \right\}$$

and the infimum is taken over all finite or countable collections of strings $\mathcal{G} \subset \mathcal{S}(\mathcal{U})$ such that $m(U) = N$ for all $U \in \mathcal{G}$ and $\mathcal{G}$ covers Z.

According to Section 10, given a set $Z \subset X$, this C-structure generates the Carathéodory dimension of Z and the lower and upper Carathéodory capacities of Z specified by the cover $\mathcal{U}$ and the map f. We denote them, respectively, by $P_Z(\varphi, \mathcal{U})$, $\underline{CP}_Z(\varphi, \mathcal{U})$, and $\overline{CP}_Z(\varphi, \mathcal{U})$. Repeating arguments in the proof of Theorem 11.1, one can show that for any $Z \subset X$ the following limits exist:

$$P_Z(\varphi) \stackrel{\text{def}}{=} \lim_{|\mathcal{U}| \to 0} P_Z(\varphi, \mathcal{U}),$$

$$\underline{CP}_Z(\varphi) \stackrel{\text{def}}{=} \lim_{|\mathcal{U}| \to 0} \underline{CP}_Z(\varphi, \mathcal{U}),$$

$$\overline{CP}_Z(\varphi) \stackrel{\text{def}}{=} \lim_{|\mathcal{U}| \to 0} \overline{CP}_Z(\varphi, \mathcal{U}).$$

We call the quantities $P_Z(\varphi)$, $\underline{CP}_Z(\varphi)$, and $\overline{CP}_Z(\varphi)$, respectively, the **non-additive topological pressure** and **non-additive lower** and **upper capacity topological pressures** of the sequence of functions φ on the set Z (with respect to f). They were introduced by Barreira in [Bar2]. We emphasize that the set Z can be arbitrary and need not be compact or invariant under the map f.

The quantities $P_Z(\varphi)$, $\underline{CP}_Z(\varphi)$, and $\overline{CP}_Z(\varphi)$ have the properties stated in Theorems 11.2 and 11.3. One can check that the functions η and ψ satisfy Condition $A4$ in Section 10. Therefore,

$$\underline{CP}_Z(\varphi) = \lim_{|\mathcal{U}| \to 0} \varliminf_{n \to \infty} \frac{1}{N} \log \Lambda(Z, \varphi, \mathcal{U}, N),$$

$$\overline{CP}_Z(\varphi) = \lim_{|\mathcal{U}| \to 0} \varlimsup_{n \to \infty} \frac{1}{N} \log \Lambda(Z, \varphi, \mathcal{U}, N),$$

where, in accordance with (2.3) and (12.2),

$$\Lambda(Z, \varphi, \mathcal{U}, N) = \inf_{\mathcal{G}} \left\{ \sum_{U \in \mathcal{G}} \exp\left(\sup_{x \in X(U)} \varphi_N(x) \right) \right\} \qquad (12.3)$$

and the infimum is taken over all finite or countable collections of strings $\mathcal{G} \subset S(\mathcal{U})$ such that $m(U) = N$ for all $U \in \mathcal{G}$ and $\mathcal{G}$ covers Z.

The use of the adjective "non-additive" is due to the following observation. A sequence of functions $\varphi = \{\varphi_n\}$ is called **additive** if

$$\varphi_{n+m}(x) = \varphi_n(x) + \varphi_m(f^n(x))$$

for any $n, m \geq 1$, and $x \in X$. One can verify that the sequence φ is additive if and only if

$$\varphi_n(x) = \sum_{k=0}^{n-1} \varphi(f^k(x)),$$

where φ is a function. It is not difficult to check that an additive sequence satisfies (12.1) if the function φ is continuous. Furthermore, in this case, for any $Z \subset X$,

$$P_Z(\varphi) = P_Z(\varphi), \quad \underline{CP}_Z(\varphi) = \underline{CP}_Z(\varphi), \quad \overline{CP}_Z(\varphi) = \overline{CP}_Z(\varphi).$$

In the second part of the book we will often deal with non-additive sequences of functions. One can naturally associate such sequences with geometric constructions of a general type in dimension theory (including Moran-like geometric constructions; see Chapter 5) as well as with dynamical systems of hyperbolic type (including smooth expanding maps and Axiom A diffeomorphisms; see Chapter 7). The non-additive topological pressure will be used as an essential tool in computing the Hausdorff dimension of the limit sets of geometric constructions as well as of invariant sets for hyperbolic dynamical systems.

We note that the sequences of functions we will deal with, being in general non-additive, often satisfy a special property which we now describe. We call a sequence of functions $\varphi = \{\varphi_n\}$ **sub-additive** if for any $n, m \geq 1$, and $x \in X$,

$$\varphi_{n+m}(x) \leq \varphi_n(x) + \varphi_m(f^n(x)). \qquad (12.4)$$

The proof of the following result is similar to the proof of Theorem 11.5.

Theorem 12.1. [Bar2]

(1) *If a set $Z \subset X$ is f-invariant and a sequence of functions φ is sub-additive then for any finite open cover $\mathcal{U}$ of X we have $\underline{CP}_Z(\varphi, \mathcal{U}) = \overline{CP}_Z(\varphi, \mathcal{U})$.*

(2) *If a set $Z \subset X$ is f-invariant and compact and a sequence of functions φ is sub-additive and satisfies*

$$\varphi_n \leq \varphi_{n+1} + K \tag{12.5}$$

for some $K > 0$ then for any finite open cover $\mathcal{U}$ of X,

$$P_Z(\varphi, \mathcal{U}) = \underline{CP}_Z(\varphi, \mathcal{U}) = \overline{CP}_Z(\varphi, \mathcal{U}) = \lim_{n \to \infty} \frac{1}{N} \log \Lambda(Z, \varphi, \mathcal{U}, N),$$

where $\Lambda(Z, \varphi, \mathcal{U}, N)$ is given by (12.2).

Appendix II

Variational Principle for Topological Pressure; Symbolic Dynamical Systems; Bowen's Equation

The mathematical foundation of the thermodynamic formalism, i.e., the formalism of equilibrium statistical physics, has been led by Ruelle [R1]. Bowen, Ruelle, and Sinai have used the thermodynamic approach to study ergodic properties of smooth hyperbolic dynamical systems (see references and discussion in [KH]). The main constituent components of the thermodynamic formalism are:

(a) the topological pressure of a continuous function φ which determines the "potential of the system";

(b) the variational principle for the topological pressure, which establishes the variational property of the "free energy" of the system (which is defined as the sum of the measure-theoretic entropy and the integral of φ with respect to a probability distribution in the phase space of the system);

(c) existence, uniqueness, and ergodic properties of equilibrium measures (which are extremes of the variational principle).

Ruelle's version of the thermodynamic formalism is based on the classical notion of topological pressure for compact invariant sets. In this appendix we outline more general versions of the variational principle by considering topological pressure on non-compact sets and non-additive topological pressure.

We use the thermodynamic formalism to describe Gibbs measures for symbolic dynamical systems. For the reader's convenience we also provide a brief description of basic notions in symbolic dynamics which are widely used in the second part of the book.

One of the main manifestations of the thermodynamic formalism in dimension theory is different versions of Bowen's equation. Its roots often provide optimal estimates (and sometimes the exact value) of the dimension of an invariant set. We describe some properties of the pressure function and study roots of Bowen's equation.

In the second part of the book the reader will find many applications of these results to dimension theory as well as to the theory of dynamical systems.

Variational Principle for Topological Pressure

Let (X, ρ) be a compact metric space with metric ρ, $f: X \to X$ a continuous map, and $\varphi: X \to \mathbb{R}$ a continuous function. Denote by $\mathfrak{M}(X)$ the set of all f-invariant Borel ergodic measures on X. Given an f-invariant (not necessarily compact) set $Z \subset X$, denote also by $\mathfrak{M}(Z) \subset \mathfrak{M}(X)$ the set of measures μ for which $\mu(Z) = 1$. For each $x \in X$ and $n \geq 0$ we define a probability measure $\mu_{x,n}$ on X by

$$\mu_{x,n} = \frac{1}{n} \sum_{k=0}^{n-1} \delta_{f^k(x)},$$

where δ_y is the δ-measure supported at the point y. Denote by $V(x)$ the set of limit measures (in the weak topology) of the sequence of measures $(\mu_{x,n})_{n\in\mathbb{N}}$. It is easy to see that $\varnothing \neq V(x) \subset \mathfrak{M}(X)$ for each $x \in X$. Put $\mathcal{L}(Z) = \{x \in Z : V(x) \cap \mathfrak{M}(Z) \neq \varnothing\}$. It is easy to check that $\mathcal{L}(Z)$ is a Borel f-invariant set.

The following statement establishes the variational principle for the topological pressure on non-compact sets. It was proved by Pesin and Pitskel' in [PP].

Theorem A2.1. *Let $Z \subset X$ be an f-invariant set. Then for any continuous function φ on X,*

$$P_{\mathcal{L}(Z)}(\varphi) = \sup_{\mu \in \mathfrak{M}(Z)} \left(h_\mu(f) + \int_Z \varphi \, d\mu \right)$$

(recall that $P_Y(\varphi)$ is the topological pressure of the function φ on the set Y and $h_\mu(f)$ is the measure-theoretic entropy of f; see Section 11).

Proof. We present a sketch of the proof following [PP]. We first show that for any f-invariant subset $Y \subset X$ (i.e., $f^{-1}(Y) = Y$) and any measure $\mu \in \mathfrak{M}(Y)$,

$$h_\mu(f) + \int_Y \varphi \, d\mu \leq P_Y(\varphi). \tag{A2.1}$$

One can prove the following statement.

Lemma 1. *For any $\varepsilon > 0$ there exists δ, $0 < \delta \leq \varepsilon$, a finite Borel partition $\xi = \{C_1, \ldots, C_m\}$, and a finite open cover $\mathcal{U} = \{U_1, \ldots, U_k\}$, $k \geq m$ of X such that*

(1) $\operatorname{diam} U_i \leq \varepsilon$, $\operatorname{diam} C_j \leq \varepsilon$, $i = 1, \ldots, k$, $j = 1, \ldots, m$;

(2) $\overline{U}_i \subset C_i$, $i = 1, \ldots, m$ *(where $\overline{A}$ denotes the closure of the subset $A \subset Y$ in the induced topology of Y);*

(3) $\mu(C_i \setminus U_i) \leq \delta$, $i = 1, \ldots, m$ *and* $\mu(\bigcup_{i=m+1}^k U_i) \leq \delta$;

(4) $2\delta \log m \leq \varepsilon$.

Given $y \in Y$, let $t_n(y)$ denote the number of those ℓ, $0 \leq \ell < n$, for which $f^\ell(y) \in U_i$ for some $i = m+1, \ldots, k$. It follows from Lemma 1 (see Statement 3) and the Birkhoff ergodic theorem that there exist $N_1 > 0$ and a set $A_1 \subset Y$ such that $\mu(A_1) \geq 1 - \delta$ and for any $y \in A_1$ and $n \geq N_1$,

$$n^{-1} t_n(y) \leq 2\delta. \tag{A2.2}$$

Set $\xi_n = \xi \vee f^{-1}\xi \vee \cdots \vee f^{-n}\xi$. It follows from the Shannon–McMillan–Breiman theorem that there exist $N_2 > 0$ and a set $A_2 \subset Y$ such that $\mu(A_2) \geq 1 - \delta$ and for any $y \in A_2$ and $n \geq N_2$,

$$\mu(C_{\xi_n}(y)) \leq \exp(-(h_\mu(f,\xi) - \delta)n), \tag{A2.3}$$

where $C_{\xi_n}(y)$ denotes the element of the partition ξ_n containing y. At last, using the Birkhoff ergodic theorem one can find $N_3 > 0$ and a set $A_3 \subset Y$ such that $\mu(A_3) \geq 1 - \delta$ and for any $y \in A_3$ and $n \geq N_3$,

$$\left| \frac{1}{n} \sum_{i=0}^{n-1} \varphi(f^i(y)) - \int_Y \varphi \, d\mu \right| \leq \delta. \tag{A2.4}$$

Set $N = \max\{N_1, N_2, N_3\}$ and $A = A_1 \cap A_2 \cap A_3$. We have that

$$\mu(A) \geq 1 - 3\delta. \tag{A2.5}$$

Choose any $n \geq N$ and any

$$\lambda < h_\mu(f,\xi) + \int_Y \varphi \, d\mu - \gamma(\mathcal{U})$$

(we recall that $\gamma(\mathcal{U}) = \sup\{|\varphi(x) - \varphi(y)| : x, y \in U_i \text{ for some } U_i \in \mathcal{U}\}$). By the definition of the topological pressure there exists a finite or countable collection of strings $\mathcal{G} \subset \mathcal{S}(\mathcal{U})$, which covers Y, (i.e., the collection of sets $\{X(\boldsymbol{U}) : \boldsymbol{U} \in \mathcal{G}\}$ covers Y) such that $m(\boldsymbol{U}) \geq N$ and

$$\left| \sum_{\boldsymbol{U} \in \mathcal{G}} \exp\left(-\lambda m(\boldsymbol{U}) + \sup_{x \in X(\boldsymbol{U})} \sum_{k=0}^{m(\boldsymbol{U})-1} \varphi(f^k(x)) \right) - M(Y, \lambda, \varphi, \mathcal{U}, N) \right| \leq \delta. \tag{A2.6}$$

Let $\mathcal{G}_\ell \subset \mathcal{G}$ be a subcollection of strings for which $m(\boldsymbol{U}) = \ell$ and $X(\boldsymbol{U}) \cap A \neq \varnothing$. Denote by P_ℓ the cardinality of $\mathcal{G}_\ell$. Set $Y_\ell = \bigcup_{\boldsymbol{U} \in \mathcal{G}_\ell} X(\boldsymbol{U})$.

Lemma 2. We have $P_\ell \geq \mu(Y_\ell \cap A) \exp\left((h_\mu(f,\xi) - \delta - 2\delta \log m)\ell \right)$.

Proof of the lemma. Let L_ℓ be the number of those elements of the partition ξ_ℓ for which

$$C_{\xi_\ell} \cap Y_\ell \cap A \neq \varnothing. \tag{A2.7}$$

It is easy to see that

$$\sum \mu(C_{\xi_\ell}) \geq \mu(Y_\ell \cap A), \tag{A2.8}$$

where the sum is taken over all elements of the partition ξ_ℓ for which Condition (A2.7) holds. Since $C_{\xi_\ell} \cap A_2 \neq \varnothing$ we obtain by (A2.3) and (A2.8) that

$$L_\ell \geq \mu(Y_\ell \cap A) \exp\left((h_\mu(f,\xi) - \delta)\ell \right). \tag{A2.9}$$

Let us fix $\boldsymbol{U} \in \mathcal{G}_\ell$. Since $X(\boldsymbol{U}) \cap A_1 \neq \varnothing$ the inequality (A2.2) implies that the number $S_\ell(\boldsymbol{U})$ of those elements C_{ξ_ℓ} of the partition ξ_ℓ, for which $X(\boldsymbol{U}) \cap C_{\xi_\ell} \cap A \neq \varnothing$, admits the following estimate:

$$S_\ell(\boldsymbol{U}) \leq m^{2\delta\ell} = \exp(2\delta\ell \log m). \tag{A2.10}$$

The lemma now follows from (A2.9) and (A2.10). ∎

We continue the proof of the inequality (A2.1). Using Lemma 2, (A2.4), and (A2.5) we obtain that

$$\sum_{U \in \mathcal{G}} \exp\left(-\lambda\, m(U) + \sup_{x \in X(U)} \sum_{k=0}^{m(U)-1} \varphi(f^k(x))\right)$$

$$\geq \sum_{\ell=N}^{\infty} \sum_{U \in \mathcal{G}_\ell} \exp\left(-\lambda\ell + \sup_{x \in X(U)} \sum_{k=0}^{m(U)-1} \varphi(f^k(x))\right)$$

$$\geq \sum_{\ell=N}^{\infty} P_\ell \exp\left(\left(-\lambda + \int_Y \varphi\, d\mu - \delta - \gamma(\mathcal{U})\right)\ell\right)$$

$$\geq \sum_{\ell=N}^{\infty} \mu(Y_\ell \cap A) \exp\left(\left(h_\mu(f,\xi) + \int_Y \varphi\, d\mu - 2\delta - 2\delta \log m - \gamma(\mathcal{U}) - \lambda\right)\ell\right)$$

$$\geq \sum_{\ell=N}^{\infty} \mu(Y_\ell \cap A) = \mu(A) \geq 1 - 3\delta.$$

We used here the fact that for sufficiently small ε

$$h_\mu(f,\xi) + \int_Y \varphi\, d\mu - 2\delta - 2\delta \log m - \gamma(\mathcal{U}) - \lambda$$

$$\geq h_\mu(f,\xi) + \int_Y \varphi\, d\mu - 2\varepsilon - 2\varepsilon \log m - \gamma(\mathcal{U}) - \lambda > 0.$$

By (A2.6) we have that $M(Y, \lambda, \varphi, \mathcal{U}, N) \geq 1 - 4\delta \geq 1/2$ if ε (and hence δ) is sufficiently small. Therefore, $P_Y(\varphi, \mathcal{U}) > \lambda$ and hence

$$P_Y(\varphi, \mathcal{U}) \geq h_\mu(f,\xi) + \int_Y \varphi\, d\mu - \gamma(\mathcal{U}).$$

Letting $\varepsilon \to 0$ yields $\gamma(\mathcal{U}) \to 0$ and $\operatorname{diam} \xi \to 0$. The latter also implies that $h_\mu(f,\xi)$ approaches $h_\mu(f)$ and (A2.1) follows.

Consider a Borel f-invariant subset $Z \subset X$. Given a measure $\mu \in \mathfrak{M}(Z)$ denote by $Z_\mu = \{x \in Z : V(x) = \{\mu\}\}$. It is easy to see that $\mu(Z_\mu) = 1$ and that $Z_\mu \subset \mathcal{L}(Z)$. Therefore, by (A2.1),

$$P_{\mathcal{L}(Z)}(\varphi) \geq P_{Z_\mu}(\varphi) \geq h_\mu(f) + \int_Z \varphi\, d\mu. \tag{A2.11}$$

We now prove that for any Borel f-invariant (not necessarily compact) subset $Y \subset X$ with the property that $V(x) \cap \mathfrak{M}(Y) \neq \varnothing$ for any $x \in Y$ we have

$$P_Y(\varphi) \leq \sup_{\mu \in \mathfrak{M}(Z)} \left(h_\mu(f) + \int_Y \varphi\, d\mu\right). \tag{A2.12}$$

Let E be a finite set and $\underline{a} = (a_0, \ldots, a_{k-1}) \in E^k$. Define the measure $\mu_{\underline{a}}$ on E by

$$\mu_{\underline{a}}(e) = \frac{1}{k}(\text{the number of those } j \text{ for which } a_j = e).$$

Set

$$H(\underline{a}) = -\sum_{e \in E} \mu_{\underline{a}}(e) \log \mu_{\underline{a}}(e).$$

Consider the set

$$R(k, h, E) = \{\underline{a} \in E^k : H(\underline{a}) \le h\}.$$

The following statement describes the asymptotic growth in k of the number of elements in the set $R(k, h, E)$. The proof is based on rather standard combinatorial arguments and is omitted.

Lemma 3. *(See Lemma 2.16 in* [Bo1]*). We have*

$$\varlimsup_{k \to \infty} \frac{1}{k} \log |R(k, h, E)| \le h.$$

Let $\mathcal{U} = \{U_1, \ldots, U_r\}$ be an open cover of X and $\varepsilon > 0$.

Lemma 4. *Given $x \in Y$ and $\mu \in V(x) \cap \mathfrak{M}(Y)$, there exists a number $m > 0$ such that for any $n > 0$ one can find $N > n$ and a string $\boldsymbol{U} \in \mathcal{S}(\mathcal{U})$ with $m(\boldsymbol{U}) = N$ satisfying:*

(1) $x \in X(\boldsymbol{U})$;

(2)

$$\sup_{x \in X(\boldsymbol{U})} \sum_{k=0}^{m(\boldsymbol{U})-1} \varphi(f^k(x)) \le N\Big(h_\mu(f, \xi) + \int_Y \varphi \, d\mu + \varepsilon\Big);$$

(3) *the string $\boldsymbol{U}$ contains a substring $\boldsymbol{U}'$ of length $m(\boldsymbol{U}') = km \ge N - m$ which, being written as $\underline{a} = (a_0, \ldots, a_{k-1})$, satisfies the inequality*

$$\frac{1}{m} H(\underline{a}) \le h_\mu(f) + \varepsilon. \tag{A2.13}$$

Proof. There exists a Borel partition $\zeta = \{C_1, \ldots, C_r\}$ of X such that $\overline{C}_i \subset U_i$, $i = 1, \ldots, r$. By the definition of measure-theoretic entropy there exists a number $m > 0$ such that

$$\frac{1}{m} H_\mu\Big(\zeta \vee \cdots \vee f^{-(m+1)}\zeta\Big) \le h_\mu(f, \zeta) + \frac{\varepsilon}{2} \le h_\mu(f) + \frac{\varepsilon}{2}.$$

Assume that μ_{x,n_j} converges to the measure μ for some subsequence n_j. For $n' > n$ we write

$$\mu_{x,n'} = \frac{n}{n'} \mu_{x,n} + \frac{n'-n}{n'} \mu_{f^n(x), n'-n}.$$

This shows that if we replace the number n_j by the closest integer which is a factor of m then the new subsequence of measures will still converge to μ. Thus, we can assume that $n_j = mk_j$.

Let $D_1, \ldots, D_t$ be the non-empty elements of the partition $\xi = \zeta \vee \cdots \vee f^{-(m+1)}\zeta$. Fix $\beta > 0$. For each D_i one can find a compact set $K_i \subset D_i$ such that $\mu(D_i \setminus K_i) \le \beta$. Each element D_i is contained in an element of the cover

$\mathcal{V} = \mathcal{U} \vee \cdots \vee f^{-(m+1)}\mathcal{U}$ which we denote by B_i. One can find disjoint open subsets V_i such that $K_i \subset V_i \subset B_i$. Moreover, there exist Borel subsets V_i^* comprising a Borel partition of X such that $V_i \subset V_i^* \subset B_i$.

Given $n_j = mk_j$, we denote by $M_i^{(j)}$ the number of those $s \in [0, n_j)$, for which $f^s(x) \in V_i^*$, and by $M_{i,q}^{(j)}$ the number of those $s \in M_i^{(j)}$, for which $s = q$ (mod m). Define

$$p_{i,q}^{(j)} = M_{i,q}^{(j)}/k_j, \quad p_i^{(j)} = M_i^{(j)}/n_j = \frac{1}{m}(p_{i,0}^{(j)} + \cdots + p_{i,m-1}^{(j)}).$$

Since μ_{x,n_j} converges to the measure μ we obtain that

$$\varliminf_{j \to \infty} p_i^{(j)} \geq \mu(K_i) \geq \mu(D_i) - \beta,$$

$$\varlimsup_{j \to \infty} p_i^{(j)} \leq \mu(K_i) + \sum_{j \neq i} \mu(D_j \setminus K_j) \leq \mu(D_i) + t\beta.$$

If j is sufficiently large and β is sufficiently small we find that

$$\frac{1}{m}\left(-\sum_i p_i^{(j)} \log p_i^{(j)}\right) \leq \frac{1}{m}\left(-\sum_i \mu(D_i) \log \mu(D_i)\right) + \frac{\varepsilon}{2} \leq h_\mu(f) + \varepsilon.$$

Since the function $g(x) = -x \log x$ is convex we obtain that

$$g(p_i^{(j)}) \geq \frac{1}{m}\sum_{q=0}^{m-1} g(p_{i,q}^{(j)})$$

and hence

$$\sum_i g(p_i^{(j)}) \geq \frac{1}{m}\sum_{q=0}^{m-1}\sum_i g(p_{i,q}^{(j)}).$$

Therefore, the inequality

$$\sum_i g(p_{i,q}^{(j)}) \leq \sum_i g(p_i^{(j)})$$

should hold for some $q \in [0, m)$. This implies that

$$\frac{1}{m}\sum_i g(p_{i,q}^{(j)}) \leq h_\mu(f) + \varepsilon.$$

Let $N = n_j + q$. For some sufficiently large j we choose a string $\boldsymbol{U} \in \mathcal{S}(\mathcal{U})$ with $m(\boldsymbol{U}) = N$ in the following way. For $s < q$ we choose $U_s \in \mathcal{U}$ which contains $f^s(x)$. Further, for every V_i^* we choose a string $\boldsymbol{U}_i = U_{0,i} \ldots U_{m-1,i}$ such that

$$V_i^* \subset U_{0,i} \cap f^{-1}(U_{1,i}) \cdots \cap f^{-m+1}(U_{m-1,i}).$$

Then, for $s \geq q$ we write $s = q + mp + e$ with $p \geq 0$ and $m > e \geq 0$ and set $U_s = U_{e,i}$, where i is chosen such that $f^{q+mp}(x) \in V_i^*$. Set $a_p = U_{0,i}U_{1,i} \ldots U_{m-1,i}$ and consider the string $U_0 \ldots U_{q-1}a_0a_1 \ldots a_{k_j-1}$. For $\underline{a} = (a_0, \ldots, a_{k_j-1})$, the measure $\mu_{\underline{a}}$ is given by probabilities $p_{i,q}^{(j)}$, $i = 1 \ldots t$ and it satisfies

$$\frac{1}{m}H(\underline{a}) = \frac{1}{m}\sum_i g(p_{i,q}^{(j)}) \leq h_\mu(f) + \varepsilon.$$

This proves the first and the third statements of the lemma. Since μ_{x,n_j} converges to the measure μ we obtain for sufficiently large N that

$$\left|\frac{1}{N}\mu_{x,N}(\varphi) - \int_Y \varphi \, d\mu\right| \leq \varepsilon.$$

This implies the second statement and completes the proof of the lemma. ∎

Given a number $m > 0$, denote by Y_m the set of points $y \in Y$ for which Lemma 4 holds for this m and some measure $\mu \in V(x) \cap \mathfrak{M}(Y)$. We have that $Y = \bigcup_{m>0} Y_m$. Denote also by $Y_{m,u}$ the set of points $y \in Y_m$ for which Lemma 3 holds for some measure $\mu \in V(x) \cap \mathfrak{M}(Y)$ satisfying $\int_Y \varphi \, d\mu \in [u - \varepsilon, u + \varepsilon]$. Set

$$c = \sup_{\mu \in \mathfrak{M}(Y)} \left(h_\mu(f) + \int_Y \varphi \, d\mu\right).$$

Note that if $x \in Y_{m,u}$ then the corresponding measure μ satisfies $h_\mu(f) \leq c - u + \varepsilon$. Let $\mathcal{G}_{m,u}$ be the collection of all strings U described in Lemma 4 that correspond to all $x \in Y_{m,u}$ and all N exceeding some number N_0. It follows from (A2.13) that for any $x \in Y_{m,u}$ the substring constructed in Lemma 3 is contained in $R(k, m(h + \varepsilon), \mathcal{U}^m)$, where $h = c - u + \varepsilon$. Therefore, the total number of the strings constructed in Lemma 4 does not exceed $b(N) = |\mathcal{U}|^m |R(k, m(h+\varepsilon), \mathcal{U}^m)|$. By Lemma 3 we obtain that

$$\overline{\lim_{N \to \infty}} \frac{\log b(N)}{N} \leq h + \varepsilon. \tag{A2.14}$$

Since the collection of strings $\mathcal{G}_{m,u}$ covers the set $Y_{m,u}$ we conclude using Lemma 4 and (A2.14) that

$$M(Y_{m,u}, \lambda, \varphi, \mathcal{U}, N_0) \leq \sum_{N=N_0}^\infty b(N) \exp\left(-\lambda m(U) + \sup_{x \in X(U)} \sum_{k=0}^{m(U)-1} \varphi(f^k(x))\right)$$

$$\leq \sum_{N=N_0}^\infty b(N) \exp\left(-\lambda m(U) + N\left(\int_Y \varphi \, d\mu + \gamma(\mathcal{U}) + \varepsilon\right)\right).$$

If N_0 is sufficiently large, we have that $b(N) \leq \exp(N(h + 2\varepsilon))$. Hence,

$$M(Y_{m,u}, \lambda, \varphi, \mathcal{U}, N_0) \leq \frac{\beta^{N_0}}{1 - \beta}, \tag{A2.15}$$

where
$$\beta = \exp\left(-\lambda + h + \int_Y \varphi\, d\mu + \gamma(\mathcal{U}) + 3\varepsilon\right).$$

It follows from (A2.15) that if $\lambda > c + \gamma(\mathcal{U}) + 4\varepsilon$ then $m_C(Y_{m,u}, \lambda) = 0$. Hence, $\lambda \geq P_{Y_{m,u}}(\varphi, \mathcal{U})$. Assume that points $u_1, \dots, u_r$ form an ε-net of the interval $[-\|\varphi\|, \|\varphi\|]$. Then
$$Y = \bigcup_{m=1}^{\infty} \bigcup_{i=1}^{r} Y_{m,u_i}.$$

We have that $\lambda \geq P_{Y_{m,u_i}}(\varphi, \mathcal{U})$ for any m and i. Therefore,
$$\lambda \geq \sup_{m,i} P_{Y_{m,u_i}}(\varphi, \mathcal{U}) = P_Y(\varphi, \mathcal{U}).$$

This implies that $c + \gamma(\mathcal{U}) + 4\varepsilon \geq P_Y(\varphi, \mathcal{U})$. Since ε can be chosen arbitrarily small it follows that $c + \gamma(\mathcal{U}) \geq P_Y(\varphi, \mathcal{U})$. Taking the limit as $|\mathcal{U}| \to 0$ yields $c \geq P_Y(\varphi)$ and the desired result follows. ∎

We state some most interesting corollaries of Theorem A2.1:

(1) *inequality for topological pressure*: for any invariant set $Z \subset X$, any continuous function φ on X, and any measure $\mu \in \mathfrak{M}(Z)$,
$$h_\mu(f) + \int_Z \varphi\, d\mu \leq P_Z(\varphi);$$

(2) *classical variational principle for topological pressure on compact sets*: for any continuous function φ on X,
$$P_X(\varphi) = \sup_{\mu \in \mathfrak{M}(X)} \left(h_\mu(f) + \int_X \varphi\, d\mu\right);$$

(3) *classical variational principle for topological entropy on compact sets*:
$$h_X(f) = \sup_{\mu \in \mathfrak{M}(X)} h_\mu(f);$$

(4) for any set $Z \subset X$, any continuous function φ on X, and any measure $\mu \in \mathfrak{M}(Z)$,
$$h_\mu(f) + \int_Z \varphi\, d\mu = P_{G_\mu}(\varphi), \qquad (A2.16)$$

where G_μ is the set of all forward generic points of the measure μ (i.e., the points for which the Birkhoff ergodic theorem holds for any continuous function on X); applying this equality to $\varphi = 0$ yields Bowen's formula for the measure-theoretic entropy of f
$$h_\mu(f) = h_{G_\mu}(f);$$

(5) let $Z \subset X$ be an invariant set; if $V(x) \cap \mathfrak{M}(Z) \neq \varnothing$ for every $x \in Z$, then
$$P_Z(\varphi) = \sup_{\mu \in \mathfrak{M}(Z)} \left(h_\mu(f) + \int_Z \varphi\, d\mu\right).$$

In [PP], Pesin and Pitskel' described an example that shows that the assumption "$V(x) \cap \mathfrak{M}(Z) \neq \varnothing$ for every $x \in Z$" is crucial and that the variational principle for the topological pressure on non-compact sets may fail otherwise. It also reveals some new and interesting phenomena associated with the topological entropy in the case of non-compact sets.

Proposition A2.1. *Let* (Σ_2, σ) *be the full shift on two-sided sequences of two symbols (the classical Bernoulli scheme) and*

$$Z = \{\omega \in \Sigma_2 : \omega \notin \tilde{G}_\mu \text{ for any measure } \mu \in \mathfrak{M}(\Sigma_2)\} = \Sigma_2 \setminus \bigcup_{\mu \in \mathfrak{M}(\Sigma_2)} \tilde{G}_\mu,$$

where $\tilde{G}_\mu$ *is the set of points which are both forward and backward generic with respect to the measure* μ. *Then* $h_Z(\sigma) = h_{\Sigma_2}(\sigma) = \log 2$.

Remark.

Note that $\mu(Z) = 0$ for any $\mu \in \mathfrak{M}(\Sigma_2)$. Hence,

$$h_Z(\sigma) = h_{\Sigma_2}(\sigma) > 0 = \sup_{\mu \in \mathfrak{M}(Z)} h_\mu(\sigma)$$

and the variational principle for the topological pressure fails (note that in this case $\mathcal{L}(Z) = \varnothing$).

The set Z is an example of so-called **metrically irregular** sets for dynamical systems which we describe in detail in Appendix IV.

Proof of the proposition. Consider a Bernoulli measure μ on Σ_2 such that $\mu(C_0) = p$ and $\mu(C_1) = 1 - p = q$, where $C_0 = \{\omega = (i_n) : i_0 = 0\}$ and $C_1 = \{\omega = (i_n) : i_0 = 1\}$ are cylinders and $p \neq q$. Given δ, $0 < \delta \leq 1$, we can choose p such that $\log 2 - h_\mu(\sigma) \leq \delta$.

Let $\{n_k\}$ be a sequence of positive integers satisfying $n_k < n_{k+1}$ and $n_k \to \infty$ as $k \to \infty$. We decompose the set of all integers into two disjoint subsets Q_1 and Q_2 in the following way: $i \in Q_1$ if $n_{2k} \leq |i| < n_{2k+1}$ and $i \in Q_2$ otherwise. Consider the map $\psi \colon \Sigma_2 \to \Sigma_2$ given as follows

$$(\psi(\omega))_n = \begin{cases} i_n & \text{if } n \in Q_1, \\ i_n + 1 \pmod 2 & \text{if } n \in Q_2 \end{cases}$$

where $\omega = (\ldots i_{-1} i_0 i_1 \ldots)$. Note that ψ is a homeomorphism (even bi-Lipschitz) but it does not commute with the shift. Set $Y = \psi(\tilde{G}_\mu)$.

Lemma 1. *If the sequence* $\{n_k\}$ *grows sufficiently fast then* $Y \subset Z$.

Proof of the lemma. Let χ be the indicator of the set C_0. If the sequence $\{n_k\}$ increases sufficiently rapidly then by the Birkhoff ergodic theorem for any $\omega \in Y$ we obtain that

$$\lim_{k \to \infty} \frac{1}{2n_{2k+1} + 1} \sum_{i=-n_{2k+1}}^{n_{2k+1}} \chi(\sigma^i(\omega)) = p, \quad \lim_{k \to \infty} \frac{1}{2n_{2k} + 1} \sum_{i=-n_{2k}}^{n_{2k}} \chi(\sigma^i(\omega)) = q.$$

Since $p \neq q$ it follows that the Birkhoff sum $\frac{1}{n} \sum_{i=0}^{n-1} \chi(\sigma^i(\omega))$ does not converge and hence $\omega \in Z$. ∎

Let ξ be the partition of Σ_2 by the sets C_0 and C_1. Given $m > 0$ and $n > 0$, denote by $\eta_m = \bigvee\limits_{j=-m}^{m} \sigma^j \xi$ and $\xi_n = (\eta_m)_n = \bigvee\limits_{j=-n}^{n} \sigma^j \eta_m$.

Lemma 2. *For every* $\omega \in \tilde{G}_\mu$

$$\lim_{n \to \infty} \left(-\frac{1}{n} \log \mu(C_{\xi_n}(\psi(x))) \right) = h_\mu(\sigma).$$

Proof of the lemma. Set $Q(i, m, n) = Q_i \cap [-m - n + 1, m + n - 1]$, where $i = 1, 2$, $m > 0$, and $n > 0$. By the law of large numbers for every $\omega \in \tilde{G}_\mu$ there exists the limit

$$-\lim_{n \to \infty} \frac{1}{|Q(1, m, n)|} \sum_{j \in Q(1, m, n)} \log \mu(C_{\sigma^j \xi}(\psi(\omega))) = h_\mu(\sigma),$$

where $|A|$ denotes the number of elements in the set A. Since the involution $0 \mapsto 1$, $1 \mapsto 0$ transfers the measure μ into a Bernoulli measure with the same entropy we obtain that

$$-\lim_{n \to \infty} \frac{1}{|Q(2, m, n)|} \sum_{j \in Q(2, m, n)} \log \mu(C_{\sigma^j \xi}(\psi(\omega))) = h_\mu(\sigma).$$

For any $n > 0$ we write

$$-\frac{1}{n} \log \mu(C_{\xi_n}(\psi(x))) = -\frac{1}{n} \log \prod_{j=-m-n+1}^{m+n-1} \mu(C_{\sigma^j \xi}(\psi(\omega)))$$

$$= -\frac{|Q(1, m, n)|}{n} \cdot \frac{1}{|Q(1, m, n)|} \sum_{j \in Q(1, m, n)} \log \mu(C_{\sigma^j \xi}(\psi(\omega)))$$

$$- \frac{|Q(2, m, n)|}{n} \cdot \frac{1}{|Q(2, m, n)|} \sum_{j \in Q(2, m, n)} \log \mu(C_{\sigma^j \xi}(\psi(\omega))).$$

The desired result follows.   ∎

We proceed with the proof of the proposition. Fix $m > 0$ and consider the partition η_m which is also a finite cover of Σ_2. It follows from Lemma 2 that for any $\gamma > 0$ there exists a set $D \subset \tilde{G}_\mu$ and a number $N > 0$ such that $\mu(D) \geq 1 - \gamma$ and for any $\omega \in D$ and $n \geq N$,

$$\mu(C_{(\eta_m)_n}(\psi(\omega))) \leq \exp\left(-n(h_\mu(\sigma) - \gamma)\right).$$

Fix $n \geq N$ and choose a collection of strings $\mathcal{G} \subset \mathcal{S}(\eta_m)$ which cover Y (i.e., the sets $\{\Sigma_2(U) : U \in \mathcal{G}\}$ cover Y) and satisfy: $m(U) \geq n$ and

$$\left| M(Y, \alpha, 0, \eta_m, n) - \sum_{U \in \mathcal{G}} \exp\left(-\alpha n\right) \right| \leq \gamma$$

Let $\mathcal{G}_\ell = \{\boldsymbol{U} \in \mathcal{G} : m(\boldsymbol{U}) = \ell\}$ and K_ℓ be the number of elements in $\mathcal{G}_\ell$. Set

$$E_\ell = \bigcup_{\boldsymbol{U} \in \mathcal{G}_\ell} \Sigma_2(\boldsymbol{U}).$$

Since the sets $\Sigma_2(\boldsymbol{U}')$ and $\Sigma_2(\boldsymbol{U}'')$ are disjoint for any distinct $\boldsymbol{U}', \boldsymbol{U}'' \in \mathcal{G}_\ell$ we obtain that

$$K_\ell \geq \frac{\mu(\psi^{-1}(E_\ell) \cap D)}{\exp\left(-\ell(h_\mu(\sigma) - \gamma)\right)}.$$

Therefore,

$$\sum_{\boldsymbol{U} \in \mathcal{G}} \exp\left(-\alpha n\right) = \sum_{\ell=n}^{\infty} \exp\left(-\alpha\ell\right) K_\ell \geq \mu(D) \exp\left((-\alpha + h_\mu(\sigma) - \gamma)\ell\right)$$

$$\sum_{\ell=n}^{\infty} \mu(\psi^{-1}(E_\ell) \cap D) \exp\left((-\alpha + h_\mu(\sigma) - \gamma)\ell\right)$$

$$\geq (1 - 2\gamma) \exp\left((-\alpha + h_\mu(\sigma) - \gamma)n\right)$$

if n is sufficiently big (such that $\mu(\psi^{-1}(E_n) \cap D) \geq (1 - 2\gamma)$). For $\alpha < h_\mu(\sigma) - \gamma$ this implies that

$$M(Y, \alpha, 0, \eta_m, n) \geq 1 - \gamma \geq \frac{1}{2}.$$

Taking the limit as $n \to \infty$ we obtain that $m_C(Z, \alpha) = \infty$ and hence

$$P_Y(0, \eta_m) \geq h_\mu(\sigma) - \gamma \geq \log 2 - \delta - \gamma.$$

Since $\operatorname{diam} \eta_m \to 0$ as $m \to \infty$ this implies that $h_Y(\sigma) = P_Y(0) \geq \log 2 - \delta - \gamma$. Taking into account that the numbers δ and γ can be chosen arbitrarily small we conclude that $h_Z(\sigma) \geq h_Y(\sigma) \geq \log 2$. This completes the proof of the proposition. ∎

Equilibrium Measures

Let φ be a continuous function on X. Given an f-invariant set $Z \subset X$ we call a measure $\mu = \mu_\varphi$ an **equilibrium measure** on Z corresponding to the function φ if $\mu_\varphi \in \mathfrak{M}(Z)$ and

$$h_{\mu_\varphi}(f) + \int_Z \varphi \, d\mu_\varphi = \sup_{\mu \in \mathfrak{M}(Z)} \left(h_\mu(f) + \int_Z \varphi \, d\mu\right).$$

The following statement establishes the existence of an equilibrium measure for a continuous function φ. Note that, in general, this measure may not be unique.

We recall that a homeomorphism f of X is called **expansive** if there exists $\varepsilon > 0$ such that for any two points $x, y \in X$ if $\rho(f^k(x), f^k(y)) \leq \varepsilon$ for all $k \in \mathbb{Z}$ then $x = y$.

Theorem A2.2. [PP] *Assume that the following conditions hold:*

(1) *f is a homeomorphism of X;*

(2) *f is expansive;*

(3) *the set $\mathfrak{M}(Z)$ is closed in $\mathfrak{M}(X)$ (in the weak*-topology).*

Then for any continuous function φ there exists an equilibrium measure μ_φ on Z.

Proof. Let ξ be a partition of X with $\operatorname{diam}\xi \leq \varepsilon$. Set $\xi_n = \bigvee_{j=-n}^{n} f^j\xi$. Since f is expansive we obtain that $\operatorname{diam}\xi_n \to 0$. Therefore, $h_\mu(f) = \lim_{n\to\infty} h_\mu(f,\xi_n)$. On the other hand, it is easy to see that $h_\mu(f,\xi_n) = h_\mu(f,\xi)$. Therefore, $h_\mu(f) = h_\mu(f,\xi)$.

We now show that the map $\mu \mapsto h_\mu(f)$ is upper semi-continuous on $\mathfrak{M}(Z)$. Then the map $\mu \mapsto h_\mu(f) + \int \varphi\,d\mu$ is also upper semi-continuous on $\mathfrak{M}(Z)$ and the desired result follows from Theorem A2.1 and the fact that an upper semi-continuous function on a compact set attains its supremum.

Fix $\mu \in \mathfrak{M}(Z)$, $\alpha > 0$, and a partition $\xi = \{C_1 \ldots C_n\}$ of X with $\operatorname{diam}\xi \leq \varepsilon$. For a sufficiently large m we have that

$$\frac{1}{m}H_\mu(\xi \vee \cdots \vee f^{-m+1}\xi) \leq h_\mu(f) + \alpha.$$

Let us fix such an m. Given β, choose compact sets

$$K_{i_0\ldots i_{m-1}} \subset \bigcap_{k=0}^{m-1} f^{-k}(C_{i_k})$$

such that

$$\mu\left(\bigcap_{k=0}^{m-1} f^{-k}(C_{i_k}) \setminus K_{i_0\ldots i_{m-1}}\right) \leq \beta.$$

This implies that

$$L_i \stackrel{\text{def}}{=} \bigcup_{j=0}^{m-1} \bigcup_{i_j-i} f^j(K_{i_0\ldots i_{m-1}}) \subset C_i.$$

Since L_i are disjoint and compact one can find a partition $\xi' = \{C'_1 \ldots C'_n\}$ of X with $\operatorname{diam}\xi' \leq \varepsilon$ such that $L_i \subset \operatorname{int}C'_i$. We have that

$$K_{i_0\ldots i_{m-1}} \subset \operatorname{int} \bigcap_{k=0}^{m-1} f^{-k}(C'_{i_k}).$$

If a measure $\nu \in \mathfrak{M}(Z)$ is close to μ in the weak*-topology then

$$\nu\left(\bigcap_{k=0}^{m-1} f^{-k}C'_{i_k}\right) \geq \mu(K_{i_0\ldots i_{m-1}}) - \beta$$

and

$$\left|\nu\left(\bigcap_{k=0}^{m-1} f^{-k}(C'_{i_k})\right) - \mu\left(\bigcap_{k=0}^{m-1} f^{-k}(C'_{i_k})\right)\right| \leq 2\beta m.$$

If β is sufficiently small it follows that

$$h_\nu(f) = h_\nu(f, \xi') \le \frac{1}{m} H_\mu(\xi' \vee \cdots \vee f^{-m+1}\xi')$$

$$\le \frac{1}{m} H_\mu(\xi \vee \cdots \vee f^{-m+1}\xi) + \alpha \le h_\mu(f) + 2\alpha.$$

This completes the proof of the theorem.   ∎

As a direct consequence of Theorem A2.2 we obtain the following statement.

Theorem A2.3. *Assume that a map f satisfies Conditions 1 and 2 of Theorem A2.2. Then for any compact f-invariant set $Z \subset X$ and any continuous function φ on Z there exists an equilibrium measure μ_φ on Z for which*

$$h_{\mu_\varphi}(f) + \int_Z \varphi \, d\mu_\varphi = P_Z(\varphi).$$

In particular, if a map f satisfies Conditions 1 and 2 of Theorem A2.2 then for any compact f-invariant set $Z \subset X$ there exists a **measure of maximal entropy**, i.e., an equilibrium measure corresponding to the function $\varphi = 0$ for which

$$h_Z(f) = \sup_{\mu \in \mathfrak{M}(Z)} h_\mu(f).$$

Let us notice that, in view of (11.14), (11.15), and Theorems 11.6 and A2.1 for any compact invariant set Z we have that

$$P_Z(\varphi) = \sup_\mu P_\mu(\varphi)$$

(provided conditions of Theorem 11.6 hold). Therefore, the variational principle for topological pressure can be viewed as a variational principle for Carathéodory dimension (where the dimension is generated by the C-structure defined by (11.1)–(11.3); see Sections 1 and 5). Moreover, any equilibrium measure is a measure of full Carathéodory dimension (with respect to the space $\mathfrak{M}$ of invariant measures; see (5.4)). In particular, any measure of maximal entropy is a measure of full Carathéodory dimension.

Non-additive Variational Principle

We now state a non-additive version of the variational principle for topological pressure established by Barreira in [Bar2]. Let $\varphi = \{\varphi_n\}$ be a sequence of continuous functions.

Theorem A2.4. *Let $Z \subset X$ be an f-invariant set. Assume that there exists a continuous function $\psi: X \to \mathbb{R}$ such that*

$$\varphi_{n+1} - \varphi_n \circ f \to \psi \tag{A2.17}$$

uniformly on Z as $n \to \infty$. Then

$$P_{\mathcal{L}(Z)}(\varphi) = \sup_{\mu \in \mathfrak{M}(Z)} \left(h_\mu(f) + \int_Z \psi \, d\mu \right).$$

As an immediate consequence of the above statement we have that *if $Z \subset X$ is an f-invariant compact set, then*

$$P_Z(\varphi) = \sup_{\mu \in \mathfrak{M}(Z)} \left(h_\mu(f) + \int_Z \psi \, d\mu \right),$$

where the function ψ satisfies (A2.17).

Falconer [F3] established another version of the variational principle for topological pressure assuming that the sequence of functions φ is sub-additive (it also should satisfy some other additional requirements) — the so-called "sub-additive" variational principle for topological pressure.

Symbolic Dynamical Systems

We briefly describe some basic concepts of symbolic dynamics which are used in the second part of the book. For each $p \in \mathbf{N}$ we denote the space of right-sided infinite sequences of p symbols by

$$\Sigma_p^+ = \left\{ \omega = (i_0 i_1 \dots) \right\} = \{1, \dots, p\}^{\mathbf{N}}.$$

We call the number i_j the j-**coordinate** of the point ω (we also use another notation ω_j). We write ω^+ for points in Σ_p^+ to stress that we are dealing with a right-sided infinite sequence.

A cylinder (or **a cylinder set**) is defined as

$$C_{i_0 \dots i_n} = \left\{ (j_0 j_1 \dots) \in \Sigma_p^+ : j_k = i_k, \ k = 0, \dots, n \right\}.$$

We also use more explicit notation $C_{i_0 \dots i_n}^+$.

Given $\beta > 1$, we endow the space Σ_p^+ with the metric

$$d_\beta(\omega, \omega') = \sum_{k=0}^{\infty} \frac{|i_k - i_k'|}{\beta^k}, \tag{A2.18}$$

where $\omega = (i_0 i_1 \dots)$ and $\omega' = (i_0' i_1' \dots)$. It induces the topology on Σ_p^+ such that the space is compact and cylinders are disjoint open (as well as closed) subsets.

The (one-sided) **shift** on Σ_p^+ is defined by

$$\sigma(\omega)_k = \omega_{k+1}$$

(we also use more explicit notation σ^+). It is easily seen to be continuous. A subset $Q \subset \Sigma_p^+$ is said to be σ-**invariant** if $\sigma(Q) = Q$. When the set $Q \subset \Sigma_p^+$ is compact and σ-invariant, the map $\sigma|Q$ is called a (one-sided) **subshift**.

Let A be a $p \times p$ matrix whose entries a_{ij} are either 0 or 1. The compact and σ-invariant subset

$$\Sigma_A^+ = \big\{ (i_0 i_1 \ldots) \in \Sigma_p^+ : a_{i_n i_{n+1}} = 1 \text{ for all } n \in \mathbb{N} \big\}$$

is called a **topological Markov chain** with the **transfer matrix** A. The map $\sigma | \Sigma_A^+$ is called a (one-sided) **subshift of finite type**. It is **topologically transitive** (i.e., for any two open subsets $U, V \subset \Sigma_A^+$ there exists $n > 0$ such that $\sigma^n(U) \cap V \neq \varnothing$) if the matrix A is **irreducible**, i.e., for each entry a_{ij} there exists a positive integer k such that $a_{ij}^k > 0$, where a_{ij}^k is the (i,j)-entry of the matrix A^k. The map $\sigma | \Sigma_A^+$ is **topologically mixing** (i.e., for any two open subsets $U, V \subset \Sigma_A^+$ there exists $N > 0$ such that $\sigma^n(U) \cap V \neq \varnothing$ for any $n > N$) if the transfer matrix A is **transitive**, i.e., $A^k > 0$ for some positive integer k.

We call (Q, σ) a **sofic system** if $Q \subset \Sigma_p^+$ is a finite factor of some topological Markov chain Σ_A^+, i.e., there exists a continuous surjective map $\zeta : \Sigma_A^+ \to Q$ such that $\sigma | Q \circ \zeta = \zeta \circ \sigma$. An example is the **even system**, i.e., the set Q of sequences of 1's and 2's, where the 2's are separated by an even number of 1's.

Similarly to the above, we consider the space of left-sided infinite sequences

$$\Sigma_p^- = \big\{ \omega = (\ldots i_{-1} i_0) \big\}.$$

We also write ω^- for points in Σ_p^-. A cylinder in Σ_p^- is denoted by $C_{i_{-n} \ldots i_0}$ (or more explicitly $C_{i_{-n} \ldots i_0}^-$). The (one-sided) **shift** is defined by

$$\sigma(\omega)_k = \omega_{k-1}.$$

(we also use more explicit notation σ^-). It is continuous.

Further, given a transfer matrix $A = (a_{ij})$, we set

$$\Sigma_A^- = \big\{ \omega = (\ldots i_{-1} i_0) \in \Sigma_p^- : a_{i_{-n} i_{-n+1}} = 1 \text{ for all } n \in \mathbb{N} \big\}.$$

The map $\sigma | \Sigma_A^-$ is a (one-sided) **subshift of finite type**.

We also consider the space of two-sided infinite sequences of p symbols

$$\Sigma_p = \{ \omega = (\ldots i_{-1} i_0 i_1 \ldots) \} = \{1, \ldots, p\}^{\mathbb{Z}}.$$

A **cylinder** (or **a cylinder set**) is defined as

$$C_{i_m \ldots i_n} = \big\{ \omega = (\ldots j_{-1} j_0 j_1 \ldots) \in \Sigma_p : j_k = i_k, \ k = m, \ldots, n \big\},$$

where $m \leq n$. Given $\beta > 1$, we endow the space Σ_p with the metric

$$d_\beta(\omega, \omega') = \sum_{k=-\infty}^{\infty} \frac{|i_k - i_k'|}{\beta^{|k|}}, \tag{A2.18'}$$

where $\omega = (\ldots i_{-1} i_0 i_1 \ldots)$ and $\omega' = (\ldots i'_{-1} i'_0 i'_1 \ldots)$. It induces the compact topology on Σ_p with cylinders to be disjoint open (and at the same time closed) subsets.

The (two-sided) **shift** $\sigma \colon \Sigma_p \to \Sigma_p$ is defined by $\sigma(\omega)_k = \omega_{k+1}$. It is an expansive homeomorphism. Given a compact σ-invariant set $Q \subset \Sigma_p$, we call the map $\sigma | Q$ a (two-sided) **subshift**.

Let A be a $p \times p$ transfer matrix with entries 0 and 1. Consider the compact σ-invariant subset

$$\Sigma_A = \left\{ \omega \in \Sigma_p : a_{\omega_n \omega_{n+1}} = 1 \text{ for all } n \in \mathbb{Z} \right\}.$$

The map $\sigma | \Sigma_A$ is called a (two-sided) **subshift of finite type**.

Let us notice that given a point $\omega \in \Sigma_A$, the set of points $\omega' \in \Sigma_A$ having the same *past* as ω (i.e., $\omega_i = \omega'_i$ for $i \leq 0$) can be identified with the cylinder $C_{i_0}^+ \subset \Sigma_A^+$. Similarly, the set of points $\omega' \in \Sigma_A$ having the same *future* as ω (i.e., $\omega_i = \omega'_i$ for $i \geq 0$) can be identified with the cylinder $C_{i_0}^- \subset \Sigma_A^-$. Thus, the cylinder $C_{i_0} \subset \Sigma_A$ can be identified with the direct product $C_{i_0}^+ \times C_{i_0}^-$.

For symbolic dynamical systems the definitions of topological pressure and lower and upper capacity topological pressure can be simplified based on the following observation (which we have already used in the proof of Proposition A2.1). Let $\mathcal{U}_n$ be the open cover of Σ_p^+ by cylinder sets $C_{i_0 \ldots i_n}$. Notice that $|\mathcal{U}_n| \to 0$ as $n \to \infty$ and for any $U \in \mathcal{S}(\mathcal{U}_n)$ the set $X(U)$ is a cylinder set. Therefore, the function $M(Z, \alpha, \varphi, \mathcal{U}_n, N)$ can be rewritten according to (11.4) as

$$M(Z, \alpha, \varphi, \mathcal{U}_n, N) =$$

$$\inf_{\mathcal{G}} \left\{ \sum_{C_{i_0 \ldots i_m} \in \mathcal{G}} \exp \left(-\alpha(m+1) + \sup_{\omega \in C_{i_0 \ldots i_m}} \sum_{k=0}^{m} \varphi(\sigma^k(\omega)) \right) \right\} \quad (A2.19)$$

and the infimum is taken over all finite or countable collections of cylinder sets $C_{i_0 \ldots i_m}$ with $m \geq N > n$ which cover Z. Furthermore, the function $R(Z, \alpha, \varphi, \mathcal{U}_n, N)$ $(N > n)$ can be rewritten according to (11.4') as

$$R(Z, \alpha, \varphi, \mathcal{U}_n, N)$$

$$= \sum_{C_{i_0 \ldots i_N} \in \mathcal{G}} \exp \left(-\alpha(N+1) + \sup_{\omega \in C_{i_0 \ldots i_N}} \sum_{k=0}^{N} \varphi(\sigma^k(\omega)) \right) \quad (A2.19')$$

and the sum is taken over the collection of all cylinder sets $C_{i_0 \ldots i_N}$ intersecting Z.

Let (Q, σ) be a symbolic dynamical system, where Q is a compact σ-invariant subset of Σ_p^+ and φ a continuous function on Q. A Borel probability measure $\mu = \mu_\varphi$ on Q is called a **Gibbs measure** (corresponding to φ) if there exist constants $D_1 > 0$ and $D_2 > 0$ such that for any $n > 0$, any cylinder set $C_{i_0 \ldots i_n}$, and any $\omega \in C_{i_0 \ldots i_n}$ we have that

$$D_1 \leq \frac{\mu(C_{i_0 \ldots i_n})}{\exp \left(-Pn + \sum\limits_{k=0}^{n} \varphi(\sigma^k(\omega)) \right)} \leq D_2, \quad (A2.20)$$

where $P = P_Q(\varphi)$. Note that if Condition (A2.20) holds for *some* number P then $P = P_Q$. Indeed, in this case for every $\varepsilon > 0$ by (A2.19) we obtain that

$$M(Z, P + \varepsilon, \varphi, \mathcal{U}_n, N) \leq D_1^{-1} \inf_{\mathcal{G}} \sum_{C_{i_0 \ldots i_m} \in \mathcal{G}} \mu(C_{i_0 \ldots i_m}) \exp(-\varepsilon + \gamma(\mathcal{U}_n))$$

$$\leq D_1^{-1} \exp(-\varepsilon + \gamma(\mathcal{U}_n)).$$

Letting $n \to \infty$ yields that $P_Q \leq P + \varepsilon$ and hence $P_q \leq P$ since ε is arbitrary. The opposite inequality can be proved in a similar fashion (I thank S. Ferleger for pointing out this argument to me).

Any Gibbs measure is an equilibrium measure but not otherwise. It is known that the *specification property* (see [KH] for definition) of a topologically mixing symbolic dynamical system (Q, σ) ensures that any equilibrium measure corresponding to a Hölder continuous function is Gibbs.

It is known that any subshift of finite type (Σ_A^+, σ) satisfies the specification property. Therefore, an equilibrium measure μ_φ, corresponding to a Hölder continuous function φ, is a Gibbs measure provided the transfer matrix A is transitive. In this case it is also a Bernoulli measure. For an arbitrary transfer matrix A, by the Perron–Frobenius theorem one can decompose the set Σ_A^+ into two shift-invariant subsets: the wandering set Q_1 (corresponding to the non-recurrent states) and the non-wandering set Q_2 (corresponding to the recurrent states). The latter can be further partitioned into finitely many shift-invariant subsets of the form $\Sigma_{A_i}^+$, where each matrix A_i is irreducible and corresponds to a class of equivalent recurrent states (see [KH] for details). Moreover, for each i there exists a number n_i such that the map σ^{n_i} is topologically mixing.

Note also that any sofic system satisfies the specification property.

We define the notion of Gibbs measures for two-sided subshifts. Let Q be a compact σ-invariant subset of Σ_p and φ a continuous function on Q. A Borel probability measure $\mu = \mu_\varphi$ on Q is called a **Gibbs measure** (corresponding to φ) if there exist constants $D_1 > 0$ and $D_2 > 0$ such that for any $m < 0$, $n > 0$, any cylinder set $C_{i_m \ldots i_n}$, and any $\omega \in C_{i_m \ldots i_n}$ we have that

$$D_1 \leq \frac{\mu(C_{i_m \ldots i_n})}{\exp\left(-P(m+n) + \sum_{k=m}^{n} \varphi(\sigma^k(\omega))\right)} \leq D_2, \tag{A2.20'}$$

where $P = P_Q(\varphi)$. Again, any Gibbs measure is an equilibrium measure and the specification property ensures otherwise.

In the case of subshifts of finite type there is a deep connection between Gibbs measures for one-sided and two-sided subshifts. In order to describe this connection consider a two-sided subshift of finite type (Σ_A, σ) and a Hölder continuous function φ on Σ_A. Choose p points $\omega^{(i)} = (\omega_j^{(i)}) \in \Sigma_A$ such that $\omega_0^{(i)} = i$ for $i = 1, \ldots, p$ and set $\Omega = (\omega^{(1)}, \ldots, \omega^{(p)})$. We now define the function r_Ω on Σ_A by

$$r_\Omega(\omega) = r_\Omega(\ldots \omega_{-2} \omega_{-1} i \, \omega_1 \omega_2 \ldots) = (\ldots \omega_{-2}^{(i)} \omega_{-1}^{(i)} i \, \omega_1 \omega_2 \ldots)$$

provided $\omega_0 = i$. Further, we define the function $\theta^{(u)} = \theta_\Omega^{(u)}$ on Σ_A by

$$\theta^{(u)}(\omega) = \varphi(r_\Omega(\omega)) + \sum_{j=0}^{\infty} \left[\varphi(\sigma^{j+1}(r_\Omega(\omega))) - \varphi(\sigma^j(r_\Omega(\sigma\omega))) \right].$$

Given a Hölder continuous function ϕ on Σ_A and a positive integer n, we define the *n-variation of the function* ϕ by

$$\mathrm{var}_n \phi = \sup\{ |\phi(\omega) - \phi(\omega')| : \omega_i = \omega'_i, \, 0 \le i < n \}.$$

If ω and ω' are two points whose i-coordinates coincide for all i between $-n$ and ∞ then $\mathrm{var}_n \phi \le C\beta^n$, where $C > 0$ is a constant and β is the coefficient in the d_β-metric on Σ_A (see (A2.18')).

Lemma A2.1. *The function $\theta^{(u)}$ is Hölder continuous, it is cohomologous to φ and hence has the same topological pressure.*

Proof. We have that

$$\theta^{(u)}(\omega) = \varphi(\omega) + \sum_{j=-1}^{\infty} \left[\varphi(\sigma^{j+1}(r_\Omega(\omega))) - \varphi(\sigma^{j+1}(\omega)) \right]$$

$$+ \sum_{j=0}^{\infty} \left[\varphi(\sigma^{j+1}((\omega))) - \varphi(\sigma^j(r_\Omega(\sigma(\omega)))) \right] = \varphi(\omega) - u(\omega) + u(\sigma(\omega)),$$

where

$$u(\omega) = \sum_{j=0}^{\infty} \left[\varphi(\sigma^j(\omega)) - \varphi(\sigma^j(r_\Omega(\omega))) \right].$$

In order to complete the proof of the lemma we need only to show that the function $u(\omega)$ is Hölder continuous. Since the i-coordinates of the points $\sigma^j(\omega)$ and $\sigma^j(r_\Omega(\omega))$ coincide for all i between $-j$ and ∞ we obtain that

$$|\varphi(\sigma^j(\omega)) - \varphi(\sigma^j(r_\Omega(\omega)))| \le \mathrm{var}_j \varphi \le C\beta^j.$$

If $\omega = (\omega_i)$ and $\omega' = (\omega'_i)$ are two points satisfying $\omega_i = \omega'_i$ for $|i| \le n$ then for any $j \in [0, n]$

$$|\varphi(\sigma^j(\omega)) - \varphi(\sigma^j(\omega'))| \le C\beta^{n-j}$$

and

$$|\varphi(\sigma^j(r_\Omega(\omega))) - \varphi(\sigma^j(r_\Omega(\omega')))| \le C\beta^{n-j}.$$

Therefore,

$$|u(\omega) - u(\omega')| \le \sum_{j=0}^{[n/2]} |\varphi(\sigma^j(\omega)) - \varphi(\sigma^j(\omega'))|$$

$$+ |\varphi(\sigma^j(r_\Omega(\omega'))) - \varphi(\sigma^j(r_\Omega(\omega)))| + 2 \sum_{j>[n/2]} C\beta^j$$

$$\le 2C \left(\sum_{j=0}^{[n/2]} \beta^{n-j} + \sum_{j>[n/2]} \beta^j \right) < \infty.$$

This completes the proof of the lemma.   ∎

It follows from the lemma that the measure μ is the Gibbs measure corresponding to $\theta^{(u)}$.

Let $\pi^+ \colon \Sigma_A \to \Sigma_A^+$ and $\pi^- \colon \Sigma_A \to \Sigma_A^-$ be the projections

$$\pi^+(\ldots i_{-1} i_0 i_1 \ldots) = (i_0 i_1 \ldots), \quad \pi^-(\ldots i_{-1} i_0 i_1 \ldots) = (\ldots i_{-1} i_0).$$

One can easily check that $\theta^{(u)}(\ldots i_{-1} i_0 i_1 \ldots) = \theta^{(u)}(\ldots i'_{-1} i'_0 i'_1 \ldots)$ whenever $i_j = i'_j$ for every $j \geq 0$. This means that there is a function $\varphi^{(u)}$ on Σ_A^+ such that $\theta^{(u)} = \varphi^{(u)} \circ \pi^+$ on Σ_A.

In a similar fashion, we define the function $\theta^{(s)} = \theta_\Omega^{(s)}$ on Σ_A, and find a function $\varphi^{(s)}$ on Σ_A^- such that $\theta^{(s)} = \varphi^{(s)} \circ \pi^-$.

The functions $\varphi^{(u)}$ and $\varphi^{(s)}$ can be shown to be Hölder continuous. We call them **stable** and **unstable parts of the function** φ. Let $\mu^{(u)}$ be the Gibbs measure for $\varphi^{(u)}$ on Σ_A^+ and $\mu^{(s)}$ the Gibbs measure for $\varphi^{(s)}$ on Σ_A^-. We have that $\mu^{(u)} = \mu \circ (\pi^+)^{-1}$ since both measures are Gibbs measures for the function $\varphi^{(u)}$, and that $\mu^{(s)} = \mu \circ (\pi^-)^{-1}$ since both measures are Gibbs measures for the function $\varphi^{(s)}$. This implies that

$$\mu^{(u)}(C_{i_0 \ldots i_n}^+) = \mu(C_{i_0 \ldots i_n}) \quad \text{and} \quad \mu^{(s)}(C_{i_0 \ldots i_n}^-) = \mu(C_{i_0 \ldots i_n}), \qquad \text{(A2.21)}$$

where $C_{i_0 \ldots i_n}^+ = \pi^+(C_{i_0 \ldots i_n})$ and $C_{i_{-n} \ldots i_0}^- = \pi^-(C_{i_{-n} \ldots i_0})$.

We want now to *normalize* the functions $\varphi^{(u)}$ and $\varphi^{(s)}$, i.e., to switch to the functions $\psi^{(u)}$ on Σ_A^+ and $\psi^{(s)}$ on Σ_A^- defined by

$$\log \psi^{(u)} = \varphi^{(u)} - P_{\Sigma_A^+}(\varphi^{(u)}), \quad \log \psi^{(s)} = \varphi^{(s)} - P_{\Sigma_A^-}(\varphi^{(s)}).$$

Clearly, $P_{\Sigma_A^+}(\log \psi^{(u)}) = P_{\Sigma_A^-}(\log \psi^{(s)}) = 0$.

Lemma A2.2. *We have that*

$$\log \psi^{(u)}(\omega^+) = \lim_{n \to \infty} \log \frac{\mu^{(u)}(C_{i_0 \ldots i_n}^+)}{\mu^{(u)}(C_{i_1 \ldots i_n}^+)} = \lim_{n \to \infty} \log \frac{\mu(C_{i_0 \ldots i_n})}{\mu(C_{i_1 \ldots i_n})}$$

for each $\omega^+ = (i_0 i_1 \ldots) \in \Sigma_A^+$ *(with uniform convergence), and*

$$\log \psi^{(s)}(\omega^-) = \lim_{n \to \infty} \log \frac{\mu^{(s)}(C_{i_{-n} \ldots i_0}^-)}{\mu^{(s)}(C_{i_{-n} \ldots i_1}^-)} = \lim_{n \to \infty} \log \frac{\mu(C_{i_{-n} \ldots i_0})}{\mu(C_{i_{-n} \ldots i_1})}$$

for each $\omega^- = (\ldots i_{-1} i_0) \in \Sigma_A^-$ *(with uniform convergence).*

Proof. The statement is an immediate corollary of the following property of Gibbs measures (see Proposition 3.2 in [PaPo]): *let* μ *be the Gibbs measure corresponding to a Hölder continuous function* ϕ *on* Σ_A^+; *then*

$$\exp(-\alpha^n |\phi|_\alpha) \leq \frac{\mu(C_{i_0 \ldots i_n}^+) \exp(-\phi(\omega))}{\mu(C_{i_1 \ldots i_n}^+)} \leq \exp(\alpha^n |\phi|_\alpha).$$

Here α is the Hölder exponent of ϕ and $|\phi|_\alpha$ is the *Hölder norm* of ϕ (i.e., $|\phi|_\alpha = \sup\{\alpha^{-n} \mathrm{var}_n \phi : n \geq 0\}$). ∎

The lemma implies that the functions $\psi^{(u)}$ and $\psi^{(s)}$ do not depend on the choice of the point Ω.

The following statement shows that the measure μ is the "direct product" of measures $\mu^{(u)}$ and $\mu^{(s)}$.

Proposition A2.2. *The following properties hold:*

(1) $P_{\Sigma_A}(\varphi) = P_{\Sigma_A^+}(\varphi^{(u)}) = P_{\Sigma_A^-}(\varphi^{(s)})$;

(2) *there exist positive constants A_1 and A_2 such that for every integers n, $m \geq 0$, and any $(\ldots i_{-1} i_0 i_1 \ldots) \in \Sigma_A$,*

$$A_1 \leq \frac{\mu(C_{i_{-m} \ldots i_n})}{\mu^{(u)}(C_{i_0 \ldots i_n}^+) \times \mu^{(s)}(C_{i_{-m} \ldots i_0}^-)} \leq A_2.$$

Proof. Since the functions $\varphi^{(s)}$ and $\varphi^{(u)}$ are cohomologous the first property follows from Lemma A2.1. The second property is an immediate consequence of identities (A2.21). ∎

Bowen's Equation

Let (X, ρ) be a compact metric space with metric ρ, $f \colon X \to X$ a continuous map, and $\varphi \colon X \to \mathbb{R}$ a continuous function. Consider the **pressure function** $\psi(t) = P_X(t\varphi)$ for $t \in \mathbb{R}$. By the continuity property of the topological pressure (see Section 11) this function is continuous. If φ is negative the function $\psi(t)$ can be shown to be strictly decreasing (see below). Since $\psi(0) = h_X(f) \geq 0$ the equation

$$P_X(t\varphi) = 0$$

has a unique root. In [Bo3], Bowen discovered this equation while studying the Hausdorff dimension of quasi-circles. This equation is known now in dimension theory as **Bowen's equation**.

If $Z \subset X$ is an arbitrary set (not necessarily invariant or compact) one can consider three **pressure functions** $\psi(t) = P_Z(t\varphi)$, $\underline{\psi}(t) = \underline{CP}_Z(t\varphi)$, and $\overline{\psi}(t) = \overline{CP}_Z(t\varphi)$.

Theorem A2.5. *Assume that the function φ is negative. Then*

(1) *the functions $\psi(t)$, $\underline{\psi}(t)$, and $\overline{\psi}(t)$ are Lipschitz continuous, convex, and strictly decreasing;*

(2) *there exist unique roots s, $\underline{s}$, and $\overline{s}$ of the equations*

$$\psi(t) = 0, \quad \underline{\psi}(t) = 0, \quad \overline{\psi}(t) = 0;$$

(3) $0 \leq s \leq \underline{s} \leq \overline{s}$ *and* $\overline{s} < \infty$ *if* $h_X(f) < \infty$;

(4) $s = 0$ *if and only if* $h_Z(f) = 0$; $\underline{s} = 0$ *if and only if* $\underline{Ch}_Z(f) = 0$; *and* $\overline{s} = 0$ *if and only if* $\overline{Ch}_Z(f) = 0$.

Proof. Lipschitz continuity of the functions $\psi(t)$, $\underline{\psi}(t)$, and $\overline{\psi}(t)$ follows from Theorem 11.4. If $t' \geq t$, we obtain that

$$-(t'-t)C_1 \leq \psi(t') - \psi(t) \leq -(t'-t)C_2,$$

where C_1 and C_2 are positive constants. This implies that the function $\psi(t)$ is strictly decreasing. Similar arguments show that the other two functions are also strictly decreasing. The proof of convexity of these functions is straightforward.

Since $\psi(0) \geq 0$, $\underline{\psi}(0) \geq 0$, and $\overline{\psi}(0) \geq 0$ the second statement follows. The last two statements are consequences of Theorems 11.2 and 11.3. ∎

We notice that by Theorem 11.5, if the set Z is invariant then $\underline{\psi}(t) = \overline{\psi}(t)$ for all t and hence $\underline{s} = \overline{s}$; if, in addition, the set Z is compact then $\psi(t) = \underline{\psi}(t) = \overline{\psi}(t)$ for all t and hence $s = \underline{s} = \overline{s}$.

Following Barreira [Bar2], we consider a non-additive version of Bowen's equation. Let $\varphi = \{\varphi_n \colon X \to \mathbb{R}\}$ be a sequence of continuous functions satisfying (12.1). We assume, in addition, that the following condition holds:

there exist negative constants B_1 and B_2 such that for any sufficiently large n

$$B_1 n \leq \varphi_n \leq B_2 n. \tag{A2.22}$$

This condition is satisfied if φ_n is additive. The following statement is an extension of Theorem A2.5 to the non-additive case.

Theorem A2.6. [Bar2] *If a sequence of functions φ satisfies Condition (A2.22) then the pressure functions $\psi(t) = P_Z(t\varphi)$, $\underline{\psi}(t) = \underline{CP}_Z(t\varphi)$, and $\overline{\psi}(t) = \overline{CP}_Z(t\varphi)$ satisfy Statements (1)–(4) of Theorem A2.5. In particular, there exist unique roots s, $\underline{s}$, and $\overline{s}$ of the equations $\psi(t) = 0$, $\underline{\psi}(t) = 0$, and $\overline{\psi}(t) = 0$ respectively, which satisfy $0 \leq s \leq \underline{s} \leq \overline{s}$.*

We notice that by Theorem 12.1, if the sequence of functions φ is sub-additive (see (12.4)) and the set Z is invariant then $\underline{\psi}(t) = \overline{\psi}(t)$ for all t and hence $\underline{s} = \overline{s}$; if, in addition, the set Z is compact and the sequence of functions $\{\varphi_n\}$ satisfies (12.5) then $\psi(t) = \underline{\psi}(t) = \overline{\psi}(t)$ for all t and hence $s = \underline{s} = \overline{s}$.

One can use the construction in Example 15.1 below to show that there exists a sequence of functions $\varphi = \{\varphi_n\}$ defined on a compact invariant set Z such that: 1) it satisfies conditions (12.1) and (A2.22) (and hence there exist unique roots $\underline{s}$ and $\overline{s}$ of the equations $\underline{\psi}(t) = 0$ and $\overline{\psi}(t) = 0$); 2) it is *not* sub-additive; 3) $\underline{s} < \overline{s}$ (see detailed description in [Bar2]).

We apply the above results to a symbolic dynamical system (Q, σ), where Q is an invariant compact subset of Σ_p^+. Assume that for any $(n+1)$-tuple $(i_0 \dots i_n)$ we have a positive number $a_{i_0 \dots i_n}$. Consider a sequence of functions $\varphi = \{\varphi_n\}$, where $\varphi_n(\omega) = \log a_{i_0 \dots i_n}$ for $\omega = (i_0 i_1 \dots)$. One can verify that the sequence of functions φ satisfies (12.1). The non-additive topological pressure and non-additive lower and upper capacity topological pressures corresponding to φ admit the following explicit description established in [Bar2].

Theorem A2.7. *For every $t \in \mathbb{R}$ we have:*

(1) *if the sequence of functions φ is sub-additive then*

$$\underline{CP}_Q(t\varphi) = \overline{CP}_Q(t\varphi) = \lim_{n \to \infty} \frac{1}{n} \log \sum_{\substack{(i_0 \ldots i_n) \\ Q\text{-admissible}}} a_{i_0 \ldots i_n}{}^t$$

and the limit exists as $n \to \infty$.

(2) *if the sequence of functions φ is sub-additive and satisfies (12.5) then*

$$P_Q(t\varphi) = \underline{CP}_Q(t\varphi) = \overline{CP}_Q(t\varphi).$$

Assume that the numbers $a_{i_0 \ldots i_n}$ satisfy the following condition:

$$e^{-\alpha n} \leq a_{i_0 \ldots i_n} \leq e^{-\beta n}, \tag{A2.23}$$

where α and β are positive constants. Clearly, (A2.23) implies (A2.22), and hence the equations

$$P_Q(t\varphi) = 0, \quad \underline{CP}_Q(t\varphi) = 0, \quad \overline{CP}_Q(t\varphi) = 0$$

have unique roots s, $\underline{s}$, and $\overline{s}$ which satisfy $0 \leq s \leq \underline{s} \leq \overline{s} < \infty$.

We consider an important particular case. Let $0 < a_1, \ldots, a_p < 1$ be numbers. Define the function φ on Σ_p^+ by $\varphi(\omega) = \log a_{i_0}$, where $\omega = (i_0 i_1 \ldots)$.

Theorem A2.8.

(1) *If $Q \subset \Sigma_p^+$ is a compact σ-invariant set, then Bowen's equation $P_Q(t\varphi) = 0$ has a unique root.*

(2) *If (Q, σ) is the full shift, i.e., $Q = \Sigma_p^+$, then Bowen's equation $P_Q(t\varphi) = 0$ is equivalent to the equation*

$$\sum_{i=1}^{p} a_i{}^t = 1.$$

(3) *If (Q, σ) is a subshift of finite type, i.e., $Q = \Sigma_A^+$ with a transfer matrix A, then Bowen's equation $P_Q(t\varphi) = 0$ is equivalent to the equation $\rho(AM_t(a)) = 1$, where $\rho(B)$ denotes the spectral radius of the matrix B and $M_t(a) = \text{diag}(a_1{}^t, \cdots, a_p{}^t)$.*

Proof. Let $f : \Sigma_A^+ \to \mathbb{R}$ be a continuous function. Consider the *transfer operator*

$$(L_\phi f)(\omega) \stackrel{\text{def}}{=} \sum_{\omega' \in \sigma^{-1}(\omega)} \exp(\varphi(\omega'))f(\omega') = \sum_k \exp(\varphi(k))f(k i_0 i_1 \ldots)A(k, i_0),$$

where $\omega = (i_0 i_1 \ldots)$ and the function $\varphi(\omega) = \log a_{i_0} = \varphi(i_0)$ depends on the first coordinate only. The eigenvalue equation (corresponding to an eigenvalue η) is

$$\sum_k \exp(\varphi(k))h(k)A(k, j) = \eta h(j).$$

According to [R1], the largest eigenvalue of L_φ is $\exp(P(\varphi))$. Hence, $\exp(P(\varphi))$ is the spectral radius of the matrix $A^*\Phi$, where Φ denotes the $(p \times p)$ diagonal matrix $\text{diag}(e^{\varphi(1)}, e^{\varphi(2)}, \cdots, e^{\varphi(p)})$. This implies the third statement. The second one is an immediate consequence of the third statement. The first statement follows from Theorem A2.5. ∎

We conclude the appendix with formulae for the topological entropy of a subshift of finite type and the Hausdorff dimension of a topological Markov chain.

Theorem A2.9. *Let A be a transfer matrix. Then*

(1) $h_{\Sigma_A^+}(\sigma) = h_{\Sigma_A^-}(\sigma) = h_{\Sigma_A}(\sigma) = \rho(A)$;

(2) $\dim_H \Sigma_A^+ = \dim_H \Sigma_A^- = \frac{\log \rho(A)}{\log \beta}$, *where β is the coefficient in the d_β-metric (see (A2.18));*

(3) $\dim_H \Sigma_A = 2\frac{\log \rho(A)}{\log \beta}$, *where β is the coefficient in the d_β-metric (see (A2.18$'$)).*

Proof. The first statement follows from the third statement of Theorem A2.8. The second and the third statements can be proved by straightforward calculation which we leave to the reader. Note that the second statement is essentially a corollary of Theorem 13.2 (see Statement 2; see also self-similar constructions in Section 13) while the third statement is a corollary of Theorem 22.2 (see also linear horseshoes in Section 23). ∎

Appendix III

An Example of Carathéodory Structure Generated by Dynamical Systems

In this Appendix we briefly discuss an example of C-structure generated by dynamical systems. This C-structure was introduced by Barreira and Schmeling [BS] in their study of metrically irregular sets for dynamical systems. See Appendix IV.

Let $f: X \to X$ be a continuous map of a compact metric space X and $\varphi: X \to \mathbb{R}$ a continuous *strictly positive* function. Consider a finite open cover $\mathcal{U}$ of X and define the collection of subsets $\mathcal{F} = \mathcal{F}(\mathcal{U})$ by (11.2) and three functions $\xi, \eta, \psi: \mathcal{S}(\mathcal{U}) \to \mathbb{R}$ as follows

$$\xi(\boldsymbol{U}) = 1, \quad \eta(\boldsymbol{U}) = \exp\big(- \sup_{x \in X(\boldsymbol{U})} S_{m(\boldsymbol{U})}\varphi(x)\big), \quad \psi(\boldsymbol{U}) = m(\boldsymbol{U})^{-1},$$

where

$$S_{m(\boldsymbol{U})}\varphi = \sum_{k=0}^{m(\boldsymbol{U})-1} \varphi \circ f^k.$$

One can directly verify that the collection of subsets $\mathcal{F}$, and the functions η, ξ, and ψ satisfy Conditions $A1, A2, A3$, and $A3'$ in Section 10 and hence determine a C-structure $\tau = \tau(\mathcal{U}) = (\mathcal{S}, \mathcal{F}, \xi, \eta, \psi)$ on X. The corresponding Carathéodory function $m_C(Z, \alpha)$ (where $Z \subset X$ and $\alpha \in \mathbb{R}$; see Section 10) depends on the cover $\mathcal{U}$ (and the function φ) and is given by

$$m_C(Z, \alpha) = \lim_{N \to \infty} M(Z, \alpha, \varphi, \mathcal{U}, N),$$

where

$$M(Z, \alpha, \varphi, \mathcal{U}, N) = \inf_{\mathcal{G}} \left\{ \sum_{U \in \mathcal{G}} \exp\big(-\alpha \sup_{x \in X(\boldsymbol{U})} S_{m(\boldsymbol{U})}\varphi(x)\big) \right\}$$

and the infimum is taken over all finite or countable collections of strings $\mathcal{G} \subset \mathcal{S}(\mathcal{U})$ such that $m(\boldsymbol{U}) \geq N$ for all $\boldsymbol{U} \in \mathcal{G}$ and $\mathcal{G}$ covers Z.

Furthermore, the Carathéodory functions $\underline{r}_C(Z, \alpha)$ and $\overline{r}_C(Z, \alpha)$ (where $Z \subset X$ and $\alpha \in \mathbb{R}$; see Section 10) depend on the cover $\mathcal{U}$ and are given by

$$\underline{r}_C(Z, \alpha) = \varliminf_{N \to \infty} R(Z, \alpha, \varphi, \mathcal{U}, N), \quad \overline{r}_C(Z, \alpha) = \varlimsup_{N \to \infty} R(Z, \alpha, \varphi, \mathcal{U}, N),$$

where

$$R(Z, \alpha, \varphi, \mathcal{U}, N) = \inf_{\mathcal{G}} \left\{ \sum_{U \in \mathcal{G}} \exp\left(-\alpha \sup_{x \in X(U)} S_N \varphi(x)\right) \right\}$$

and the infimum is taken over all finite or countable collections of strings $\mathcal{G} \subset S(\mathcal{U})$ such that $m(U) = N$ for all $U \in \mathcal{G}$ and $\mathcal{G}$ covers Z.

According to Section 10, given a set $Z \subset X$, the C-structure τ generates the Carathéodory dimension of Z and lower and upper Carathéodory capacities of Z specified by the cover $\mathcal{U}$ and the map f. We denote them by $BS_{\varphi,\mathcal{U}}(Z)$, $\underline{BS}_{\varphi,\mathcal{U}}(Z)$, and $\overline{BS}_{\varphi,\mathcal{U}}(Z)$ respectively. Repeating arguments in the proof of Theorem 11.1 one can show that *for any set $Z \subset X$ the following limits exist:*

$$BS_{\varphi}(Z) \stackrel{\text{def}}{=} \lim_{|\mathcal{U}| \to 0} BS_{\varphi,\mathcal{U}}(Z),$$

$$\underline{BS}_{\varphi}(Z) \stackrel{\text{def}}{=} \lim_{|\mathcal{U}| \to 0} \underline{BS}_{\varphi,\mathcal{U}}(Z),$$

$$\overline{BS}_{\varphi}(Z) \stackrel{\text{def}}{=} \lim_{|\mathcal{U}| \to 0} \overline{BS}_{\varphi,\mathcal{U}}(Z).$$

We call the quantities $BS_{\varphi}(Z)$, $\underline{BS}_{\varphi}(Z)$, and $\overline{BS}_{\varphi}(Z)$, respectively the BS-**dimension of a set** and **lower** and **upper** BS-**box dimensions of the set** Z (specified by the function φ and the map f; after Barreira and Schmeling). We emphasize that the set Z can be arbitrary and need not be compact or invariant under the map f.

The main properties of BS-dimension and lower and upper BS-box dimensions are described below. They are immediate corollaries of the definitions and Theorems 1.1 and 2.1.

Theorem A3.1.

(1) $BS_{\varphi}\varnothing = 0$; $BS_{\varphi}(Z) \geq 0$ *for any non-empty set $Z \subset X$.*

(2) $BS_{\varphi}(Z_1) \leq BS_{\varphi}(Z_2)$ *if $Z_1 \subset Z_2 \subset X$.*

(3) $BS_{\varphi}(Z) = \sup_{i \geq 1} BS_{\varphi}(Z_i)$, *where $Z = \bigcup_{i \geq 1} Z_i$ and $Z_i \subset X$, $i = 1, 2, \ldots$.*

(4) $\underline{BS}_{\varphi}\varnothing = \overline{BS}_{\varphi}\varnothing = 0$; $\underline{BS}_{\varphi}(Z) \geq 0$ *and $\overline{BS}_{\varphi}(Z) \geq 0$ for any non-empty set $Z \subset X$.*

(5) $\underline{BS}_{\varphi}(Z_1) \leq \underline{BS}_{\varphi}(Z_2)$ *and $\overline{BS}_{\varphi}(Z_1) \leq \overline{BS}_{\varphi}(Z_2)$ if $Z_1 \subset Z_2 \subset X$.*

(6) $\underline{BS}_{\varphi}(Z) \geq \sup_{i \geq 1} \underline{BS}_{\varphi}(Z_i)$ *and $\overline{BS}_{\varphi}(Z) \geq \sup_{i \geq 1} \overline{BS}_{\varphi}(Z_i)$, where $Z = \bigcup_{i \geq 1} Z_i$ and $Z_i \subset X$, $i = 1, 2, \ldots$.*

(7) *If $h: X \to X$ is a homeomorphism which commutes with f (i.e., $f \circ h = h \circ f$) then*

$$BS_{\varphi}(Z) = BS_{\varphi \circ h^{-1}}(h(Z)),$$

$$\underline{BS}_{\varphi}(Z) = \underline{BS}_{\varphi \circ h^{-1}}(h(Z)), \quad \overline{BS}_{\varphi}(Z) = \overline{BS}_{\varphi \circ h^{-1}}(h(Z)).$$

(8) *For any two continuous functions φ and ψ on X*

$$|BS_{\varphi}(Z) - BS_{\psi}(Z)| \leq \|\varphi - \psi\|,$$

$$|\underline{BS}_{\varphi}(Z) - \underline{BS}_{\psi}(Z)| \leq \|\varphi - \psi\|,$$

$$|\overline{BS}_{\varphi}(Z) - \overline{BS}_{\psi}(Z)| \leq \|\varphi - \psi\|,$$

where $\|\cdot\|$ denotes the supremum norm in the space of continuous functions on X.

Remarks.

(1) In the case $\varphi = 1$, the BS-dimension of a set $Z \subset X$ and lower and upper BS-box dimensions of Z coincide with the topological entropy of f on Z and lower and upper capacity topological entropies of f on Z respectively.

(2) Compare the definitions of topological pressure and BS-dimension one can obtain that for any set $Z \subset X$ the BS-dimension of Z is a unique root of Bowen's equation $P_Z(-s\varphi) = 0$, i.e., $s = BS_\varphi(Z)$.

Let μ be a Borel probability measure on X which is invariant under f. According to Section 10, the C-structure $\tau = (\mathcal{S}, \mathcal{F}, \xi, \eta, \psi)$ on X generates the Carathéodory dimension of μ and lower and upper Carathéodory capacities of μ specified by the cover $\mathcal{U}$ and the map f. We denote them by $BS_{\varphi,\mathcal{U}}(\mu)$, $\underline{BS}_{\varphi,\mathcal{U}}(\mu)$, and $\overline{BS}_{\varphi,\mathcal{U}}(\mu)$ respectively. It follows from what was said above that there exist the limits

$$BS_\varphi(\mu) \overset{\text{def}}{=} \lim_{|\mathcal{U}| \to 0} BS_{\varphi,\mathcal{U}}(\mu),$$

$$\underline{BS}_\varphi(\mu) \overset{\text{def}}{=} \lim_{|\mathcal{U}| \to 0} \underline{BS}_{\varphi,\mathcal{U}}(\mu),$$

$$\overline{BS}_\varphi(\mu) \overset{\text{def}}{=} \lim_{|\mathcal{U}| \to 0} \overline{BS}_{\varphi,\mathcal{U}}(\mu).$$

We call these quantities the BS-**dimension of a measure** and **lower** and **upper** BS-**box dimensions of the measure** μ (specified by the function φ and the map f). Given a point $x \in X$, we set in accordance with (10.1)

$$\underline{\mathcal{D}}_{C,\mu,\alpha}(x, \varphi, \mathcal{U}) = \varliminf_{N \to \infty} \inf_{\boldsymbol{U}} \frac{\log \mu(X(\boldsymbol{U}))}{- \sup\limits_{y \in X(\boldsymbol{U})} S_N \varphi(y)},$$

$$\overline{\mathcal{D}}_{C,\mu,\alpha}(x, \varphi, \mathcal{U}) = \varlimsup_{N \to \infty} \sup_{\boldsymbol{U}} \frac{\log \mu(X(\boldsymbol{U}))}{- \sup\limits_{y \in X(\boldsymbol{U})} S_N \varphi(y)},$$

where the infimum and supremum are taken over all strings $\boldsymbol{U}$ with $x \in X(\boldsymbol{U})$ and $m(\boldsymbol{U}) = N$.

We leave the proof of the following statement to the reader (compare to Proposition 11.1).

Theorem A3.2. *If μ is a Borel probability measure on X invariant under the map f and ergodic, then*

(1) *for every $\alpha \in \mathbb{R}$ and μ-almost every $x \in X$,*

$$\lim_{|\mathcal{U}| \to 0} \underline{\mathcal{D}}_{C,\mu,\alpha}(x, \varphi, \mathcal{U}) = \lim_{|\mathcal{U}| \to 0} \overline{\mathcal{D}}_{C,\mu,\alpha}(x, \varphi, \mathcal{U}) = \frac{h_\mu(f)}{\int_X \varphi \, d\mu} \overset{\text{def}}{=} d,$$

where $h_\mu(f)$ is the measure-theoretic entropy of f.

(2) $BS_\varphi(\mu) = \underline{BS}_\varphi(\mu) = \overline{BS}_\varphi(\mu) = d$.

As a consequence of Theorem A3.2 we establish the *variational principle* for the BS-dimension:

$$BS_\varphi(G) = \sup_{\mu \in \mathfrak{M}(X)} BS_\varphi(\mu), \qquad (A3.1)$$

where $G = \bigcup_{\mu \in \mathfrak{M}(X)} G_\mu$ and G_μ is the set of all forward generic points of the measure μ (see (A2.16)). Indeed, by Theorem A3.2 and (A2.16),

$$0 = h_\mu(f) - BS_\varphi(\mu) \int_X \varphi \, d\mu = P_G(-BS_\varphi(\mu)\varphi).$$

This implies that $BS_\varphi(\mu) = BS_\varphi(G_\mu)$ and hence

$$BS_\varphi(G) = \sup_{\mu \in \mathfrak{M}(X)} BS_\varphi(G_\mu) = \sup_{\mu \in \mathfrak{M}(X)} BS_\varphi(\mu).$$

Part II

**Applications to Dimension
Theory and Dynamical Systems**

Chapter 5

Dimension of Cantor-like Sets and Symbolic Dynamics

It is now the prevailing opinion among experts in dimension theory that the coincidence of the Hausdorff dimension and lower and upper box dimensions of a set is a "rare" phenomenon and can occur only if a set has a "rigid" geometric structure. Nevertheless, there exists a broad class of subsets in $\mathbb{R}^m$, known as Cantor-like sets, for which the coincidence usually takes place. These sets were a traditional object of study in dimension theory for a long time and also served as a touchstone for different techniques. Computing their Hausdorff dimension and box dimension was a great challenge to experts in the field who had to create a number of highly non-trivial methods of study. These sets often were the source for exciting examples that demonstrated different phenomena in dimension theory.

The Cantor-like sets are defined by geometric constructions of different types. We begin with the most basic geometric construction. Starting from p arbitrary closed subsets $\Delta_1, \ldots, \Delta_p$ of $\mathbb{R}^m$ we define a Cantor-like set by

$$F = \bigcap_{n=0}^{\infty} \bigcup_{(i_0 \ldots i_n)} \Delta_{i_0 \ldots i_n},$$

where the **basic sets** on the nth step of the geometric construction, $\Delta_{i_0 \ldots i_n}$, $i_k = 1, \ldots, p$ $(n \geq 0)$ are closed and satisfy the following conditions:

(CG1) $\Delta_{i_0 \ldots i_n j} \subset \Delta_{i_0 \ldots i_n}$ for $j = 1, \ldots, p$;
(CG2) $\operatorname{diam} \Delta_{i_0 \ldots i_n} \to 0$ as $n \to \infty$;
(CG3) (**separation condition**): $\Delta_{i_0 \ldots i_n} \cap \Delta_{j_0 \ldots j_n} \cap F = \varnothing$ for any $(i_0 \ldots i_n) \neq (j_0 \ldots j_n)$.

The set F is perfect, nowhere dense, and totally disconnected. See Figure 1.

The description of a geometric construction includes the description of its symbolic representation and its geometry.

Symbolic representation. Let Σ_p^+ be the set of all one-sided infinite sequences $(i_0 i_1 \ldots)$ of p symbols (p is the number of basic sets on the first step of the construction) endowed with the d_β metric given by (A2.18) (see Appendix II). Given $\omega = (i_0 i_1 \ldots) \in \Sigma_p^+$ the intersection $\bigcap_{n=0}^{\infty} \Delta_{i_0 \ldots i_n}$ is non-empty and by Condition (CG2), consists of only one point x. Thus, the formula $\chi(\omega) = x$ defines correctly a continuous map $\chi \colon \Sigma_p^+ \to F$ which we call the **coding map**. By the separation condition this map is one-to-one and hence (since Σ_p^+ is compact) it is a homeomorphism.

117

We consider a geometric construction modeled by a general symbolic dynamical system (Q, σ), where $Q \subset \Sigma_p^+$ is a compact set invariant under the shift map σ, i.e., $\sigma(Q) = Q$. Namely, we allow only basic sets that correspond to admissible n-tuples $(i_0 \ldots i_n)$ with respect to Q, i.e., there exists $(j_0 j_1 \ldots) \in Q$ such that $j_0 = i_0$, $j_1 = i_1, \ldots, j_n = i_n$. Such constructions are called **symbolic geometric constructions**. The **limit set** F is defined by

$$F = \bigcap_{n=0}^{\infty} \bigcup_{\substack{(i_0 \ldots i_n) \\ Q\text{-admissible}}} \Delta_{i_0 \ldots i_n}.$$

One can classify geometric constructions according to their symbolic representation. A construction is called a **simple geometric construction** if it is modeled by the full shift, i.e., $Q = \Sigma_p^+$ (see Figure 1). Another important class of geometric constructions are constructions modeled by subshifts of finite type. Namely, the geometric construction is called a **Markov geometric construction** if $Q = \Sigma_A^+$ (we remind the reader that Σ_A^+ consists of all sequences $(i_0 i_1 \ldots)$ admissible with respect to the transfer matrix A with entries $A(i, j) = 0$ or 1, i.e., $A(i_j, i_{j+1}) = 1$ for $j = 0, 1 \ldots$; see Appendix II; see also Figure 2 in Section 13). Finally, a geometric construction is called a **sofic geometric construction** if (Q, σ) is a sofic system (i.e., a finite factor of a subshift of finite type).

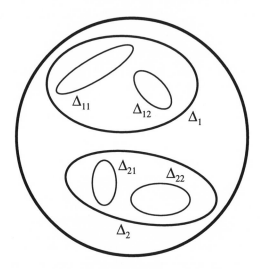

Figure 1. A SIMPLE GEOMETRIC CONSTRUCTION.

Geometry of the construction. This includes the information on the placement of the basic sets, their geometric shapes, and "sizes". If one has strong control over the shape and sizes of basic sets, then the placement can be fairly arbitrary, and vice-versa. To illustrate this we first consider constructions whose geometry seems to be most simple where one has complete control over the shape and sizes of basic sets which are *balls*. Denote by $r_{i_0...i_n}$ the radius and by $x_{i_0...i_n} \in \mathbb{R}^m$ the center of the basic set (ball) $\Delta_{i_0...i_n}$ on the nth step of the construction. The collection of numbers $r_{i_0...i_n}$ and coordinates of points $x_{i_0...i_n}$ provide complete information on the geometry of the construction and, in particular, is sufficient to compute the Hausdorff dimension and box dimension of the limit set. Surprisingly, one can often use only the numbers $r_{i_0...i_n}$ to obtain refined estimates for the Hausdorff dimension (from below) and the upper box dimension (from above) of the limit set (see Section 15).

The main tool to study this geometric construction is the thermodynamic formalism, developed in Chapter 4 and Appendix II (in particular its non-additive version), applied to the underlying symbolic dynamical system. Namely, consider the sequence of functions $\varphi_n(\omega) = \log r_{i_0...i_n}$ on Q, where $\omega = (i_0 i_1 ...)$. Clearly it satisfies Condition (12.1). According to Appendix II (see Theorem A2.6) the equations

$$P_Q(s\{\varphi_n\}) = 0, \quad \overline{CP}_Q(s\{\varphi_n\}) = 0$$

(where P_Q and $\overline{CP}_Q$ denote the non-additive topological pressure and non-additive upper capacity topological pressure specified by the sequence of functions $\{\varphi_n\}$) have unique roots provided the sequence of radii admits the asymptotic estimate (A2.23). We denote these roots by $\underline{s}$ and $\overline{s}$ respectively. Clearly, $\underline{s} \leq \overline{s}$. In Section 15 we will show that, under some mild additional assumptions, the number $\underline{s}$ provides a lower bound for the Hausdorff dimension and the number $\overline{s}$ provides an upper bound for the upper box dimension of the limit set (see Theorem 15.1).

We emphasize that the above approach works well for an arbitrary collection of numbers $r_{i_0...i_n}$ which may depend on the whole "past" and may not admit any asymptotic behavior as $n \to \infty$. Yet, the Hausdorff dimension and lower and upper box dimensions may not coincide and their exact values may depend on the placement of basic sets (see Example 15.1). However, if, for instance, the sequence of functions φ_n is sub-additive (see Condition (12.4)), then the Hausdorff dimension and lower and upper box dimensions coincide and are completely determined by the sizes of basic sets only, i.e., the numbers $r_{i_0...i_n}$. In this case $\underline{s} = \overline{s} \stackrel{\text{def}}{=} s$ and regardless of the placement of basic sets, the Hausdorff dimension and lower and upper box dimensions of the limit set coincide and are equal to s. Moreover, s is the unique root of the equation

$$P_Q(s\{\varphi_n\}) = 0.$$

An equation of this type was first discovered by Bowen in his study of the Hausdorff dimension of quasi-circles (see [Bo2]) and now bears his name. Bowen's equation is one of the main manifestations of the general Carathéodory construction developed in Chapter 1: it expresses intimate relations between two Carathéodory dimension characteristics: the Hausdorff dimension of the limit set of a geometric construction and non-additive topological pressure specified by the sequence of functions $\{\varphi_n\}$ for the underlying symbolic dynamical system

(Q, σ). Bowen's equation contains all necessary information on the geometric construction that matters in computing the Hausdorff dimension of the limit set: the information on the underlying symbolic dynamical system and its realization into the Euclidean space by basic sets.

Bowen's equation seems to be "universal" in dimension theory: we will demonstrate that the Hausdorff dimension of limit sets for various classes of geometric constructions as well as invariant sets for various classes of dynamical systems can be computed as roots of Bowen equations written with respect to the underlying symbolic dynamical systems.

In 1946 Moran, in his seminal paper [Mo], introduced and studied geometric constructions with basic sets satisfying the following conditions:

(CM1) each basic set $\Delta_{i_0...i_n}$ is the closure of its interior;

(CM2) $\Delta_{i_0...i_n j} \subset \Delta_{i_0...i_n}$ for $j = 1, \ldots, p$;

(CM3) each basic set $\Delta_{i_0...i_n j}$ is geometrically similar to the basic set $\Delta_{i_0...i_n}$ for every j;

(CM4) for any $(i_0 \ldots i_n) \neq (j_0 \ldots j_n)$ the basic sets $\Delta_{i_0...i_n}$ and $\Delta_{j_0...j_n}$ do not overlap, i.e., their interiors are disjoint;

(CM5) $\operatorname{diam} \Delta_{i_0...i_n j} = \lambda_j \operatorname{diam} \Delta_{i_0...i_n}$, where $0 < \lambda_j < 1$ for $j = 1, \ldots, p$ are constants.

Such constructions are called **Moran geometric constructions** (the name coined by Cawley and Mauldin [CM]). The numbers λ_i are known as **ratio coefficients** of the construction since they determine the rate of decreasing the sizes of basic sets. We stress that basic sets of a Moran geometric construction at the same step may not be disjoint (they may intersect along boundaries; compare to Condition (CG3)). Therefore, the limit set F may not be a Cantor-like set and the coding map χ may not be injective (but it is still surjective).

Moran considered only geometric constructions modeled by the full shift. His remarkable observation was that an "optimal" cover, which can be used to obtain the exact values of the Hausdorff dimension of the limit set (we call it a **Moran cover**) is completely determined by the symbolic representation of the geometric construction. In particular, regardless of the placement of basic sets, the Hausdorff dimension of the limit set is completely determined by the ratio coefficients. Namely, Moran discovered that it is a unique root of the equation

$$\sum_{i=1}^{p} \lambda_i{}^t = 1$$

which is a particular case of Bowen's equation corresponding to the full shift. Moran constructions modeled by subshifts of finite type or sofic systems were studied in [MW].

In [PW1], Pesin and Weiss demonstrated that Moran's approach can be greatly extended to a much more general class of **Moran-like geometric constructions with stationary (constant) ratio coefficients**. They discovered that the only property of basic sets that matters in constructing Moran covers, is the following: there exist closed balls $B_{i_0...i_n}$ of radii $r_{i_0...i_n} = C \prod_{j=0}^{n} \lambda_{i_j}$ $(C > 0$ is a constant) with *disjoint* interiors such that

$$B_{i_0...i_n} \subset \Delta_{i_0...i_n}.$$

The topology and geometry of basic sets of these constructions may be quite complicated (for example, they may not be connected and their boundary may be fractal). Moreover, the basic sets at step n of the construction need not be geometrically similar to the basic sets at step $n - 1$ (see Condition (CM3) in the definition of Moran geometric constructions). Another important feature is that basic sets $\Delta_{i_0...i_n}$ at step n of the construction *may* intersect each other (although the interiors of the balls $B_{i_0...i_n}$ *must* be disjoint).

In [PW1], Pesin and Weiss also showed that a slight modification of Moran's approach allows one to deal with geometric constructions modeled by an arbitrary symbolic dynamical system (Q, σ). The main tool of study, as explained above, is the thermodynamic formalism described in Chapter 4 and Appendix II. In particular, regardless of the placement of basic sets the Hausdorff dimension and lower and upper box dimensions of the limit set F coincide and the common value is a unique root of Bowen's equation

$$P_Q(s\varphi) = 0,$$

where $\varphi(\omega) = \log \lambda_{i_0}$ for any $\omega = (i_0 i_1 \ldots) \in Q$. Moreover, the push forward to F by the coding map of an equilibrium measure μ on Q corresponding to the function $s\varphi$ is an invariant measure m of full dimension, i.e., $\dim_H F = \dim_H m$ (see discussion in Section 5). It is worth emphasizing that if μ_φ is a Gibbs measure (this happens, for example, if $Q = \Sigma_A^+$ for some transitive transfer matrix A) the Hausdorff measure of F is equivalent to m_φ; in particular, it is positive and finite.

In [St], Stella considered a particular case of geometric constructions modeled by subshifts of finite type assuming that basic sets do not overlap and satisfy Conditions (CM1) and (CM4). In [Bar2], Barreira pointed out that using a non-additive version of the thermodynamic formalism, one can generalize results by Pesin and Weiss to **Moran-like geometric constructions with non-stationary ratio coefficients** where the radii of balls $B_{i_0...i_n}$ may depend on the whole past $(i_0 \ldots i_n)$ (see Section 15).

Moran-like geometric constructions with stationary or non-stationary ratio coefficients can serve as models to approximate geometric constructions with arbitrary geometry of basic sets. For example, one can choose $B_{i_0...i_n}$ to be inscribed balls and use their radii to estimate the Hausdorff dimension (from below) of the limit set. We note that Moran-like geometric constructions are isotropic, i.e., the ratio coefficients do not depend on the directions in $\mathbb{R}^m$. There are geometric constructions which do not satisfy this assumption and have different rates of contraction in different directions in $\mathbb{R}^m$. Examples include geometric constructions in $\mathbb{R}^2$ with basic sets to be rectangles or ellipsis (see Sections 14 and 16).

Nevertheless, the idea of approximating geometric constructions (CG1–CG2) with complicated geometry of basic sets (which may not even satisfy separation condition (CG3)) by Moran-like geometric constructions can be fruitfully used to some extent. We call a geometric construction **regular** if it admits such an approximation. We stress that basic sets of a regular geometric construction need not resemble balls in any geometrical sense. Roughly speaking the regularity of a geometric construction means that its basic sets can be effectively replaced by balls such that the new geometric construction admits a Moran cover. In Section 14 we give a formal mathematical definition of the notion of regularity and study some properties of regular constructions. In particular, we obtain a general lower bound for the Hausdorff dimension of the limit set of the construction (see Theorem 14.1).

One can further use the approach, based on an effective replacement of the basic set of a geometric construction (CG1–CG2) by disjoint balls, to obtain a general upper bound for the upper box dimension of its limit set (see Theorem 14.5).

In general, it may be extremely difficult to find an effective approximation of a given geometric construction; for instance, inscribed balls often fail to produce a reasonable approximation (see Example 14.1). A more sophisticated approach to the description of geometric constructions with complicated geometry of basic sets is based upon the study of the **induced map** G on the limit set F generated by the symbolic dynamical system. We notice that if basic sets of a geometric construction are disjoint the coding map is injective. Therefore, the shift σ induces a map G on the limit set F given by $G = \chi \circ \sigma \circ \chi^{-1}$.

The dynamics of G bears the complete information on the symbolic representation of the construction (via the map $\sigma|Q$) as well as on its geometry (via the coding map χ, i.e., the embedding of the symbolic dynamics into $\mathbb{R}^m$ by χ). It seems quite plausible that those characteristics of the dynamics of G that are connected to the instability of its trajectories are relevant for computing the Hausdorff dimension and box dimension of F. This can be clearly seen when G is an expanding map. In Section 15 we introduce appropriate characteristics of instability and demonstrate how to use them in estimating the Hausdorff dimension and box dimension of the limit set.

In dimension theory there is a popular class of geometric constructions, known as **self-similar geometric constructions**, which are effected by a finite collection of **similarity maps**. This means that the basic sets $\Delta_{i_0 \ldots i_n}$ are given by

$$\Delta_{i_0 \ldots i_n} = h_{i_0} \circ h_{i_1} \circ \cdots \circ h_{i_n}(D),$$

where $h_1, \ldots, h_p \colon D \to D$ are conformal affine maps, i.e., they satisfy the property: $\operatorname{dist}(h_i(x), h_i(y)) = \lambda_i \operatorname{dist}(x, y)$ for any $x, y \in D$ (where D is the unit ball in $\mathbb{R}^m$) with fixed $0 < \lambda_i < 1$ (for detailed description of self-similar constructions and related results see [F1] where further references can also be found). These constructions are a very special type of Moran geometric constructions (CM1–CM5) where not only the sizes of basic sets but also the gaps between them are strongly controlled.

A more general class of geometric constructions is formed by **geometric constructions with contraction maps** where the maps h_i are bi-Lipschitz contraction maps. This means that for any $x, y \in D$,

$$\underline{\lambda}_i \operatorname{dist}(x, y) \leq \operatorname{dist}(h_i(x), h_i(y)) \leq \overline{\lambda}_i \operatorname{dist}(x, y),$$

where $0 < \underline{\lambda}_i \leq \overline{\lambda}_i < 1$. These constructions are regular (see Section 14).

13. Moran-like Geometric Constructions with Stationary (Constant) Ratio Coefficients

We begin with a geometric construction with the simplest geometry of basic sets modeled by a symbolic dynamical system (Q, σ), where $Q \subset \Sigma_p^+$ is a compact shift invariant set. We assume that basic sets are closed and satisfy:

(CPW1) $\Delta_{i_0 \ldots i_n j} \subset \Delta_{i_0 \ldots i_n}$ for $j = 1, \ldots, p$;

(CPW2) $\underline{B}_{i_0 \ldots i_n} \subset \Delta_{i_0 \ldots i_n} \subset \overline{B}_{i_0 \ldots i_n}$, where $\underline{B}_{i_0 \ldots i_n}$ and $\overline{B}_{i_0 \ldots i_n}$ are closed balls of radii $\underline{r}_{i_0 \ldots i_n}$ and $\overline{r}_{i_0 \ldots i_n}$;

(CPW3)

$$\underline{r}_{i_0 \ldots i_n} = K_1 \prod_{j=0}^{n} \lambda_{i_j}, \quad \overline{r}_{i_0 \ldots i_n} = K_2 \prod_{j=0}^{n} \lambda_{i_j}, \tag{13.1}$$

where $0 < \lambda_j < 1$ for $j = 1, \ldots, p$, $K_1 > 0$, and $K_2 > 0$ are constants;

(CPW4) $\operatorname{int} \underline{B}_{i_0 \ldots i_n} \cap \operatorname{int} \underline{B}_{j_0 \ldots j_m} = \varnothing$ for any $(i_0 \ldots i_n) \neq (j_0, \ldots, j_m)$ and $m \geq n$.

This class of geometric constructions was introduced by Pesin and Weiss in [PW1]. Basic sets of these constructions are *essentially* balls, although their topology and geometry may be quite complicated. Furthermore, basic sets $\Delta_{i_0 \ldots i_n}$ at step n of the construction may *intersect* each other. The numbers λ_i are called **ratio coefficients**. They are fixed and do not depend on the basic sets.

Self-similar constructions (see below) or more general Moran geometric constructions (CM1–CM5) (see the introduction to this chapter) are particular examples of geometric constructions (CPW1–CPW4).

The geometric simplicity of the geometric constructions (CPW1–CPW4) will allow us to illustrate better the role of symbolic dynamics. Given a p-tuple $\lambda = (\lambda_1, \ldots, \lambda_p)$ such that $0 < \lambda_i < 1$, there exists a uniquely defined nonnegative number s_λ such that

$$P_Q(s_\lambda \log \lambda_{i_0}) = 0$$

(we remind the reader that P_Q denotes the topological pressure on Q with respect to the shift σ; see Section 11 and Appendix II). Denote by $\chi: Q \to F$ the coding map (see definition in the introduction to this chapter). Note that it is Hölder continuous. To see this let $\omega_1 = (i_0 i_1 \ldots i_n j \ldots)$ and $\omega_2 = (i_0 i_1 \ldots i_n k \ldots)$ be two points in Q with $j \neq k$. We have

$$\|\chi(\omega_1) - \chi(\omega_2)\| \leq \prod_{j=0}^{n} \lambda_{i_j} \leq (\max_{1 \leq i \leq p} \lambda_i)^n \leq (\beta^n)^\alpha \leq C d_\beta(\omega_1, \omega_2)^\alpha,$$

where $C > 0$, $0 < \alpha < 1$ are constants and d_β is the metric in Σ_p^+ (see (A2.18) in Appendix II). This implies that any Hölder continuous function on F pulls back by χ to a Hölder continuous function on Q. Notice that, in general, the coding map χ is not invertible (since basic sets at the same level of the construction may intersect each other) and even if it is invertible it is not necessarily Hölder continuous (this depends on the placement of the basic sets, i.e., the gaps between them).

Let μ_λ be an equilibrium measure for the function $(i_0 i_1 \ldots) \mapsto s_\lambda \log \lambda_{i_0}$ on Q, and m_λ the push forward measure to F under χ (i.e., $m_\lambda(Z) = \mu_\lambda(\chi^{-1}(Z))$ for any Borel set $Z \subset F$).

We describe a special cover of the limit set F for a symbolic geometric construction (CPW1–CPW4) which allows one to build an *optimal* cover to be used to compute the Hausdorff dimension and box dimension of F. We call this cover a **Moran cover**.

Let $\lambda = (\lambda_1, \ldots, \lambda_p)$ be a vector of numbers with $0 < \lambda_i < 1$, $i = 1, \ldots, p$. Fix $0 < r < 1$. Given a point $\omega = (i_0 i_1 \ldots) \in Q$, let $n(\omega) = n(\omega, r, \lambda)$ denote the unique positive integer such that

$$\lambda_{i_0} \lambda_{i_1} \ldots \lambda_{i_{n(\omega)}} > r, \quad \lambda_{i_0} \lambda_{i_1} \ldots \lambda_{i_{n(\omega)+1}} \leq r. \tag{13.2}$$

It is easy to see that $n(\omega) \to \infty$ as $r \to 0$ uniformly in ω. Fix $\omega \in Q$ and consider the cylinder set $C_{i_0 \ldots i_{n(\omega)}} \subset Q$. We have that $\omega \in C_{i_0 \ldots i_{n(\omega)}}$. Furthermore, if $\omega' \in C_{i_0 \ldots i_{n(\omega)}}$ and $n(\omega') \leq n(\omega)$ then

$$C_{i_0 \ldots i_{n(\omega)}} \subset C_{i_0 \ldots i_{n(\omega')}}.$$

Let $C(\omega)$ be the largest cylinder set containing ω with the property that $C(\omega) = C_{i_0 \ldots i_{n(\omega'')}}$ for some $\omega'' \in C(\omega)$ and $C_{i_0 \ldots i_{n(\omega')}} \subset C(\omega)$ for any $\omega' \in C(\omega)$. The sets $C(\omega)$ corresponding to different $\omega \in Q$ either coincide or are disjoint. We denote these sets by $C^{(j)}$, $j = 1, \ldots, N_r$. There exist points $\omega_j \in Q$ such that $C^{(j)} = C_{i_0 \ldots i_{n(\omega_j)}}$. These sets form a *disjoint* cover of Q which we denote by $\mathfrak{U}_r = \mathfrak{U}_{r,Q}(\lambda)$. The sets $\Delta^{(j)} = \chi(C^{(j)})$, $j = 1, \ldots, N_r$ are not necessarily disjoint and comprise a cover of F (which we will denote by the same symbol $\mathfrak{U}_r$ if it does not cause any confusion). We have that $\Delta^{(j)} = \Delta_{i_0 \ldots i_{n(x_j)}}$ for some $x_j \in F$.

Given a subset $R \subset Q$ (not necessarily invariant) one can repeat the above arguments to construct a Moran cover of R which we denote by $\mathfrak{U}_{r,R}(\lambda)$. It consists of sets $C^{(j)} = C_{i_0 \ldots i_{n(\omega_j)}}$, where $\omega_j \in R$ and the intersection $C^{(j)} \cap C^{(k)} \cap R$ is empty for any $j \neq k$ (while the intersection $C^{(j)} \cap C^{(k)}$ may *not* be empty).

The crucial role that Moran covers play in studying the Hausdorff dimension and box dimension of F can be understood in view of the following observation. Given a point $x \in F$ and a number $r > 0$, there exists a number $M > 0$ which does *not* depend on x, r, and λ and satisfies the following property: *the number of basic sets $\Delta^{(j)}$ in a Moran cover $\mathfrak{U}_{r,Q}(\lambda)$ that have non-empty intersection with the ball $B(x, r)$ is bounded from above by M.* We call M a **Moran multiplicity factor**. Moreover, given a subset $R \subset Q$, a point $x \in F$, and a number $r > 0$, the number of basic sets $\Delta^{(j)}$ in a Moran cover $\mathfrak{U}_{r,R}(\lambda)$ that have non-empty intersection with the ball $B(x, r)$ is bounded from above by M.

Theorem 13.1. [PW1] *Let F be the limit set for a geometric construction (CPW1–CPW4) modeled by a symbolic dynamical system (Q, σ). Then*

(1) $\dim_H F = \underline{\dim}_B F = \overline{\dim}_B F = s_\lambda$;

(2) $\dim_H m_\lambda = s_\lambda$;

(3)

$$s_\lambda = -\frac{h_{\mu_\lambda}(\sigma | Q)}{\int_Q \log \lambda_{i_0} \, d\mu_\lambda}.$$

Proof. Set $s = s_\lambda$ and $d = \dim_H F$. We first show that $s \leq d$. Fix $\varepsilon > 0$. By the definition of Hausdorff dimension there exists a number $r > 0$ and a cover of F by balls B_ℓ, $\ell = 1, 2, \dots$ of radius $r_\ell \leq r$ such that

$$\sum_\ell r_\ell^{d+\varepsilon} \leq 1. \tag{13.3}$$

For every $\ell > 0$ consider a Moran cover $\mathfrak{U}_{r_\ell}$ of F and choose those basic sets from the cover that intersect B_ℓ. Denote them by $\Delta_\ell^{(1)}, \dots, \Delta_\ell^{(m(\ell))}$. Note that $\Delta_\ell^{(j)} = \Delta_{i_0 \dots i_{n(\ell,j)}}$ for some $(i_0 \dots i_{n(\ell,j)})$. By (13.1) and (13.2) it follows that

$$K_1 \prod_{k=0}^{n(\ell,j)} \lambda_{i_k} \leq \underline{r}_{i_0 \dots i_{n(l,j)}} \leq \overline{r}_{i_0 \dots i_{n(l,j)}} \leq C_1 r_\ell, \tag{13.4}$$

where $C_1 > 0$ is a constant independent of ℓ and j. The property of the Moran cover implies that $m(\ell) \leq M$, where $M > 0$ is a Moran multiplicity factor (which is independent of ℓ).

The sets $\{\Delta_\ell^{(j)}, \ j = 1, \dots, m(\ell), \ \ell = 1, 2, \dots\}$ comprise a cover $\mathcal{G}$ of F, and the corresponding cylinder sets $C_\ell^{(j)} = C_{i_0 \dots i_{n(\ell,j)}}$ comprise a cover of Q. By (13.3) and (13.4)

$$\sum_{\Delta_\ell^{(j)} \in \mathcal{G}} \prod_{k=0}^{n(\ell,j)} \lambda_{i_k}^{d+\varepsilon} \leq \sum_\ell \sum_{j=1}^{m(\ell)} \prod_{k=0}^{n(\ell,j)} \lambda_{i_k}^{d+\varepsilon}$$

$$\leq M \left(\frac{C_1}{K_1} \right)^{d+\varepsilon} \sum_\ell r_\ell^{d+\varepsilon} \leq M \left(\frac{C_1}{K_1} \right)^{d+\varepsilon}.$$

Given a number $N > 0$, choose r so small that $n(\ell, j) \geq N$ for all ℓ and j. We now have that for any $n > 0$ and $N > n$,

$$M(Q, 0, \varphi, \mathcal{U}_n, N) \leq \sum_{\Delta_\ell^{(j)} \in \mathcal{G}} \exp \left(\sup_{\omega \in C_\ell^{(j)}} \sum_{k=0}^{n(\ell,j)} \varphi(\sigma^k(\omega)) \right)$$

$$= \sum_{\Delta^{(j)} \in \mathcal{G}} \prod_{k=0}^{n(\ell,j)} \lambda_{i_k}^{d+\varepsilon} \leq M \left(\frac{C_1}{K_1} \right)^{d+\varepsilon},$$

where $M(Q, 0, \varphi, \mathcal{U}_n, N)$ is defined by (A2.19) (see Appendix II) with

$$\varphi(\omega) = (d + \varepsilon) \log \lambda_{i_0}$$

(and $\alpha = 0$). This implies that

$$P_Q((d + \varepsilon) \log \lambda_{i_0}) \le 0.$$

Hence, by Theorem A2.5 (see Appendix II), $s \le d + \varepsilon$. Since this inequality holds for all ε we conclude that $s \le d$.

Denote $\bar{d} = \overline{\dim}_B F$. We now show that $\bar{d} \le s$. Fix $\varepsilon > 0$. By the definition of the upper box dimension (see Section 6) there exists a number $r = r(\varepsilon) > 0$ such that $N(F, r) \ge r^{\varepsilon - \bar{d}}$ (recall that $N(F, r)$ is the least number of balls of radius r needed to cover the set F). Consider a Moran cover $\mathfrak{U}_r$ of Q by basic sets $C^{(j)} = C_{i_0 \ldots i_{n(\omega_j)}}, j = 1, \ldots, N_r$. Let $\Delta^{(j)} = \chi(C^{(j)}) = \Delta_{i_0 \ldots i_{n(x_j)}}$, where $x_j = \chi(\omega_j)$. Note that this cover need not be optimal, i.e., $N_r \ge N(F, r)$. By (13.2) there exists $A > 1$ such that for $j = 1, \ldots, N_r$,

$$\frac{r}{A} \le \prod_{k=0}^{n(\omega_j)+1} \lambda_{i_k} \le r$$

and hence

$$C_2 \log \frac{1}{r} - 1 \le n(\omega_j) \le C_3 \log \frac{A}{r} + 1,$$

where $C_2 > 0$ and $C_3 > 0$ are constants. This implies that $n(\omega_j)$ can take on at most $B \stackrel{\text{def}}{=} C_3 \log \frac{A}{r} - C_2 \log \frac{1}{r} + 2$ possible values.

We now think of having N_r balls and B baskets. Then there exists a basket containing at least $\frac{N_r}{B}$ balls. This implies that there exists a positive integer $N \in [C_2 \log \frac{1}{r} - 1, C_3 \log \frac{A}{r} + 1]$ such that

$$\text{card } \{j : n(\omega_j) = N\} \ge \frac{N_r}{B} \ge \frac{N(F, r)}{B} \ge \frac{r^{\varepsilon - \bar{d}}}{C_3 \log \frac{A}{r}},$$

where card denotes the cardinality of the corresponding set. If r is sufficiently small we obtain that

$$\text{card } \{j : n(\omega_j) = N\} \ge r^{2\varepsilon - \bar{d}}.$$

Consider an arbitrary cover $\mathcal{G}$ of Q by cylinder sets $C_{i_0 \ldots i_N}$. It follows that

$$\sum_{C_{i_0 \ldots i_N} \in \mathcal{G}} \prod_{k=0}^{N} \lambda_{i_k}^{\bar{d}-2\varepsilon} \ge \sum_{j : n(x_j)=N} \prod_{k=0}^{n(x_j)} \lambda_{i_k}^{\bar{d}-2\varepsilon}$$

$$\ge \sum_{j : n(x_j)=N} \left(\frac{r}{A}\right)^{\bar{d}-2\varepsilon} \ge A^{2\varepsilon-\bar{d}} r^{\bar{d}-2\varepsilon} r^{2\varepsilon-\bar{d}} \ge C_4,$$

where $C_4 > 0$ is a constant. We now have that for any $n > 0$ and $N > n$,

$$R(Q, 0, \varphi, \mathcal{U}_n, N) = \sum_{C_{i_0 \ldots i_N} \in \mathcal{G}} \exp \left(\sup_{\omega \in C_{i_0 \ldots i_N}} \sum_{k=0}^{N} \varphi(\sigma^k(\omega)) \right)$$

$$= \sum_{C_{i_0 \ldots i_N} \in \mathcal{G}} \prod_{k=0}^{N} \lambda_{i_k}{}^{\bar{d}-2\varepsilon} \geq C_4,$$

where $R(Q, 0, \varphi, \mathcal{U}_n, N)$ is defined by (A2.19$'$) (see Appendix II) with $\alpha = 0$ and

$$\varphi(\omega) = (\bar{d} - 2\varepsilon) \log \lambda_{i_0}.$$

By Theorem 11.5 this implies that

$$\overline{CP}_Q \left((\bar{d} - 2\varepsilon) \log \lambda_{i_0} \right) = P_Q \left((\bar{d} - 2\varepsilon) \log \lambda_{i_0} \right) \geq 0$$

and hence $\bar{d} - 2\varepsilon \leq s$ (see Theorem A2.5 in Appendix II). Since this inequality holds for all ε we conclude that $\bar{d} \leq s$. This completes the proof of the first statement.

In order to prove the second statement we need only to establish that $s \leq \dim_H m_\lambda$. Assume first that the measure μ_λ is a Gibbs measure corresponding to the function $s \log \lambda_{i_0}$. By (A2.20) (see Appendix II) there exist positive constants D_1 and D_2 such that for $j = 1, \ldots, N_r$

$$D_1 \leq \frac{m_\lambda(\Delta_{i_0 \ldots i_{n(x_j)}})}{\prod_{k=0}^{n(x_j)} \lambda_{i_k}{}^{s_\lambda}} \leq D_2. \tag{13.5}$$

Consider the open Euclidean ball $B(x, r)$ of radius r centered at a point x. Let $N(x, r)$ denote the number of sets $\Delta^{(j)}$ that have non-empty intersection with $B(x, r)$. It follows from the property of the Moran cover that $N(x, r) \leq M$, where M is a Moran multiplicity factor. By (13.5) and (13.2) we obtain that for every x and every $r > 0$,

$$m_\lambda(B(x, r)) \leq \sum_{j=1}^{N(x,r)} m_\lambda(\Delta^{(j)}) \leq \sum_{j=1}^{N(x,r)} D_2 \prod_{k=0}^{n(x_j)} \lambda_{i_k}{}^{s}$$

$$\leq C_5 \sum_{j=1}^{N(x,r)} \prod_{k=0}^{n(x_j)+1} \lambda_{i_k}{}^{s} \leq C_5 N(x, r) r^s \leq C_5 M r^s, \tag{13.6}$$

where $C_5 > 0$ is a constant. It follows that the measure m_λ satisfies the uniform mass distribution principle (see Section 7) and hence $\dim_H m_\lambda \geq s$.

We turn to the general case when μ_λ is just an equilibrium measure. By definition

$$h_{\mu_\lambda}(\sigma | Q) + s \int_Q \log \lambda_{i_0} d\mu_\lambda = 0, \tag{13.7}$$

where $h_\mu(\sigma|Q) \overset{\text{def}}{=} h$ is the measure-theoretic entropy. Let us first assume that μ_λ is ergodic. Fix $\varepsilon > 0$. It follows from the Shannon–McMillan–Breiman theorem that for μ_λ-almost every $\omega \in Q$ one can find $N_1(\omega) > 0$ such that for any $n \geq N_1(\omega)$,

$$\mu_\lambda(C_{i_0\dots i_n}(\omega)) \leq \exp(-(h-\varepsilon)n), \tag{13.8}$$

where $C_{i_0\dots i_n}(\omega)$ is the cylinder set containing ω.

If the measure μ_λ is ergodic it follows from the Birkhoff ergodic theorem, applied to the function $s \log \lambda_{i_0}$, that for μ_λ-almost every $\omega \in Q$ there exists $N_2(\omega)$ such that for any $n \geq N_2(\omega)$,

$$s \int_Q \log \lambda_{i_0} d\mu_\lambda \leq \frac{1}{n} \log \prod_{j=0}^n \lambda_{i_j}^{\ s} + \varepsilon. \tag{13.9}$$

Combining (13.7), (13.8), and (13.9) we obtain that for μ_λ-almost every $\omega \in Q$ and $n \geq \max\{N_1(\omega), N_2(\omega)\}$,

$$\mu_\lambda(C_{i_0\dots i_n}(\omega)) \leq \prod_{j=0}^n \lambda_{i_j}^{\ s} \exp(2\varepsilon n) \leq \prod_{j=0}^n \lambda_{i_j}^{\ s-\alpha},$$

where $\alpha = 2\varepsilon / \min_j \log(1/\lambda_j) > 0$. This implies that for μ_λ-almost every $\omega \in Q$ and any $n \geq \max\{N_1(\omega), N_2(\omega)\}$,

$$\mu_\lambda(C_{i_0\dots i_n}(\omega)) \leq \prod_{j=0}^n \lambda_{i_j}^{\ s-\alpha}. \tag{13.10}$$

If μ_λ is not ergodic, then (13.10) is still valid and can be shown by decomposing μ_λ into its ergodic components.

Given $\ell > 0$, denote by $Q_\ell = \{\omega \in Q : N_1(\omega) \leq \ell \text{ and } N_2(\omega) \leq \ell\}$. It is easy to see that $Q_\ell \subset Q_{\ell+1}$ and $Q = \bigcup_{\ell=1}^\infty Q_\ell \pmod 0$. Thus, there exists $\ell_0 > 0$ such that $\mu_\lambda(Q_\ell) > 0$ if $\ell \geq \ell_0$. Let us choose $\ell \geq \ell_0$.

Given $0 < r < 1$, consider a Moran cover $\mathfrak{U}_{r,Q_\ell}$ of the set Q_ℓ. It consists of sets $C_\ell^{(j)}$, $j = 1, \dots, N_{r,\ell}$ for which there exist points $\omega_j \in Q$ such that $C_\ell^{(j)} = C_{i_0\dots i_{n(\omega_j)}}$. Set $\Delta_\ell^{(j)} = \chi(C_\ell^{(j)})$.

Consider the open Euclidean ball $B(x,r)$ of radius r centered at a point x. Let $N = N(x,r,\ell)$ denote the number of sets $\Delta_\ell^{(j)}$ that have non-empty intersection with $B(x,r)$. By the property of the Moran cover we have that $N \leq M$, where M is a Moran multiplicity factor. It now follows from (13.10) and (13.2) that

$$m_\lambda(B(x,r) \cap \chi(Q_\ell)) \leq \sum_{j=1}^N m_\lambda(\Delta_\ell^{(j)}) \leq \sum_{j=1}^N \prod_{k=0}^{n(x_j)} \lambda_{i_k}^{\ s-\alpha}$$

$$\leq K_2 N(x,r,\ell) r^{s-\alpha} \leq K_2 M r^{s-\alpha}.$$

Since $m_\lambda(\chi(Q_\ell)) > 0$ by the Borel Density Lemma (see Appendix V) for m_λ-almost every $x \in \chi(Q_\ell)$ there exists a number $r_0 = r_0(x)$ such that for every $0 < r \leq r_0$ we have

$$m_\lambda(B(x,r)) \leq 2m_\lambda(B(x,r) \cap \chi(Q_\ell)).$$

This implies that for any $\ell > \ell_0$ and m_λ-almost every $x \in \chi(Q_\ell)$,

$$\underline{d}_{m_\lambda}(x) = \varliminf_{r \to 0} \frac{\log m_\lambda(B(x,r))}{\log r} \geq \varliminf_{r \to 0} \frac{\log m_\lambda(B(x,r) \cap \chi(Q_\ell))}{\log r} \geq s_\lambda - \alpha.$$

Since the sets Q_ℓ are nested and exhaust the set Q (mod 0) we obtain that $\underline{d}_{m_\lambda}(x) \geq s_\lambda - \alpha$ for m_λ-almost every $x \in F$. This implies that $\dim_H m_\lambda \geq s_\lambda - \alpha$. Since α can be arbitrarily small this proves that $\dim_H F \geq \dim_H m_\lambda \geq s_\lambda$.

Note that μ_λ is an equilibrium measure corresponding to the function $s_\lambda \log \lambda_{i_0}$. Therefore,

$$0 = P_Q(s_\lambda \log \lambda_{i_0}) = h_{\mu_\lambda}(\sigma|Q) + s_\lambda \int_Q \log \lambda_{i_0} \, d\mu_\lambda$$

and the third statement follows.   ∎

In view of Statement 2 of Theorem 13.1 the measure m_λ is an invariant measure of full Carathéodory dimension (see (5.4)). We call it simply **measure of full dimension**.

The next statement provides an upper estimate for the number s_λ.

Theorem 13.2. *Let F be the limit set for a geometric construction (CPW1–CPW4) modeled by a symbolic dynamical system (Q, σ). Then*

(1) [PW1]

$$s_\lambda \leq \frac{h_Q(\sigma)}{-\log \lambda_{\max}},$$

where $\lambda_{\max} = \max\{\lambda_k : 1 \leq k \leq p\}$ and $h_Q(\sigma)$ is the topological entropy of σ on Q; equality occurs if $\lambda_i = \lambda$ for $i = 1, \ldots, p$; in particular, if $h_Q(\sigma) = 0$, then

$$\dim_H F = \underline{\dim}_B F = \overline{\dim}_B F = 0;$$

(2) [Fu] *if $\lambda_i = \lambda$ for $i = 1, \ldots, p$, then*

$$\dim_H F = \underline{\dim}_B F = \overline{\dim}_B F = s_\lambda = \frac{h_Q(\sigma)}{-\log \lambda}.$$

Proof. It follows from Statement 3 of Theorem 13.1 that

$$s_\lambda = \frac{h_{\mu_\lambda}(\sigma \,|\, Q)}{-\int \log \lambda_{i_0} d\mu_\lambda} \leq \frac{h_Q(\sigma)}{-\log \lambda_{\max}}.$$

The case of equality is obvious. If $\lambda_i = \lambda$ for $i = 1, \ldots, p$, then μ_λ is a measure of maximal entropy (since the function $\varphi(\omega) = -s \log \lambda$ is constant). Thus, $h_{\mu_\lambda}(\sigma \,|\, Q) = h_Q(\sigma)$. This proves the desired results.   ∎

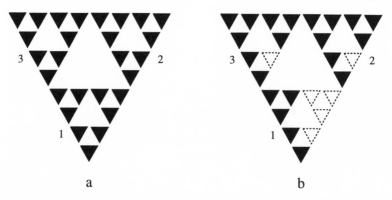

Figure 2. SIERPIŃSKI GASKETS:
a) A Simple Construction, b) A Markov Construction.

We consider two special cases of symbolic geometric constructions — simple geometric constructions and Markov geometric constructions, specified by the full shift and a subshift of finite type respectively. Examples are shown on Figure 2. This is the well-known Sierpiński gasket — the limit set for a Moran geometric construction (CM1–CM5) on the plane with $\lambda_1 = \lambda_2 = \lambda_3 < \frac{1}{2}$ and $p = 3$. The case of a simple construction is shown on Figure 2a, and the case of a Markov construction on Figure 2b, where we forbid all configurations whose codings contain a 1 followed by a 2. The corresponding transfer matrix is $\begin{pmatrix} 1 & 0 & 1 \\ 1 & 1 & 1 \\ 1 & 1 & 1 \end{pmatrix}$.

We consider the particular case of a subshift of finite type (Σ_A^+, σ). Given p numbers $0 < \lambda_1, \ldots, \lambda_p < 1$, we define a $p \times p$ diagonal matrix $M_t(\lambda) = \mathrm{diag}(\lambda_1{}^t, \cdots, \lambda_p{}^t)$. Let $\rho(B)$ denote the spectral radius of the matrix B.

Theorem 13.3.

(1) Let F be the limit set for a geometric construction (CPW1–CPW4) modeled by a subshift of finite type (Σ_A^+, σ). Then

$$\dim_H F = \underline{\dim}_B F = \overline{\dim}_B F = s_\lambda,$$

where s_λ is the unique root of the equation $\rho(AM_t(\lambda)) = 1$.

(2) Let F be the limit set for a geometric construction (CPW1–CPW4) modeled by the full shift. Then

$$\dim_H F = \underline{\dim}_B F = \overline{\dim}_B F = s_\lambda,$$

where s_λ is the unique root of the equation

$$\sum_{i=1}^{p} \lambda_i{}^t = 1.$$

Proof. The desired result follows from Theorem 13.1 and Theorem A2.8 (see Appendix II). ∎

For instance, for the Sierpiński gaskets shown on Figures 2a and 2b with $\lambda = \frac{1}{2}$ we have respectively that $s_\lambda = \log 3 (\log 2)^{-1}$ and $s_\lambda = \log \frac{3+\sqrt{5}}{2} (\log 2)^{-1}$.

A more delicate question in dimension theory is whether the Hausdorff measure of the limit set at dimension is finite. In general, the answer is negative. Below we will show that it is positive provided the measure μ_λ is a *Gibbs measure* (see Condition (A2.20) in Appendix II). If a geometric construction is Markov then the measure μ_λ is Gibbs provided the transfer matrix is transitive, i.e., the subshift is topologically mixing (the case of an arbitrary transfer matrix can be reduced, in a sense, to the case of transitive transfer matrix; see Appendix II). A more general class of geometric constructions, for which μ_λ is still a Gibbs measure, includes geometric constructions modeled by mixing sofic systems or by mixing subshifts satisfying specification property (see the definition in [KH]). The latter form a much broader class of geometric constructions than the class of geometric constructions modeled by subshifts of finite type.

Theorem 13.4. [PW1] *Let F be the limit set for a geometric construction (CPW1-CPW4) modeled by a symbolic dynamical system (Q, σ). Assume that the measure m_λ is a Gibbs measure then*

(1) *the measure μ_λ satisfies the uniform mass distribution principle;*
(2) *the Hausdorff measure $m_H(\cdot, s_\lambda)$ is equivalent to the measure m_λ; in particular, $0 < m_H(F, s_\lambda) < \infty$;*
(3) *$\underline{d}_{m_\lambda}(x) = \overline{d}_{m_\lambda}(x) = s_\lambda$ for every $x \in F$;*
(4) *$\dim_H(F \cap U) = s_\lambda$ for any open set U which has non-empty intersection with F.*

Remark.

If μ is an arbitrary Gibbs measure on Q, then by Theorem 15.4 below its push forward measure $m = \chi_* \mu$ is exact dimensional (see definition of exact dimensional measures in Section 7) and the pointwise dimension of m is constant **almost** everywhere (and is equal to $h_\mu(\sigma|Q) / \int_Q \log \lambda_{i_0} d\mu$).

Proof of the theorem. Since μ_λ is a Gibbs measure one can repeat arguments in the proof of Theorem 13.1 (see (13.5) and (13.6)) and conclude that it satisfies the uniform mass distribution principle. This proves the first statement. Moreover, (13.6) implies that $s_\lambda \leq \underline{d}_{m_\lambda}(x)$ for *every* $x \in F$. It follows that $s_\lambda \leq \dim_H F$ and $m_H(F, s_\lambda) > 0$.

We now prove that $m_\lambda(\cdot) \leq \text{const} \times m_H(\cdot, s_\lambda)$. Given $\delta > 0$ and a Borel subset $Z \subset F$, there exists $\varepsilon > 0$ and a cover of Z by balls B_k of radius $r_k \leq \varepsilon$ satisfying

$$\sum_k (r_k)^{s_\lambda} \leq m_H(F, s_\lambda) + \delta.$$

It follows from (13.6) that

$$m_\lambda(Z) \leq \sum_k m_\lambda(B_k) \leq C_1 M \sum_k (r_k)^{s_\lambda} \leq C_1 M m_H(F, s_\lambda) + C_1 M \delta,$$

where $C_1 > 0$ is a constant. Since δ is chosen arbitrarily this implies that $m_\lambda(Z) \leq C_1 M m_H(F, s_\lambda)$.

We now show that $m_H(\cdot, s_\lambda) \leq \text{const} \times m_\lambda(\cdot)$. Let $Z \subset F$ be a closed subset. Given $\delta > 0$, there exists $\varepsilon > 0$ such that for any cover $\mathcal{U}$ of Z by open sets whose diameter $\leq \varepsilon$ we have

$$m_H(Z, s_\lambda) \leq \sum_{U \in \mathcal{U}} (\text{diam}\, U)^{s_\lambda} + \delta. \tag{13.11}$$

Note that one can choose a cover $\mathcal{U}$ of Z by basic sets $\Delta^{(k)} = \Delta_{i_0 \dots i_{n(k)}}$ satisfying $\text{diam}\, \Delta^{(k)} \leq \varepsilon$ and

$$\sum_{\Delta^{(k)} \in \mathcal{U}} m_\lambda(\Delta^{(k)}) \leq m_\lambda(Z) + \delta.$$

We can apply (13.11) to this cover $\mathcal{U}$ and obtain using (13.1) and (13.5) that

$$m_H(Z, s_\lambda) \leq \sum_{\Delta^{(k)} \in \mathcal{U}} (\text{diam}\, \Delta^{(k)})^{s_\lambda} + \delta \leq K_2 \sum_{\Delta^{(k)} \in \mathcal{U}} \prod_{j=0}^{n(k)} \lambda_{i_j}^{s_\lambda} + \delta$$

$$\leq K_2 D_1^{-1} \sum_{\Delta^{(k)} \in \mathcal{U}} m_\lambda(\Delta^{(k)}) + \delta \leq K_2 D_1^{-1} m_\lambda(Z) + (K_2 D_1^{-1} + 1)\delta.$$

Since δ is chosen arbitrarily this implies the second statement.

Fix $0 < r < 1$. For each $\omega \in Q$ choose $n(\omega)$ according to (13.2). It follows from (13.1) that $\Delta_{i_0 \dots i_{n(\omega)+1}} \subset B(x, K_2 r)$, where $x = \chi(\omega)$. By virtue of (13.5) for all $\omega \in Q$,

$$m_\lambda(B(x, K_2 r)) \geq m_\lambda(\Delta_{i_0 \dots i_{n(x)+1}}) \geq D_1 \prod_{k=0}^{n(\omega)+1} \lambda_{i_k}^{s_\lambda} \geq C_2 r^{s_\lambda},$$

where $C_2 > 0$ is a constant. It follows that for all $x \in F$,

$$\overline{d}_{m_\lambda}(x) = \varliminf_{r \to 0} \frac{\log m_\lambda(B(x, r))}{\log r} \leq s_\lambda.$$

This implies the third statement.

We now prove the last statement. It follows from the second statement that $m_H(F \cap \Delta_{i_0 \dots i_n}, s_\lambda) > 0$. Thus, $\dim_H(F \cap \Delta_{i_0 \dots i_n}) \geq s_\lambda$. Now, let U be any open set with $F \cap U \neq \varnothing$. If $x = \chi(i_0 i_1 \dots) \in F \cap U$ and $n > 0$ is sufficiently large then $\Delta_{i_0 \dots i_n} \subset U$. Therefore,

$$s_\lambda \geq \dim_H(F \cap U) \geq \dim_H(F \cap \Delta_{i_0 \dots i_n}) \geq s_\lambda.$$

This completes the proof of the theorem.   ∎

Self-similar Constructions

There is a special class of geometric constructions of type (CG1–CG2) which are most studied in the literature (see for example, [F1]) — **self-similar geometric constructions**. They are geometric constructions with basic sets $\Delta_{i_0 \ldots i_n}$ given as follows:

$$\Delta_{i_0 \ldots i_n} = h_{i_0} \circ h_{i_1} \circ \cdots \circ h_{i_n}(D),$$

where $h_1, \ldots, h_p \colon D \to D$ are **conformal affine maps** (i.e., maps that satisfy $\mathrm{dist}(h_i(x), h_i(y)) = \lambda_i \mathrm{dist}(x, y)$ for any $x, y \in D$, where D is the unit ball in $\mathbb{R}^m$). Here $0 < \lambda_i < 1$ are **ratio coefficients**. These geometric constructions can be modeled by an arbitrary symbolic dynamical system (Q, σ).

Clearly, self-similar constructions are a particular case of Moran geometric constructions with stationary ratio coefficients (CPW1–CPW4). Therefore, by Theorem 13.1, the Hausdorff dimension and lower and upper box dimensions of the limit set of a self-similar construction coincide. The common value is the unique root of Bowen's equation $P_Q(s\varphi) = 0$, where $\varphi(\omega) = \log \lambda_{i_0}$ for $\omega = (i_0 i_1 \ldots)$ (if a self-similar construction is modeled by a subshift of finite type or the full shift the number s can be computed as stated in Theorem 13.3).

If we further assume that basic sets at the same level of a self-similar construction do not overlap (i.e., their interiors are disjoint) then the geometric construction is a Moran geometric construction (CM1–CM5).

Moreover, if basic sets at the same level are disjoint the coding map is a homeomorphism. Thus, the induced map G on the limit set of the self-similar construction (which in this case is a smooth map) provides a *smooth realization* of the subshift (Q, σ). We also remark that if the ratio coefficients of the maps h_i are *equal* (say, to a number λ) then the coding map is an isometry between the limit set of the geometric construction and Q (endowed with the metric $d_{\lambda^{-1}}$; see (A2.18) in Appendix II). Thus, it preserves the Hausdorff dimension and box dimension which can be computed by Statement 2 of Theorem 13.2 (compare to Theorem A2.9 in Appendix II).

14. Regular Geometric Constructions

In this section we follow Pesin and Weiss [PW1] and introduce a class of geometric constructions that admit approximations by Moran-like geometric constructions with stationary ratio coefficients. This allows us to obtain effective lower bounds for the Hausdorff dimension of the limit sets.

We will control the geometry of basic sets by numbers $\gamma_1, \ldots, \gamma_p$ such that one can replace the basic sets $\Delta_{i_0 \ldots i_n}$ by balls of radius $\prod_{j=1}^{n} \gamma_{i_j}$. In some cases these balls coincide with the largest balls that can be inscribed in the basic sets. However, this is not always the case and below we present an example where the "optimal" numbers $\gamma_1, \ldots, \gamma_p$ are completely independent of the radii of the largest inscribed balls (see Example 14.1 below).

Consider a geometric construction (CG1–CG2) modeled by a symbolic dynamical system (Q, σ) (see the introduction to this chapter; it is worth emphasizing that we do not require the separation condition (CG3)). Given $0 < r < 1$ and a vector of numbers $\gamma = (\gamma_1, \ldots, \gamma_p)$, $0 < \gamma_i < 1$, $i = 1, \ldots, p$, consider a

Moran cover $\mathfrak{U}_r = \mathfrak{U}_r(\gamma) = \{\Delta^{(j)}\}$ of the limit set F constructed in Section 13. Given an open Euclidean ball $B(x, r)$ of radius r centered at x, denote by $R(x, r)$ the number of sets $\Delta^{(j)}$ that have non-empty intersection with $B(x, r)$. We call a vector γ **estimating** if

$$R(x, r) \leq \text{constant} \tag{14.1}$$

uniformly in x and r. We call a symbolic geometric construction (CG1–CG2) **regular** if it admits an estimating vector. If $\gamma = (\gamma_1, \ldots, \gamma_p)$ is an estimating vector for a regular geometric construction, then any vector $\tilde{\gamma} = (\tilde{\gamma}_1, \ldots, \tilde{\gamma}_p)$ for which $\gamma_i \geq \tilde{\gamma}_i$, $i = 1, \ldots, p$ is also estimating.

We provide an example of a regular geometric construction on the plane that illustrates how the choice of the estimating vector can be made.

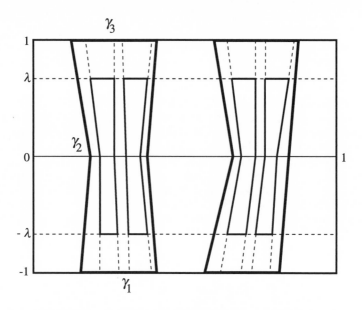

Figure 3. A REGULAR GEOMETRIC CONSTRUCTION.

Example 14.1 [PW1] Let $\gamma_1, \gamma_2, \gamma_3$, and λ be any numbers in $(0, 1)$. Given a number $i = 1, 2, 3$ consider a simple geometric construction (CG1–CG2) on the interval $[0, 1] \times \{i - 2\}$ with 2^n basic sets of size $\gamma_i{}^n$ at step n. We denote this construction by $CG(\gamma_i)$. Since the 2^n intervals at step n in each of these constructions are clearly ordered we may refer to the ith subinterval at step n, $1 \leq i \leq 2^n$ of these constructions. Consider the 2^n polygons in $[0, 1] \times [-1, 1]$ having six vertices which consist of the two endpoints of the ith subinterval at step n for all three constructions. We define the 2^n basic sets at step n by intersecting these 2^n polygons with the rectangle $[0, 1] \times [-\lambda^n, \lambda^n]$. This produces a simple geometric construction (CG1–CG2) on the plane. See Figure 3.

It is easy to see that the limit set F of this geometric construction coincides with the limit set of the construction $CG(\gamma_2)$. Hence, by Theorem 13.3, $\dim_H F = \frac{\log 2}{-\log \gamma_2}$ and does not depend on γ_1, γ_3, or λ. Let us choose numbers $\gamma_1, \gamma_2, \gamma_3$, and λ such that $\gamma_2 < \gamma_1 = \gamma_3 < \lambda$ and $\gamma_2 < \lambda\gamma_1$ or $\gamma_2 < \lambda\gamma_3$. One can see that the inscribed and circumscribed balls of the basic sets at step n have radii which are bounded from below and above by $C_1\gamma_1^n$ and $C_2\lambda^n$, respectively, where C_1 and C_2 are positive constants which are independent of n. Thus, these balls cannot be used to determine the Hausdorff dimension of the limit set.   ∎

Consider the positive number s_γ such that $P_Q(s_\gamma \log \gamma_{i_0}) = 0$, where P_Q denotes the topological pressure with respect to the shift σ on Q (see Section 11). Let μ_γ denote an equilibrium measure for the function $(i_0 i_1 \dots) \mapsto s_\gamma \log \gamma_{i_0}$ on Q, and let m_γ be the push forward measure on F under the coding map χ (i.e., $m_\gamma(Z) = \mu_\gamma(\chi^{-1}(Z))$ for any Borel set $Z \subset F$). The following result provides a lower bound for the Hausdorff dimension of the limit set. Its proof is quite similar to the proof of Statement 2 of Theorem 13.1.

Theorem 14.1. [PW1] *Let F be the limit set for a regular symbolic geometric construction. Then $\dim_H F \geq s_\gamma$ for any estimating vector γ. Hence, $\dim_H F \geq \sup s_\gamma$, where the supremum is taken over all estimating vectors γ.*

In the case when the measure μ_γ is Gibbs one can strengthen Theorem 14.1 and prove a statement that is similar to Theorem 13.4.

Theorem 14.2. [PW1] *Let F be the limit set of a regular symbolic geometric construction and γ an estimating vector. Assume that the measure μ_γ is a Gibbs measure. Then*

(1) *the measure m_γ satisfies the uniform mass distribution principle;*
(2) $0 < m_H(F, s_\gamma)$; *moreover, $m_\gamma(Z) \leq C\, m_H(Z, s_\gamma)$ for any measurable set $Z \subset F$, where $C > 0$ is a constant;*
(3) $s_\gamma \leq \underline{d}_{m_\gamma}(x)$ *for every $x \in F$;*
(4) $\dim_H(F \cap U) \geq s_\gamma > 0$ *for any open set U which has non-empty intersection with F.*

The second statement of Theorem 14.2 is non-trivial only when $s_\gamma = \dim_H F$. Otherwise, $m_H(F, s_\gamma) = \infty$. If $s_\gamma < s = \dim_H F$, then the s-Hausdorff measure may be zero or infinite. Theorem 14.2 holds for simple geometric constructions or Markov geometric constructions with transitive transfer matrix.

In [Bar1], Barreira gave sufficient conditions for a geometric construction to be regular. Roughly speaking, it requires that the basic sets contain sufficiently large open balls. We begin with geometric constructions on the line.

Theorem 14.3. *Assume that each basic set $\Delta_{i_0 \dots i_n}$ of a symbolic geometric construction (CG1–CG2) on the line contains an interval $I_{i_0 \dots i_n}$ of length $0 < \lambda_{i_0 \dots i_n} < 1$ such that $I_{i_0 \dots i_n} \cap I_{j_0 \dots j_n} = \varnothing$ for any $(i_0 \dots i_n) \neq (j_0 \dots j_n)$. Assume also that there exists $0 < \gamma < 1$ such that*

$$\varliminf_{n \to \infty} \min \frac{1}{n} \log \lambda_{i_0 \dots i_n} \geq \log \gamma,$$

where the minimum is taken over all Q-admissible n-tuples $(i_0 \ldots i_n)$. Then the geometric construction is regular with the estimating vector $(\gamma e^{-\varepsilon}, \ldots, \gamma e^{-\varepsilon})$ for any $\varepsilon > 0$.

Proof. Given $\varepsilon > 0$, we have $\lambda_{i_0 \ldots i_n} > (\gamma e^{-\varepsilon})^n$ for every $(i_0 i_1 \ldots) \in Q$ and any sufficiently large n. Given $r > 0$, one can find a unique number $n = n(r) > 0$ such that

$$(\gamma e^{-\varepsilon})^{n+1} \leq r < (\gamma e^{-\varepsilon})^n.$$

For any interval I of length r there exist at most two basic sets of length $\geq (\gamma e^{-\varepsilon})^n$ intersecting I. Therefore, for every point x in the limit set the number $R(x, r)$ in the definition of regular geometric constructions (see (14.1)) we obtain that $R(x, r) \leq 2$ for all sufficiently small r. Hence, the construction is regular with the estimating vector $(\gamma e^{-\varepsilon}, \ldots, \gamma e^{-\varepsilon})$. ∎

We now formulate a criterion of regularity for a geometric construction in $\mathbb{R}^m$ with $m > 1$. Fix a point $x \in F$ and a number $\lambda > 0$. Given $n > 0$, consider two basic sets $\Delta_{i_0 \ldots i_n}$ and $\Delta_{j_0 \ldots j_n}$ intersecting the ball $B(x, \lambda^n)$. Denote by $\alpha(\Delta_{i_0 \ldots i_n}, \Delta_{j_0 \ldots j_n})$ the minimum angle of spherical sectors centered at x which contain both $\Delta_{i_0 \ldots i_n}$ and $\Delta_{j_0 \ldots j_n}$. Let $\alpha_n(x, \lambda)$ be the minimum of all the angles $\alpha(\Delta_{i_0 \ldots i_n}, \Delta_{j_0 \ldots j_n})$.

Theorem 14.4. *Assume that each basic set of a symbolic geometric construction (CG1–CG2) in $\mathbb{R}^m$ with $m > 1$ contains a ball $B_{i_0 \ldots i_n} \subset \Delta_{i_0 \ldots i_n}$ of radius λ^n with $0 < \lambda < 1$. Assume also that there exists $\delta > 0$ such that $\alpha_n(x, \lambda) \geq \delta \lambda^n$ for all $x \in F$ and $n \geq 1$ and that $B_{i_0 \ldots i_n} \cap B_{j_0 \ldots j_n} = \varnothing$ for any $(i_0 \ldots i_n) \neq (j_0 \ldots j_n)$. Then the geometric construction is regular with the estimating vector $(\lambda, \ldots, \lambda)$.*

Proof. By elementary geometry there exists a universal constant $C = C(\delta)$ such that the maximum number of sets $\Delta_{i_0 \ldots i_n}$ intersecting a ball $B(\lambda^n)$ does not exceed C. This proves the result. ∎

Consider a simple regular geometric construction in $\mathbb{R}^m$ with the limit set F. It follows from Theorem 14.2 that the Hausdorff dimension of any open set U which has non-empty intersection with F satisfies $\dim_H(F \cap U) \geq s$ for some $s > 0$. The following example shows that the converse statement may not be true.

Example 14.2. [Bar1] *For each $s \in (0, 1)$, there exists a geometric construction (CG1–CG2) on the line modeled by the full shift (Σ_2^+, σ) such that*

(1) $\dim_H(F \cap U) = s$ *for any open set U with $F \cap U \neq \varnothing$;*
(2) *the construction is non-regular.*

Proof. Define the function $m: Q \to \mathbb{N} \cup \{+\infty\}$ by

$$m(\omega) = m(i_0 i_1 \ldots) = \begin{cases} +\infty, & \omega = 0 \\ \text{least } j \in \mathbb{N} \text{ with } i_j = 1, & \omega \neq 0. \end{cases}$$

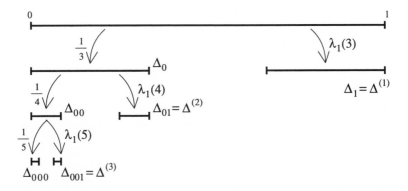

Figure 4. A NON-REGULAR GEOMETRIC CONSTRUCTION.

Define also the numbers $\lambda_{i_0\ldots i_n}$ by

$$
\lambda_{i_0\ldots i_n} = \begin{cases} \prod_{j=0}^{n} \lambda_{i_j}(j+2), & n < \ell \\ \left(\prod_{j=0}^{\ell} \lambda_{i_j}(j+2)\right) \times \left(\prod_{k=\ell+1}^{n} \lambda_{i_k}(\ell+2)\right), & n \geq \ell \end{cases}
$$

where $\ell = m(i_0\ldots i_n\, 0\, \ldots)$, $\lambda_0(j) = \frac{1}{j}$, and $\lambda_1(j) = \left(1 - (\frac{1}{j})^s\right)^{1/s}$. Notice that

$$\lambda_0(j)^s + \lambda_1(j)^s = 1 \tag{14.2}$$

for each $j > 0$. We consider basic sets spaced as shown on Figure 4. They have the following property: if $\Delta_{i_0\ldots i_n} = [a_n, b_n]$ then $a_n \in \Delta_{i_0\ldots i_n 0}$ and $b_n \in \Delta_{i_0\ldots i_n 1}$.

Define intervals $\Delta^{(j)} = \Delta_{i_0\ldots i_j}$, where $(i_0\ldots i_j) = (0\ldots 01)$. Inside each $\Delta^{(j)}$ we have a sub-construction modeled by (Σ_2^+, σ) with rates $\lambda_{0\ldots 01 i_0\ldots i_n}/\lambda_{0\ldots 01} = \prod_{k=0}^{n} \lambda_{i_k}(j+2)$. Therefore, the Hausdorff dimension of $F \cap \Delta^{(j)}$ is equal to s, where s is the unique root of equation (14.2), with $j+2$ instead of j (see Theorem 13.3). Hence, $\dim_H(F \cap \Delta^{(j)}) = s$. Since $F = \{0\} \cup \bigcup_{j>0}(F \cap \Delta^{(j)})$ it follows that $\dim_H F = \sup_{j>0} \dim_H(F \cap \Delta^{(j)}) = s$. Now, if $F \cap U \neq \varnothing$, there exists $x = \chi(i_0 i_1 \ldots) \in (F \cap U) \setminus \{0\}$, and $n > 0$ such that $\Delta_{i_0\ldots i_n} \subset U$. Hence, $s = \dim_H F \geq \dim_H(F \cap U) \geq \dim_H(F \cap \Delta_{i_0\ldots i_n}) = s$. This proves the first statement.

Consider a vector (γ_1, γ_2). For each $n > 0$, set $r_n = 2/(n-1)!$. Select now the smallest positive integer k such that $\prod_{j=0}^{k} \gamma_{i_j} > r_n$ and $\prod_{j=0}^{k+1} \gamma_{i_j} \leq r_n$ for some $(i_0 i_1 \ldots) \in \Sigma_2^+$. Set $\gamma = \min\{\gamma_1, \gamma_2\}$ and observe that $\gamma^{k+1} \leq r_n$. Hence, $k \geq \log r_n / \log \gamma - 1$. Therefore, for $k_n = \log r_n / \log \gamma - 1$ we get $R(0, r_n) \geq 2^{k_n - n}$ (where $R(0, r_n)$ is defined in (14.1)). We also have

$$
k_n - n \geq \frac{\log 2 - \log(n-1)!}{\log \gamma} - 1 - n \underset{n\to\infty}{\approx} \frac{\log n!}{-\log \gamma} - n \underset{n\to\infty}{\approx} \frac{\log n!}{-\log \gamma}.
$$

Therefore, there exists $D > 0$ such that $k_n - n \geq D \log n!$ for all $n > 0$, and we obtain

$$R(0, r_n) \geq 2^{D \log n!} = n!^{D \log 2} > \left(\frac{2}{r_n} \right)^{D \log 2}.$$

As the sequence r_n decreases monotonically to 0 and (γ_1, γ_2) is arbitrary we proved that the construction is non-regular.   ∎

If a geometric construction is regular one can effectively replace its basic sets by balls to obtain a lower bound for the Hausdorff dimension of its limit set (see Theorem 14.1). We further exploit this approach and show that, under some mild assumption, a geometric construction (CG1–CG2) (whose basic sets are, in general, *arbitrary* close subsets and are *possibly intersecting*) can be effectively compared with a geometric construction whose basic sets are *disjoint balls*.

Theorem 14.5. *Let F be the limit set of a geometric construction (CG1–CG2) modeled by a symbolic dynamical system (Q, σ). Assume that there exist numbers $\lambda_1, \ldots, \lambda_p$, $0 < \lambda_i < 1$ such that for any admissible n-tuple $(i_0 \ldots i_n)$, $i_j = 1, \ldots, p$, we have*

$$diam \, \Delta_{i_0 \ldots i_n} \leq C \prod_{j=0}^{n} \lambda_{i_j}, \qquad (14.3)$$

where $C > 0$ is a constant. Then

(1) *there exists a self-similar geometric construction modeled by a symbolic dynamical system (Q, σ) satisfying: a) its basic sets $B_{i_0 \ldots i_n}$ are disjoint balls, b) $diam \, B_{i_0 \ldots i_n} = 2C \prod_{j=0}^{n} \lambda_{i_j}$, and c) there is a Lipschitz continuous map ψ from its limit set $\tilde{F}$ onto F;*

(2) *$\overline{\dim}_B F \leq s_\lambda$.*

Proof. Consider a self-similar geometric construction with ratio coefficients $\lambda_1, \ldots, \lambda_p$ modeled by a symbolic dynamical system (Q, σ) whose basic sets $B_{i_0 \ldots i_n}$ are disjoint balls of radii $C \prod_{j=0}^{n} \lambda_{i_j}$. Denote its basic set by $\tilde{F}$. Let $\chi: Q \to F$ and $\tilde{\chi}: Q \to \tilde{F}$ be the coding maps. Consider the map $\psi = \chi \circ \tilde{\chi}^{-1}: \tilde{F} \to F$. We shall show that ψ is a (locally) Lipschitz continuous map. Choose $x, y \in \tilde{F}$ with $\rho(x, y) \leq \varepsilon$. We have that $\tilde{\chi}^{-1}(x) = (i_0 i_1 \ldots)$, $\tilde{\chi}^{-1}(y) = (j_0 j_1 \ldots)$, and $i_0 = j_0, \ldots, i_n = j_n$, $i_{n+1} \neq j_{n+1}$ for some $n > 0$. Therefore, $\|x - y\| \geq C_1 \prod_{j=0}^{n} \lambda_{i_j}$, where $C_1 > 0$ is a constant. One can also see that $\psi(x), \psi(y) \in \Delta_{i_0 \ldots i_n}$. Hence, by (14.3), $\|\psi(x) - \psi(y)\| \leq C \prod_{j=0}^{n} \lambda_{i_j}$. Since the map ψ is onto this proves the first statement. The second statement follows immediately from the first one. ∎

Geometric Constructions with Ellipsis

We describe a special class of regular geometric constructions. We say that a geometric construction (CG1–CG3) in $\mathbb{R}^2$ is a **construction with ellipsis** (see Figure 5) if each basic set $\Delta_{i_0 \ldots i_n}$ is an ellipse with axes $\underline{\lambda}^n/2$ and $\overline{\lambda}^n/2$, for some $0 < \underline{\lambda} < \overline{\lambda} < 1$ (we stress that we require the separation condition (CG3)). Such constructions were studied by Barreira in [Bar1].

Theorem 14.6. *Let F be the limit set of a geometric construction with ellipsis. Then the construction is regular with the estimating vector (λ, λ), where λ is any number in the interval $(0, \underline{\lambda}^2/\overline{\lambda})$.*

Proof. For each $\lambda \in (0,1)$ define the function $g_\lambda : (0,1) \times (0,1) \to \mathbb{R}$ by

$$g_\lambda(\underline{\lambda}, \overline{\lambda}) = \begin{cases} \left[(\overline{\lambda}/\underline{\lambda}^2)^t \underline{\lambda}\right]^{1/2}, & \lambda < \overline{\lambda} \\ (\overline{\lambda}/\underline{\lambda})^t, & \lambda = \overline{\lambda} \\ \underline{\lambda}^{1-t}, & \lambda > \overline{\lambda} \end{cases}$$

where $t = \log \underline{\lambda}/\log \lambda$. Consider an ellipse E_n with axes $\underline{\lambda}^{\alpha n}/2$ and $\overline{\lambda}^{\alpha n}/2$, where $\alpha = \log \underline{\lambda}/\log \overline{\lambda}$. We assume that it is located outside the ball $B(0, \underline{\lambda}^n)$, is tangent to this ball at a point, and that the major axis of the ellipse points towards 0. Denote by $2\beta(n)$ the smallest angular sector centered at 0 that contains E_n.

The desired result follows immediately from Theorem 14.4 and the following lemma.

Lemma. *For each $\lambda \in (0, \overline{\lambda})$, there exists a number $C > 0$ such that $\tan \beta(n) \sim C g_\lambda(\underline{\lambda}, \overline{\lambda})^{-n}$ as $n \to \infty$. In particular, $\beta(n)$ decreases exponentially with the rate $g_\lambda(\underline{\lambda}, \overline{\lambda})^{-1}$ when $\underline{\lambda}^2/\overline{\lambda} < \lambda < \overline{\lambda}$ and is uniformly bounded away from 0 if $0 < \lambda \leq \underline{\lambda}^2/\overline{\lambda}$.*

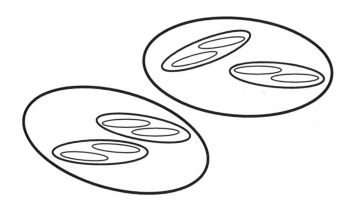

Figure 5. A Construction with Ellipsis.

Proof of the lemma. Consider an orthogonal coordinate system centered at 0 with the x-axis directed along the major axis of E_n. If $m(n) = \tan \beta(n)$ is the slope of a line starting at 0 and tangent to E_n, the points of tangency are solutions of the equation

$$\left(\frac{x - \underline{\lambda}^n - \overline{\lambda}^{\alpha n}/2}{\overline{\lambda}^{\alpha n}/2}\right)^2 + \left(\frac{m(n)x}{\underline{\lambda}^{\alpha n}/2}\right)^2 = 1,$$

provided that the discriminant of this equation is zero:

$$4\left(\frac{\underline{\lambda}^n + \overline{\lambda}^{\alpha n}/2}{\overline{\lambda}^{2\alpha n}/4}\right)^2 - 4\left(\frac{1}{\overline{\lambda}^{2\alpha n}/4} + \frac{m(n)^2}{\underline{\lambda}^{2\alpha n}/4}\right) \times \left(\left(\frac{\underline{\lambda}^n + \overline{\lambda}^{\alpha n}/2}{\overline{\lambda}^{\alpha n}/2}\right)^2 - 1\right) = 0.$$

Set $b_\alpha = \underline{\lambda}/\overline{\lambda}^\alpha$. One can see that $4\,m(n)^2 = b_1^{2\alpha n}/[b_\alpha^n(b_\alpha^n + 1)]$. We consider the following cases:

(a) $\lambda > \overline{\lambda}$; then $b_\alpha > 1$ and $2\,m(n) \sim (b_1{}^{2\alpha}/b_\alpha{}^2)^{n/2} = (\underline{\lambda}^{\alpha-1})^n = g_\lambda(\underline{\lambda}, \overline{\lambda})^{-n}$;

(b) $\lambda = \overline{\lambda}$; then $b_\alpha = 1$ and $2\sqrt{2}\,m(n) \sim (b_1{}^\alpha)^n = g_\lambda(\underline{\lambda}, \overline{\lambda})^{-n}$;

(c) $\lambda < \overline{\lambda}$; then $b_\alpha < 1$ and $2\,m(n) \sim (b_1{}^{2\alpha}/b_\alpha)^{n/2} = g_\lambda(\underline{\lambda}, \overline{\lambda})^{-n}$.

It is easy to check that if $\underline{\lambda}^2/\overline{\lambda} < \lambda < \overline{\lambda}$, we have $g_\lambda(\underline{\lambda}, \overline{\lambda}) > 1$ and if $0 < \lambda \le \underline{\lambda}^2/\overline{\lambda}$, we have $g_\lambda(\underline{\lambda}, \overline{\lambda}) \le 1$. Since $\tan x \sim x$ as $x \to 0$ the desired result follows. ∎

15. Moran-like Geometric Constructions with Non-stationary Ratio Coefficients

In this section we study Moran-like geometric constructions with ratio coefficients at step n depending on all the previous steps. This class of geometric constructions was introduced by Barreira in [Bar2]. Consider a geometric construction modeled by a symbolic dynamical system (Q, σ). We assume that $\sigma|Q$ is topologically mixing and the following conditions hold:

(CB1) $\Delta_{i_0\ldots i_n j} \subset \Delta_{i_0\ldots i_n}$ for $j = 1, \ldots, p$;

(CB2)

$$\underline{B}_{i_0\ldots i_n} \subset \Delta_{i_0\ldots i_n} \subset \overline{B}_{i_0\ldots i_n},$$

where $\underline{B}_{i_0\ldots i_n}$ and $\overline{B}_{i_0\ldots i_n}$ are closed balls with radii $K_1 r_{i_0\ldots i_n}$ and $K_2 r_{i_0\ldots i_n}$ and $0 < K_1 \le K_2$;

(CB3) $\operatorname{int} \underline{B}_{i_0\ldots i_n} \cap \operatorname{int} \underline{B}_{j_0\ldots j_m} = \varnothing$ for any $(i_0 \ldots i_n) \ne (j_0 \ldots j_n)$ and $m \ge n$.

The class of geometric constructions (CB1–CB3) is quite broad and includes geometric constructions (CPW1–CPW4) (see Section 13), geometric constructions with contraction maps (see below), and more general geometric constructions with quasi-conformal expanding induced maps (see Theorem 15.5 below). The study of geometric constructions (CB1–CB3) is based upon the non-additive version of the thermodynamic formalism (see Section 12 and Appendix II).

We define the sequence of functions $\varphi = \{\varphi_n\colon Q \to \mathbb{R}\}$ by

$$\varphi_n(\omega) = \log r_{i_0\ldots i_n} \tag{15.1}$$

for each $\omega = (i_0 i_1 \ldots) \in Q$. We wish to consider Bowen's equations

$$P_Q(s\,\varphi) = 0, \quad \overline{CP}_Q(s\,\varphi) = 0 \tag{15.2}$$

and to show that they have unique roots. We shall use these equations to produce sharp estimates and sometimes obtain exact values of the Hausdorff dimension and box dimension of the limit set of a geometric construction (CB1–CB3). We need the following two assumptions on the radii of the basic sets.

The **first assumption** is the following:

there exist $L_1 > 1$ and $L_2 > 0$ such that for any $(i_0 i_1 \ldots) \in Q$ and $n \geq 0$,

$$r_{i_0 \ldots i_n} \leq L_1^{-n}, \quad r_{i_0 \ldots i_n i_{n+1}} \geq L_2 r_{i_0 \ldots i_n}. \tag{15.3}$$

The role of Condition (15.3) is the following. First of all, one can easily derive from (15.3) that for any $(i_0 i_1 \ldots) \in Q$ and any $n \geq 1$,

$$e^{-\alpha n} \leq r_{i_0 \ldots i_n} \leq e^{-\beta n}, \tag{15.4}$$

where α and β are positive constants, i.e., the radii of basic sets decay exponentially. Therefore, the sequence of functions $\{\varphi_n\}$ (see (15.1)) satisfies Condition (A2.22) (see Appendix II). By Theorem A2.6 Bowen's equations (15.2) have respectively unique roots $\underline{s}$ and $\overline{s}$ that satisfy $0 \leq \underline{s} \leq \overline{s} < \infty$.

Furthermore, Condition (15.3) implies that for every $n \geq 0$,

$$\varphi_n + \log L_2 \leq \varphi_{n+1}.$$

Thus, Condition (12.5) holds.

At last, given $0 < r < 1$, one can construct, using Condition (15.3), a **Moran cover** $\mathfrak{U}_r$ of the limit set F. It consists of basic sets $\Delta^{(j)} = \Delta_{i_0 \ldots i_{n(x_j)}}$ such that:

(a) $x_j \in F$; $r_{i_0 \ldots i_{n(x_j)}} > r$ and $r_{i_0 \ldots i_{n(x_j)+1}} \leq r$;

(b) the corresponding cylinder sets $C^{(j)} = C_{i_0 \ldots i_{n(x_j)}} \subset Q$ are disjoint.

Moran covers have the following crucial property: given a point $x \in F$ and a number $r > 0$ the number of basic sets $\Delta_{i_0 \ldots i_{n(x_j)}} = \chi(C_{i_0 \ldots i_{n(x_j)}})$ in a Moran cover $\mathfrak{U}_r$ that have non-empty intersection with the ball $B(x, r)$ is bounded from above by a number M, which is independent of x and r (a **Moran multiplicity factor**).

Repeating arguments in the proof of Statement 1 of Theorem 13.1 one can prove the following result.

Theorem 15.1. [Bar2] *Let F be the limit set of a geometric construction (CB1–CB3) modeled by a symbolic dynamical system (Q, σ). Assume that the numbers $r_{i_0 \ldots i_n}$ satisfy Condition (15.3). Then*

$$\underline{s} \leq \dim_H F \leq \underline{\dim}_B F \leq \overline{\dim}_B F \leq \overline{s}.$$

The **second assumption** we need is the following:

the sequence φ is sub-additive, i.e., $\varphi_{n+m} \leq \varphi_n + \varphi_m \circ \sigma^n$ on Q (see (12.3)).

By virtue of (15.1) the sub-additivity of the sequence φ is equivalent to the following condition:

$$r_{i_0 \ldots i_{n+m}} \leq r_{i_0 \ldots i_n} \times r_{i_{n+1} \ldots i_{n+m}}. \tag{15.5}$$

We now establish the remarkable result that under assumptions (15.3) and (15.5) the Hausdorff dimension and box dimension of the limit set coincide regardless of the placement of the basic sets. The common value is the unique root of Bowen's equation $P_Q(s\varphi) = 0$. The proof uses Theorem A2.7 (see Appendix II) and is a modification of the proof of Theorem 13.1.

Theorem 15.2. [Bar2] *Let F be the limit set of a geometric construction (CB1–CB3) modeled by a symbolic dynamical system (Q, σ). Assume that the numbers $r_{i_0 \ldots i_n}$ satisfy Conditions (15.3) and (15.5). Then*

$$\dim_H F = \underline{\dim}_B F = \overline{\dim}_B F = \underline{s} = \overline{s} = s,$$

where s is the unique root of Bowen's equations

$$P_Q(s\varphi) = \overline{CP}_Q(s\varphi) = \lim_{n \to \infty} \frac{1}{n} \log \sum_{\substack{(i_0 \ldots i_n) \\ Q\text{-admissible}}} r_{i_0 \ldots i_n}{}^s = 0.$$

The assumption that the sequence φ is sub-additive is crucial. The following example demonstrates that the Hausdorff dimension of the limit set of a geometric construction (CB1–CB3) may not agree with the upper box dimension if the sequence φ fails to be sub-additive.

Example 15.1. [Bar2] *There exists a geometric construction (CB1–CB3) on the line modeled by the full shift (Σ_2^+, σ) such that*

(1) *each basic set $\Delta_{i_0 \ldots i_n}$ is a closed interval of length depending only on n;*
(2) *the sequence φ is not sub-additive;*
(3) $\underline{s} = \dim_H F = \underline{\dim}_B F < \overline{\dim}_B F = \overline{s}$.

Proof. Fix numbers $b > a > 0$ and choose $\delta > 0$ such that $\delta < \frac{1}{2}(b - a)$. Let m_j be a sequence of positive integers. Denote

$$n_k(a, b) = b \sum_{\substack{j \leq k \\ j \text{ is odd}}} m_j + \sum_{\substack{j \leq k \\ j \text{ is even}}} m_j, \qquad n_k = \sum_{j=1}^{k} m_j.$$

We define inductively the sequence m_j. Set $m_1 = 1$ and choose m_k such that $|n_k(a, b)/n_k - b| < \delta/k$, if k is odd, and $|n_k(a, b)/n_k - a| < \delta/k$, if k is even. We introduce now the sequence of functions $\varphi_n \colon \Sigma_2^+ \to \mathbb{R}$ by

$$\varphi_n(\omega) = -n_k(a, b) - \begin{cases} b(n - n_k) & \text{if } k \text{ is odd} \\ a(n - n_k) & \text{if } k \text{ is even} \end{cases}$$

where k is the largest positive integer such that $n_k < n$. Note that

$$\varphi_n(\omega) = \text{const} \stackrel{\text{def}}{=} \lambda_n \tag{15.6}$$

and $\varphi_{n_k} > n_k \varphi_0$. This implies that $\varphi = \{\varphi_n\}$ is not sub-additive.

One can build a geometric construction (CB1–CB3) on the line modeled by Σ_2^+ with basic sets $\Delta_{i_0 \ldots i_n}$ to be closed intervals of length $\exp \lambda_n$. Let F be the limit set. Clearly, $N(F, e^{\lambda_n}) \leq 2^n$. On the other hand, given an interval of length $\exp \lambda_n$ there exist at most two basic sets which intersect it and at least one, if the interval intersects F. Therefore, $2N(F, e^{\lambda_n}) \geq 2^n$. Since $\lambda_{n+1} - \lambda_n \geq -b$ by Theorem 6.4 we have

$$\underline{\dim}_B F = \lim_{n \to \infty} \frac{\log N(F, e^{\lambda_n})}{-\lambda_n} = \log 2 \times \lim_{n \to \infty} \frac{n}{-\lambda_n} = \frac{\log 2}{b},$$

$$\overline{\dim}_B F = \overline{\lim_{n \to \infty}} \frac{\log N(F, e^{\lambda_n})}{-\lambda_n} = \log 2 \times \overline{\lim_{n \to \infty}} \frac{n}{-\lambda_n} = \frac{\log 2}{a}.$$

By Theorem 15.1 we obtain that $\dim_H F \geq \underline{s}$, where $\underline{s}$ is the unique root of the first equation in (15.2). Since $\lambda_n \geq -bn$ for all $n > 0$ we conclude that

$$\frac{\log 2}{b} = \underline{s} = \dim_H F = \underline{\dim}_B F < \overline{\dim}_B F = \frac{\log 2}{a}.$$

We need only to show that $\overline{s} = \log 2/a$. Observe that for every $s \in \mathbb{R}$,

$$\frac{1}{n} \log \sum_{(i_0 \ldots i_n)} \exp \left(s \sup_{C_{i_0 \ldots i_n}} \varphi_n \right) = \frac{1}{n} \log(2^n \exp(s\lambda_n)) = \log 2 + \frac{s\lambda_n}{n},$$

where $C_{i_0 \ldots i_n}$ are cylinder sets. By definition $\lambda_{n_k} = -n_k(a, b)$. Therefore,

$$\overline{CP}_{\Sigma_2^+}(s\varphi) = \log 2 + s \lim_{n \to \infty} \frac{\lambda_n}{n} \geq \log 2 - s \lim_{k \to \infty} \frac{n_k(a, b)}{n_k} \geq \log 2 - sa.$$

Moreover, we have that $-bn \leq \lambda_n \leq -an$ and thus, $\overline{CP}_{\Sigma_2^+}(s\varphi) = \log 2 - sa$. This implies that $\overline{s} = \log 2/a$. ∎

Pointwise Dimension of Measures on Limit Sets of Moran-like Geometric Constructions

Let ν be a Borel probability measure on the limit set F of a geometric construction (CB1–CB3) modeled by a symbolic dynamical system (Q, σ). We formulate a criterion that allows one to estimate the lower and upper pointwise dimensions of ν. Given $x \in F$, set

$$\underline{d}(x) = \inf \lim_{n \to \infty} \frac{\log \nu(\Delta_{i_0 \ldots i_n})}{\log |\Delta_{i_0 \ldots i_n}|},$$

$$\overline{d}(x) = \inf \overline{\lim_{n \to \infty}} \frac{\log \nu(\Delta_{i_0 \ldots i_n})}{\log |\Delta_{i_0 \ldots i_n}|},$$

where $|\Delta_{i_0 \ldots i_n}|$ denotes the diameter of the basic set $\Delta_{i_0 \ldots i_n}$ and the infimum is taken over all $\omega = (i_0 i_1 \ldots) \in Q$ such that $x = \chi(\omega)$.

Theorem 15.3. *Assume that a geometric construction (CB1–CB3) satisfies Condition (15.3). Then*

(1) $\overline{d}_\nu(x) \leq \overline{d}(x)$ *for all $x \in F$;*

(2) $\underline{d}(x) \leq \underline{d}_\nu(x)$ *for ν-almost all $x \in F$;*

(3) *if $\underline{d}(x) = \overline{d}(x) \overset{\text{def}}{=} d(x)$ for ν-almost every $x \in F$, then $\underline{d}_\nu(x) = \overline{d}_\nu(x) = d(x)$ for ν-almost every $x \in F$.*

Proof. Fix $x \in F$ and choose $\omega = (i_0 i_1 \ldots) \in Q$ such that $\chi(\omega) = x$. Given $r > 0$, choose $n = n(r, \omega)$ such that $|\Delta_{i_0 \ldots i_n}| < r$ and $|\Delta_{i_0 \ldots i_{n-1}}| \geq r$. Since $x \in \Delta_{i_0 \ldots i_n}$ we have that $\Delta_{i_0 \ldots i_n} \subset B(x, r)$. This implies

$$\frac{\log \nu(B(x, r))}{\log r} \leq \frac{\log \nu(\Delta_{i_0 \ldots i_n})}{\log r}.$$

We also have $r \leq \frac{1}{L_2} |\Delta_{i_0 \ldots i_n}|$, where L_2 is the constant in (15.3). It follows that

$$\underline{d}_\nu(x) = \varliminf_{r \to 0} \frac{\log \nu(B(x, r))}{\log r} \leq \varliminf_{n = n(r, \omega) \to \infty} \frac{\log \nu(\Delta_{i_0 \ldots i_n})}{\log(\frac{1}{L_2} |\Delta_{i_0 \ldots i_n}|)} \leq \varliminf_{m \to \infty} \frac{\log \nu(\Delta_{i_0 \ldots i_m})}{\log |\Delta_{i_0 \ldots i_m}|}.$$

Since this is true for every $\omega = (i_0 i_1 \ldots) \in Q$ with $\chi(\omega) = x$ the first statement follows.

We now prove the second statement. Given $\alpha > 0$ and $C > 0$, define

$$Q_{\alpha, C} = \{\omega = (i_0 i_1 \ldots) \in Q : \nu(\Delta_{i_0 \ldots i_n}) \leq C |\Delta_{i_0 \ldots i_n}|^\alpha \text{ for all } n \geq 0\}.$$

Set $F_{\alpha, C} = \chi(Q_{\alpha, C})$. If $\omega = (i_0 i_1 \ldots) \in Q_{\alpha, C}$ and $\chi(\omega) = x$ then

$$\underline{d}(x) = \inf \varliminf_{n \to \infty} \frac{\log \nu(\Delta_{i_0 \ldots i_n})}{\log |\Delta_{i_0 \ldots i_n}|} \geq \alpha.$$

Moreover, the set $F_\alpha = \cup_{C > 0} F_{\alpha, C}$ coincides with the set of points for which $\underline{d}(x) \geq \alpha$. We will show that $\underline{d}_\nu(x) \geq \alpha$ for almost every $x \in F_\alpha$. This will imply Statement 2. Indeed, if $\underline{d}(x) > \underline{d}_\nu(x)$ on a set of positive measure then there exists α such that $\underline{d}(x) > \alpha > \underline{d}_\nu(x)$ on a set of positive measure.

Fix $x \in F_{\alpha, C}$ and $r > 0$. Consider a Moran cover $\mathfrak{U}_r$ of the set $F_{\alpha, C}$ and choose those basic sets in the cover that have non-empty intersection with the ball $B(x, r)$. By the property of the Moran cover there are points $x_j \in F_{\alpha, C}$, $j = 1, \ldots, M$ (where M is a Moran multiplicity factor which is independent of x and r) and basic sets $\Delta^{(j)}$ such that $x_j \in \Delta^{(j)} = \Delta_{i_0 \ldots i_{n(x_j)}}$, $\operatorname{diam} \Delta^{(j)} \leq r$, and

$$B(x, r) \cap F_{\alpha, C} \subset \bigcup_{j=1}^{M} (\Delta^{(j)} \cap F_{\alpha, C}).$$

It follows that

$$\nu(B(x, r) \cap F_{\alpha, C}) \leq \sum_{j=1}^{M} \nu(\Delta^{(j)} \cap F).$$

Assume that $\nu(F_{\alpha, C}) > 0$. By the Borel Density Lemma (see Appendix V) for ν-almost every $x \in F_{\alpha, C}$ there exists a number $r_0 = r_0(x)$ such that for every $0 < r \leq r_0$ we have

$$\nu(B(x, r) \cap F) \leq 2\nu(B(x, r) \cap F_{\alpha, C}).$$

This implies that for ν-almost every $x \in F_{\alpha, C}$,

$$\underline{d}_\nu(x) = \varliminf_{r \to 0} \frac{\log \nu(B(x, r))}{\log r} \geq \varliminf_{r \to 0} \frac{\log \sum_{j=1}^{M} \nu(\Delta^{(j)} \cap F)}{\log r}$$

$$\geq \varliminf_{r \to 0} \frac{\log \sum_{j=1}^{M} C |\Delta_{i_0 \ldots i_{n(x_j)}}|^\alpha}{\log r} \geq \varliminf_{r \to 0} \frac{\log \sum_{j=1}^{M} C_1 r^\alpha}{\log r} = \alpha,$$

where $C_1 > 0$ is a constant.

The last statement is a direct consequence of the preceding statements. ∎

Let ν be a Borel probability measure on the limit set F of a geometric construction (CB1–CB3). Even if the pointwise dimension $d_\nu(x)$ of ν exists almost everywhere it may *not* be constant and may essentially depend on x. As Example 25.2 shows, the pointwise dimension may not be constant even if ν is a Gibbs measure. This phenomenon is caused by the non-stationarity of geometric constructions (CB1–CB3). The situation is different for geometric constructions (CPW1–CPW4) where ratio coefficients do not depend on the step of the construction.

Theorem 15.4. *Let F be the limit set of a geometric construction (CPW1–CPW4) modeled by a symbolic dynamical system (Q, σ) and μ an ergodic measure on Q. Let also m be the push forward measure of μ to F. Then m is exact dimensional (see Section 7) and for μ-almost every $\omega = (i_0 i_1 \dots) \in Q$ we have*

$$\underline{d}_m(x) = \overline{d}_m(x) = \frac{h_\mu(\sigma|Q)}{-\int_Q \log \lambda_{i_0} \, d\mu},$$

where $x = \chi(\omega)$.

Proof. Since μ is ergodic by the Birkhoff ergodic theorem applied to the function $\omega \mapsto \log \lambda_{i_0}$ (where $\omega = (i_0 i_1 \dots) \in Q$) we have that for μ-almost every ω the following limit exists:

$$\lim_{n \to \infty} \frac{1}{n} \sum_{k=0}^{n} \log \lambda_{i_k} = \int_Q \log \lambda_{i_0} \, d\mu.$$

Exploiting again the fact that μ is an ergodic measure, by the Shannon–McMillan–Breiman theorem we obtain that for μ-almost every $\omega = (i_0 i_1 \dots) \in Q$,

$$\lim_{n \to \infty} \frac{1}{n} \log \mu(C_{i_0 \dots i_n}) = -h_\mu(\sigma|Q).$$

It follows from Condition (CPW3) that for μ-almost every $\omega = (i_0 i_1 \dots) \in Q$,

$$\lim_{n \to \infty} \frac{1}{n} \log \operatorname{diam} \Delta_{i_0 \dots i_n} = \lim_{n \to \infty} \frac{1}{n} \sum_{k=0}^{n} \log \lambda_{i_k} = \int_Q \log \lambda_{i_0} \, d\mu.$$

The desired result follows now from Theorem 15.3.   ∎

Geometric Constructions with Quasi-conformal Induced Map

Let F be the limit set of a geometric construction (CG1–CG3) in $\mathbb{R}^m$ modeled by a subshift (Q, σ). Since we require the separation condition (CG3) the coding map $\chi\colon Q \to F$ is a homeomorphism and the **induced map** $G\colon F \to F$ is well defined by $G = \chi \circ \sigma \circ \chi^{-1}$. We have the following commutative diagram

$$
\begin{array}{ccc}
Q & \xrightarrow{\ \sigma\ } & Q \\
\chi \downarrow & & \downarrow \chi \\
F & \xrightarrow{\ G\ } & F
\end{array}
\ .
$$

It is easy to see that G is a continuous endomorphism onto F. By the result of Parry [Pa] it is a local homeomorphism if and only if the subshift is a subshift of finite type, i.e. $Q = \Sigma_A^+$, where A is a transfer matrix. From now on we consider this case and assume that A is transitive, i.e., the shift is topologically mixing (see Appendix II).

The induced map encodes information about the sizes, shapes, and placement of the basic sets of the geometric construction and hence can be used to control the geometry of the construction. In order to illustrate this let us fix a number $k > 0$. For each $\omega = (i_0 i_1 \ldots) \in \Sigma_A^+$ and $n \geq 0$, we define numbers

$$
\begin{aligned}
\underline{\lambda}(\omega, n) &= \underline{\lambda}_k(\omega, n) = \inf \left\{ \frac{\|G^n(x) - G^n(y)\|}{\|x - y\|} \right\}, \\
\overline{\lambda}(\omega, n) &= \overline{\lambda}_k(\omega, n) = \sup \left\{ \frac{\|G^n(x) - G^n(y)\|}{\|x - y\|} \right\},
\end{aligned}
\tag{15.7}
$$

where the infimum and the supremum are taken over all distinct $x, y \in F \cap \Delta_{i_0 \ldots i_{n+k}}$. It may happen that $\underline{\lambda}(\omega, n) = 0$ or $\overline{\lambda}(\omega, n) = \infty$ for all sufficiently large n. In the case

$$
0 < \underline{\lambda}(\omega, n) \leq \overline{\lambda}(\omega, n) < \infty,
$$

consider the limits

$$
\underline{\lambda}(x) = \lim_{n \to \infty} \frac{1}{n} \log \underline{\lambda}(\omega, n), \quad \overline{\lambda}(x) = \lim_{n \to \infty} \frac{1}{n} \log \overline{\lambda}(\omega, n),
$$

where $x = \chi(\omega)$. Notice that by the multiplicative ergodic theorem (see for example [KH]), the limits exist for almost all x with respect to any Borel G-invariant measure μ on F provided that

$$
\int_F \log^- \underline{\lambda}(x, 1) \, d\mu < \infty, \quad \int_F \log^+ \overline{\lambda}(x, 1) \, d\mu < \infty.
\tag{15.8}
$$

If the map G is smooth the numbers $\underline{\lambda}(x)$ and $\overline{\lambda}(x)$ coincide with the largest and smallest Lyapunov exponents of G at x (see definition of Lyapunov exponents in Section 26). When G is continuous these numbers can serve as a substitution for the Lyapunov exponents (see [Ki]).

We consider the case when the trajectories of the induced map are strongly unstable. More precisely, we call the induced map **expanding** if there exist constants $b \geq a > 1$ and $r_0 > 0$, such that for each $x \in F$ and $0 < r < r_0$ we have

$$
B(G(x), ar) \subset G\big(B(x, r)\big) \subset B(G(x), br).
\tag{15.9}
$$

Note that if the induced map G is expanding then it is (locally) bi-Lipschitz. Furthermore, if the induced map G is expanding then the placement of basic sets of the geometric construction cannot be arbitrary (see Theorem 15.5 below).

We now specify the choice of the number k in (15.7). Namely, we assume that k is so large that

$$
\operatorname{diam} \Delta_{i_0 \ldots i_{n+k}} \leq r_0.
\tag{15.10}
$$

We say that an expanding induced map G is **quasi-conformal** if there exist numbers $C > 0$ and $k > 0$ (satisfying (15.10)) such that for each $\omega \in \Sigma_A^+$ and $n \geq 0$,

$$\overline{\lambda}(\omega, n) \leq C \underline{\lambda}(\omega, n). \tag{15.11}$$

As the following example shows geometry of constructions (CG1–CG3) with expanding quasi-conformal induced maps (i.e., the placement of basic sets and their "sizes") is sufficiently "rigid".

Theorem 15.5. *Let F be the limit set of a geometric construction (CG1–CG3) in $\mathbb{R}^m$ modeled by a subshift of finite type (Σ_A^+, σ). Assume that the induced map G is quasi-conformal and that the basic sets on the first step of the construction have non-empty interiors. Then*

(1) *the construction satisfies Conditions (CB1–CB3), i.e., it is a Moran-like geometric construction with non-stationary ratio coefficients; moreover, it also satisfies Conditions (15.3) and (15.5);*

(2) $\dim_H F = \underline{\dim}_B F = \overline{\dim}_B F = s$, *where s is a unique number satisfying*

$$\lim_{n \to \infty} \frac{1}{n} \log \sum_{\substack{(i_0 \ldots i_n) \\ \Sigma_A^+\text{-admissible}}} \left(diam\left(F \cap \Delta_{i_0 \ldots i_n}\right)\right)^s = 0.$$

Proof. We outline the proof of the theorem. Since the basic sets on the first step of the construction have non-empty interiors we observe that each basic set $\Delta_{i_0 \ldots i_n}$ of the geometric construction satisfies

$$\underline{B}_{i_0 \ldots i_n} \subset \Delta_{i_0 \ldots i_n} \subset \overline{B}_{i_0 \ldots i_n},$$

where $\underline{B}_{i_0 \ldots i_n}$ and $\overline{B}_{i_0 \ldots i_n}$ are closed balls with radii $\underline{r}_{i_0 \ldots i_n}$ and $\overline{r}_{i_0 \ldots i_n}$. Moreover, there exist numbers $C_1 > 0$ and $C_2 > 0$ such that

$$\frac{C_1}{\overline{\lambda}(\omega, n)} \leq \underline{r}_{i_0 \ldots i_n} \leq \overline{r}_{i_0 \ldots i_n} \leq \frac{C_2}{\underline{\lambda}(\omega, n)}.$$

Since the induced map G is quasi-conformal this implies that the geometric construction satisfies Condition (CB2). Clearly, Conditions (CB1) and (CB3) hold and thus, the construction is a Moran-like geometric construction with non-stationary ratio coefficients.

By straightforward calculations one can show that given $\omega = (i_0 i_1 \ldots) \in \Sigma_A^+$ and $n, m \geq 1$,

$$\underline{\lambda}(\omega, n + m) \geq \underline{\lambda}(\omega, n) \times \underline{\lambda}(\sigma^n(\omega), m),$$

and similarly,

$$\overline{\lambda}(\omega, n + m) \leq \overline{\lambda}(\omega, n) \times \overline{\lambda}(\sigma^n(\omega), m).$$

Since the induced map G is quasi-conformal the above inequalities imply Condition (15.5). It follows from (15.9) and (15.10) that

$$a \leq \underline{\lambda}(\omega, 1) \leq \overline{\lambda}(\omega, 1) \leq b.$$

Thus, the first inequality in (15.3) holds. Similar arguments show that the second inequality also holds. This implies the first statement. The second statement follows from the first one and Theorem 15.2. ∎

One can build a geometric construction (CPW1–CPW4) for which the induced map on the limit set is not expanding: whether it is expanding depend on the placement of basic sets on each step of the construction (see below). We present now more sophisticated examples which illustrate properties of induced maps.

Example 15.2. [Bar2] *There exists a geometric construction (CG1–CG3) on the line modeled by the full shift* (Σ_3^+, σ) *such that:*

(1) *each basic set* $\Delta_{i_0 \dots i_n}$ *is a closed interval;*
(2) *there exists a point* $\omega \in \Sigma_3^+$ *such that* $\underline{\lambda}(\omega, n) = 0$ *and* $\overline{\lambda}(\omega, n) = \infty$ *for each* $n \geq 1$; *hence, the induced map* G *is not expanding.*

Proof. Consider a geometric construction on the line modeled by the full shift on 3 symbols for which $\Delta_{i_0 \dots i_n} = [a_{jn}, b_{jn}]$ for each $(i_0 \dots i_n) = (0 \dots 0)$ and $j = 0, 1, 2$. One can choose the basic sets such that:

(a) for each $n \geq 0$ the points b_{0n}, a_{1n}, b_{1n}, and b_{2n} lie in the limit set F;
(b) the difference $a_{1n} - b_{0n}$ is $e^{-a(n+1)}$ for n even and $e^{-b(n+1)}$ for n odd;
(c) the difference $a_{2n} - b_{1n}$ is $e^{-a(n+1)}$ for $[n/2]$ even and $e^{-b(n+1)}$ for $[n/2]$ odd.

Here a and b are positive distinct constants (see Figure 6a where $a = \log 5$, $b = \log 6$, and the intervals $\Delta_{i_0 \dots i_n}$ are of length 5^{-n}). This implies that $\underline{\lambda}(\omega, n) = 0$ and $\overline{\lambda}(\omega, n) = \infty$ for the point $(i_0 i_1 \dots) = (00 \dots)$ and all $n \geq 1$. ∎

The following example shows that there are geometric constructions (CG1–CG3) for which the induced map may be expanding but not quasi-conformal. It also illustrates that in this case Theorem 15.5 may fail.

Example 15.3. [Bar2] *There exists a geometric construction (CG1–CG3) on the line modeled by the full shift* (Σ_2^+, σ) *such that*

(1) *each basic set* $\Delta_{i_0 \dots i_n}$ *is a closed interval of length depending only on n;*
(2) *the induced map* G *is expanding but not quasi-conformal;*
(3) $\dim_H F = \underline{\dim}_B F < \overline{\dim}_B F$.

Proof. (See Figure 6b.) Let λ_n be numbers defined by (15.6). Consider a geometric construction (CG1–CG3) on the line modeled by the full shift on 2 symbols with the basic sets $\Delta_{i_0 \dots i_n}$ to be closed intervals of length $\exp \lambda_n$. Given a basic set $\Delta_{i_0 \dots i_n}$, we require that the sets $\Delta_{i_0 \dots i_n 0}$ and $\Delta_{i_0 \dots i_n 1}$ be attached respectively to the left and right end-points of the interval $\Delta_{i_0 \dots i_n}$. This guarantees that for each $x, y \in F \cap \Delta_{i_0 \dots i_n}$ such that $x \neq y$, the ratio $\|G^n x - G^n y\| / \|x - y\|$ is of the form

$$\frac{e^{\lambda_{m-n}} + \sum_{j=1}^{\infty} k_j e^{\lambda_{m-n+j}}}{e^{\lambda_m} + \sum_{j=1}^{\infty} k_j e^{\lambda_{m+j}}},$$

for some $m \geq 0$, and k_j taking on one of the values $\{-2, -1, 0, 1, 2\}$ for each $j \geq 0$. Notice that not all sequences k_n are admissible. Since

$$e^{an} \leq e^{\lambda_{m-n}} / e^{\lambda_m} \leq e^{bn},$$

we obtain

$$\overline{\lambda}(\omega, n) \leq \sup_{m \geq 0} \frac{e^{\lambda_{m-n}}}{e^{\lambda_m}} \times \frac{1 + 2\sum_{j=1}^{\infty} e^{-aj}}{1 - 2\sum_{j=1}^{\infty} e^{-aj}} \leq \frac{e^{-bn}(1 + e^{-a})}{1 - 3e^{-a}} < \infty,$$

$$\underline{\lambda}(\omega, n) \geq \inf_{m \geq 0} \frac{e^{\lambda_{m-n}}}{e^{\lambda_m}} \times \frac{1 - 2\sum_{j=1}^{\infty} e^{-aj}}{1 + 2\sum_{j=1}^{\infty} e^{-aj}} \geq \frac{e^{-an}(1 - 3e^{-a})}{1 + e^{-a}} > 0,$$

for all $b \geq a > \log 3$. If a is sufficiently large we have

$$1 < \frac{e^{-a}(1 - 3e^{-a})}{1 + e^{-a}} \leq \frac{e^{-b}(1 + e^{-a})}{1 - 3e^{-a}} < \infty.$$

It follows that $\alpha \leq \underline{\lambda}(\omega, 1) \leq \overline{\lambda}(\omega, 1) \leq \beta$, where α and β are some positive constants. Therefore, for a sufficiently small r_0 and any $x, y \in F$ we have that $\alpha\|x - y\| \leq \|G(x) - G(y)\| \leq \beta\|x - y\|$ provided $\|x - y\| \leq r_0$. Since G is a local homeomorphism one can easily derive from here that G is expanding.

Notice that by the construction of the numbers λ_n, we have that

$$\sup_{m \geq 0} (\lambda_m - \lambda_{m-n}) = -an, \quad \inf_{m \geq 0} (\lambda_m - \lambda_{m-n}) = -bn \qquad (15.12)$$

for any $n \geq 0$. Hence,

$$\overline{\lambda}(\omega, n) \geq \sup_{m \geq 0} \frac{e^{\lambda_{m-n}}}{e^{\lambda_m}} \times \frac{1 - 2\sum_{j=1}^{\infty} e^{-aj}}{1 + 2\sum_{j=1}^{\infty} e^{-aj}} \geq \frac{e^{bn}(1 - 3e^{-a})}{1 + e^{-a}},$$

$$\underline{\lambda}(\omega, n) \leq \inf_{m \geq 0} \frac{e^{\lambda_{m-n}}}{e^{\lambda_m}} \times \frac{1 + 2\sum_{j=1}^{\infty} e^{-aj}}{1 - 2\sum_{j=1}^{\infty} e^{-aj}} \leq \frac{e^{an}(1 + e^{-a})}{1 - 3e^{-a}}.$$

This proves that the induced map G is not quasi-conformal.

One can see that $2^{n-1} \leq N(F, e^{\lambda_n}) \leq 2^n$ (recall that $N(F, r)$ is the smallest number of balls of radius r needed to cover F). Therefore, (15.12) implies that

$$\underline{\dim}_B F = \frac{\log 2}{b}, \quad \overline{\dim}_B F = \frac{\log 2}{a}.$$

Notice that our construction is a geometric construction (CB1–CB3) with basic sets satisfying (15.3). By Theorem 15.1 we have that $\dim_H F \geq s$, where s is the unique root of the equation $P_{\Sigma_2^+}(s\varphi) = 0$. Since $\lambda_n \geq -bn$ for all $n \geq 0$ we conclude that $s \geq \log 2/b$. Hence, $\dim_H F = \underline{\dim}_B F = \log 2/b$. ∎

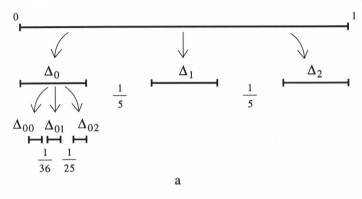

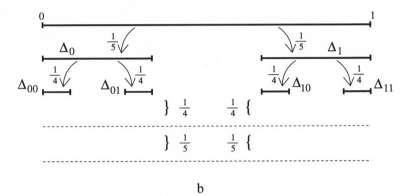

Figure 6. GEOMETRIC CONSTRUCTIONS WITH INDUCED MAPS:
a) Non-expanding, b) Non-conformal.

Geometric Constructions with Contraction Maps

There is a special class of geometric constructions (CG1–CG2) that are well-known in the literature (see for example, [F1]) — **constructions with contraction maps**. Their basic sets $\Delta_{i_0 \ldots i_n}$ are given as follows

$$\Delta_{i_0 \ldots i_n} = h_{i_0} \circ h_{i_1} \circ \cdots \circ h_{i_n}(D).$$

Here D is the unit ball in $\mathbb{R}^m$ and $h_1, \ldots, h_p \colon D \to D$ are bi-Lipschitz contraction maps, i.e., for any $x, y \in D$,

$$\underline{\lambda}_i \operatorname{dist}(x, y) \leq \operatorname{dist}(h_i(x), h_i(y)) \leq \overline{\lambda}_i \operatorname{dist}(x, y),$$

where $0 < \underline{\lambda}_i \leq \overline{\lambda}_i < 1$. They are modeled by a subshift (Q, σ). It is easy to see that these geometric constructions are Moran-like geometric constructions with non-stationary ratio coefficients of type (CB1–CB3) and hence can be treated accordingly.

If we require the separation condition (CG3) then the coding map is a home-omorphism and we can consider the induced map G on the limit set F of the construction which acts as follows: $G(x) = h_i(x)^{-1}$ for each $x \in F \cap \Delta_i$ and $i = 1, \ldots, p$. Hence, G is expanding, i.e., it satisfies (15.9). As we mentioned above G is a local homeomorphism if and only if Q is a topological Markov chain, i.e., $Q = \Sigma_A^+$ for some transfer matrix A (which we assume to be transitive).

One can also verify that for each $\omega = (i_0 i_1 \ldots) \in \Sigma_A^+$ and $n > 0$,

$$\prod_{j=0}^{n} \underline{\lambda}_{i_j} \leq \underline{\lambda}(\omega, n) \leq \overline{\lambda}(\omega, n) \leq \prod_{j=0}^{n} \overline{\lambda}_{i_j}.$$

By Theorem 15.2 the Hausdorff dimension and lower and upper box dimensions of the limit set of a construction with contraction maps admit the following estimates:

$$\underline{d} \leq \dim_H F \leq \underline{\dim}_B F \leq \overline{\dim}_B F \leq \overline{d},$$

where $\underline{d}$ and $\overline{d}$ are unique roots of Bowen's equations $P_{\Sigma_A^+}(d\,\underline{\varphi}) = 0$ and $P_{\Sigma_A^+}(d\,\overline{\varphi}) = 0$ with $\underline{\varphi}(\omega) = -\log \underline{\lambda}_{i_0}$ and $\overline{\varphi}(\omega) = -\log \overline{\lambda}_{i_0}$.

Notice that, in general, $\underline{d} \leq \underline{s}$ and $\overline{d} \geq \overline{s}$, where $\underline{s}$ and $\overline{s}$ are roots of Bowen's equations (15.2) (see Theorem 15.1). The inequalities can be strict (see [Bar2]). This illustrates that one can obtain more refined estimates of the Hausdorff dimension and upper box dimension using the numbers $\underline{\lambda}(\omega, n)$ and $\overline{\lambda}(\omega, n)$ than using the Lipschitz constants of the contraction maps (i.e., the numbers $\underline{\lambda}_i$ and $\overline{\lambda}_i$).

If $\underline{\lambda}_i = \overline{\lambda}_i$ for any $i = 1, \ldots, p$ then the the contraction maps are affine and conformal. The induced map G is quasi-conformal (and in fact, is smooth and conformal; see Section 20) and the Hausdorff dimension and lower and upper box dimensions of the limit set coincide. The common value is given by Theorem 13.3. In particular, if the geometric construction is simple (i.e., is modeled by the full shift) then s is the unique root of the equation

$$\lambda_1{}^s + \cdots + \lambda_p{}^s = 1.$$

Geometric Constructions Associated with Schottky Groups

A **Kleinian group** is a discrete subgroup of the group of all linear fractional transformations $z \mapsto \frac{az+b}{cz+d}$ of the complex plane $\widehat{\mathbb{C}}$ with the determinant $ad - bc = 1$ or -1. A linear fractional transformation g is said to be hyperbolic if $\mathrm{tr}^2 g = (a+d)^2 > 4$ and loxodromic if $\mathrm{tr}^2 g \in \widehat{\mathbb{C}} \setminus [0, 4]$.

A **(classical) Schottky group** Γ is a Kleinian group with finitely many generators $g_1, \ldots g_p$, $p \geq 1$ which act in the following way: there exist $2p$ disjoint circles $\gamma_1, \gamma_1', \ldots, \gamma_p, \gamma_p'$ bounding a $2p$-connected region D for which $g_j(D) \cap D = \varnothing$ and $g_j(\gamma_j) = \gamma_j'$ for $j = 1, \ldots, p$. The group Γ is known to be free and purely loxodromic, i.e., all non-trivial elements of Γ have either hyperbolic or loxodromic type (see [Mas], [Kr]).

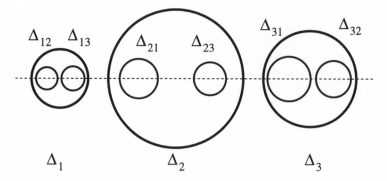

Figure 7. A GEOMETRIC CONSTRUCTION CORRESPONDING TO
A REFLECTION GROUP WITH THREE GENERATORS.

The group Γ is said to act discontinuously at a point $z \in \widehat{\mathbb{C}}$ if the stabilizer
$\Gamma_z = \{g \in \Gamma : g(z) = z\}$ of z in Γ is finite, and z has a neighborhood U_z such
that $g(U_z) \cap U_z = \varnothing$ for all $g \in \Gamma/\Gamma_z$ and $g(U_z) = U_z$ for $g \in \Gamma_z$. The set $\Omega(\Gamma)$ of
points $z \in \widehat{\mathbb{C}}$ at which Γ acts discontinuously is called the region of discontinuity
of the group Γ. It is known that if Γ is a classical Schottky group then the factor
$\Omega(\Gamma)/\Gamma$ is a closed surface of genus p.

The limit set for a classical Schottky group can be viewed as the limit set
for a geometric construction with maps which are conformal but, in general, are
neither affine nor contracting. This construction is modeled by a (one-sided)
subshift of finite type. In the special case $\gamma_j = \gamma'_j$ Schottky group is called
the **reflection group** (see Figure 7 where $p = 3$). One can show that the
maps, generating the geometric construction associated with a reflection group,
are contracting and the transfer matrix has 0s along main diagonal while other
entries are equal to 1. Note that the induced map G on the limit set of the
geometric construction is smooth conformal and expanding.

Geometric Constructions (CPW1–CPW4) with Disjoint Basic Sets

We consider a geometric construction (CPW1–CPW4) (see Section 13) mod-
eled by a subshift of finite type (Σ_A^+, σ). We assume that the basic sets of the
construction are disjoint. One can verify that for each $\omega = (i_0 i_1 \dots) \in \Sigma_A^+$,
$n > 0$,

$$K_1 \prod_{j=0}^n \lambda_{i_j} \leq \underline{\lambda}(\omega, n) \leq \overline{\lambda}(\omega, n) \leq K_2 \prod_{j=0}^n \lambda_{i_j}.$$

One can construct a geometric construction (CPW1–CPW4) for which the induced map G is not expanding (we leave this as an easy exercise for the reader).
However, if it is expanding it is quasi-conformal (this follows from the above
inequalities and Condition (13.1)). Thus, $\underline{s} = \overline{s} = s_\lambda$, where s_λ is the number in
Theorem 13.1 (i.e., the Hausdorff dimension and lower and upper box dimensions
of the limit set coincide and are equal to s_λ).

16. Geometric Constructions with Rectangles; Non-coincidence of Box Dimension and Hausdorff Dimension of Sets

A crucial feature of Moran-like geometric constructions with stationary or non-stationary ratio coefficients is that they are, so to speak, *isotropic*, i.e., the ratio coefficients do not depend on the directions in the space. This is a very strong requirement and that is why the placement of the basic sets may be fairly arbitrary.

A simple example of *anisotropic* geometric constructions is provided by constructions with rectangles. As in the case of Moran-like geometric constructions one still has a complete control over the sizes and shapes of the basic sets but needs two collections of ratio coefficients to control length and width of rectangles on the nth step. In this section we consider the simplest case when the ratio coefficients are constant and do not depend on the step of the construction. Even so, we will see that the Hausdorff and box dimensions of the limit set may depend on the placement of the basic sets and may not agree. This can happen even if a geometric construction is "most close" to a self-similar construction, i.e., it is given by finitely many affine maps (the so-called general Sierpiński carpets; discussed later in Section 16).

In this section we present examples which illustrate how the equality between the Hausdorff dimension and box dimension can be destroyed. These examples are also a source for understanding some "pathological" properties of the pointwise dimension (see Section 25). A surprising phenomenon is that the Hausdorff dimension of the limit set for constructions with rectangles may also depend on some delicate number-theoretic properties of ratio coefficients corresponding to different directions.

Geometric Constructions with Rectangles

We call a symbolic geometric construction (CG1–CG3) on the plane modeled by a symbolic dynamical system (Q, σ) a **construction with rectangles** if there exist $2p$ numbers $\underline{\lambda}_i$ and $\overline{\lambda}_i$, $i = 1, \ldots, p$, $0 < \underline{\lambda}_i \leq \overline{\lambda}_i < 1$ such that the basic set $\Delta_{i_0 \ldots i_n}$ is a rectangle with the largest side not exceeding $K_1 \prod_{j=1}^{n} \overline{\lambda}_{i_j}$ and the smallest side not less than $K_2 \prod_{j=1}^{n} \underline{\lambda}_{i_j}$, where K_1 and K_2 are positive constants. See Figure 8. We stress that we require the separation condition (CG3).

Let $s_{\underline{\lambda}}$ and $s_{\overline{\lambda}}$ be the unique roots of Bowen's equations $P_Q(s_{\underline{\lambda}} \log \underline{\lambda}_{i_0}) = 0$ and $P_Q(s_{\overline{\lambda}} \log \overline{\lambda}_{i_0}) = 0$ respectively and $m_{\underline{\lambda}}$ and $m_{\overline{\lambda}}$ be the push forward measures by the coding map of equilibrium measures $\mu_{\underline{\lambda}}$ and $\mu_{\overline{\lambda}}$ corresponding to the functions $(i_0 i_1 \ldots) \mapsto s_{\underline{\lambda}} \log \underline{\lambda}_{i_0}$ and $(i_0 i_1 \ldots) \mapsto s_{\overline{\lambda}} \log \overline{\lambda}_{i_0}$.

Theorem 16.1. *Let F be the limit set for a symbolic geometric construction with rectangles. Then*

(1) $s_{\underline{\lambda}} \leq \dim_H F \leq \underline{\dim}_B F \leq \overline{\dim}_B F \leq s_{\overline{\lambda}}$;

(2) *If the measures $\mu_{\underline{\lambda}}$ and $\mu_{\overline{\lambda}}$ on Q are Gibbs, then $s_{\underline{\lambda}} \leq \underline{d}_{m_{\underline{\lambda}}}(x)$ and $\overline{d}_{m_{\overline{\lambda}}}(x) \leq s_{\overline{\lambda}}$.*

Proof. Notice that the geometric construction with rectangle is regular with the estimating vector $(\underline{\lambda}_1, \ldots, \underline{\lambda}_p)$. The result follows from Theorems 14.1, 14.2, and 14.5. ∎

In [Bar1], Barreira gave a description of geometric constructions with rectangles. We follow his approach and assume for simplicity that $\underline{\lambda}_i = \underline{\lambda}$ and $\overline{\lambda}_i = \overline{\lambda}$ for $i = 1, \ldots, p$ for some $0 < \underline{\lambda} < \overline{\lambda} < 1$.

Given $\omega = (i_0 i_1 \ldots) \in Q$ and $n > 0$ denote by $\gamma_n(\omega) \in \mathbb{S}^1 \equiv [0, 2\pi]/\{0, 2\pi\}$ the direction of the longest sides of $\Delta_{i_0 \ldots i_n}$, i.e., the sides of length $\overline{\lambda}^n$. We first show that $\gamma_n(\omega)$ converges uniformly when $n \to \infty$ with an exponential rate.

Proposition 16.1.

(1) *For each $\omega \in Q$, there exists the limit $\gamma(\omega) \overset{\text{def}}{=} \lim\limits_{n \to \infty} \gamma_n(\omega)$.*

(2) *There exists $C > 0$ such that for all $\omega \in Q$,*

$$|\gamma(\omega) - \gamma_n(\omega)| < C(\underline{\lambda}/\overline{\lambda})^n. \tag{16.1}$$

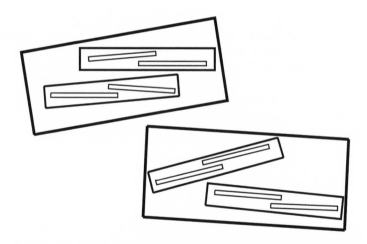

Figure 8. A Geometric Construction with Rectangles.

Proof. Fix $\omega = (i_0 i_1 \ldots) \in Q$. We first consider the "worst" possible placement of rectangles on the nth step of the construction (see Figure 9): a pair of diametrically opposed vertices of a rectangle $\Delta_{i_0 \ldots i_{n+1}}$ touches the two longest sides of a rectangle $\Delta_{i_0 \ldots i_n}$.

Let θ_n be the angle between the longest sides of $\Delta_{i_0 \ldots i_n}$ and $\Delta_{i_0 \ldots i_{n+1}}$ (see Figure 9). We notice that θ_n does not depend on ω and satisfies the equation

$$\overline{\lambda}^{n+1} \sin \theta_n + \underline{\lambda}^{n+1} \cos \theta_n = \underline{\lambda}^n.$$

Setting $a = \overline{\lambda}/\underline{\lambda}$ we obtain that $\sin\theta_n$ satisfies the following equation:

$$\left(a^{n+1}\sin\theta_n - 1/\underline{\lambda}\right)^2 = 1 - \sin^2\theta_n.$$

The formal roots of this equation are

$$\sin\theta_n = \frac{1}{1 + a^{2(n+1)}}\left(\frac{1}{\underline{\lambda}}a^{n+1} \pm \sqrt{1 - \frac{1}{\underline{\lambda}^2} + a^{2(n+1)}}\right).$$

Since $\cos\theta_n > 0$ we have to choose the root with minus in front of the square root. The discriminant of this equation is non-negative if and only if $|\Delta_{i_0\dots i_n}| \equiv \left[\underline{\lambda}^{2(n+1)} + \overline{\lambda}^{2(n+1)}\right]^{1/2} < \underline{\lambda}^n$. This takes place if n is sufficiently large, i.e., $|\Delta_{i_0\dots i_{n+1}}| < \underline{\lambda}^n$. One can also see that $\sin\theta_n$ is asymptotically equivalent to $(1/\underline{\lambda} - 1)\, a^{-(n+1)}$ as $n \to \infty$. Since $\underline{\lambda} < 1$ there exist $C_1 > 0$ and $C_2 > 0$ such that $C_1 a^{-n} < \sin\theta_n < C_2 a^{-n}$ for all sufficiently large n. One can now choose C_1 and C_2 to obtain in addition that $C_1 a^{-n} < \theta_n < C_2 a^{-n}$. Therefore, if $n_0 > 0$ is sufficiently large and $n > m \geq n_0$ one has

$$\frac{C_1 a^{-(m+1)}}{1 - a^{-1}}\left[1 - a^{-(n-m)}\right] < \sum_{k=m+1}^{n}\theta_k < \frac{C_2 a^{-(m+1)}}{1 - a^{-1}}\left[1 - a^{-(n-m)}\right]. \qquad (16.2)$$

We now consider a general placement of the basic sets. Given $\omega \in Q$, there exist angles $\widetilde{\theta}_k(\omega)$ (for each $k > 0$) with $\left|\widetilde{\theta}_k(\omega)\right| \leq \theta_k$ such that $\gamma_n(\omega) = \sum_{k=1}^{n}\widetilde{\theta}_k$. Therefore, if $n > m \geq n_0$, using (16.2) we have

$$\left|\gamma_n(\omega) - \gamma_m(\omega)\right| = \left|\sum_{k=m+1}^{n}\widetilde{\theta}_k(\omega)\right| \leq \sum_{k=m+1}^{n}\theta_k < C_2 a^{-(m+1)}.$$

This shows that the sequence $\left(\gamma_n(\omega)\right)_{n\in\mathbb{N}}$ converges and satisfies (16.1).   ■

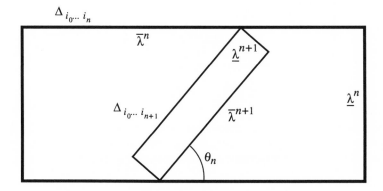

Figure 9. The "Worst" Disposition of Rectangles.

Proposition 16.2. *The vector field* $\gamma \circ \chi^{-1} \colon F \to \mathbb{S}^1$ *is uniformly continuous.*

Proof. Let $\omega = (i_0 i_1 \dots) \in Q$. Consider a sequence $\omega_n \to \omega$ as $n \to \infty$. There exists a sequence of numbers $m_n \to \infty$ as $n \to \infty$ such that $\chi(\omega) \in \Delta_{i_0 \dots i_n}$ for each $n > 0$ and $k \geq m_n$. Therefore, $\gamma_n(\omega_k) = \gamma_n(\omega)$ for $n \in \mathbb{N}$ and $k \geq m_n$. Using (16.1) we conclude that

$$\begin{aligned}
\big|\gamma(\omega_k) - \gamma(\omega)\big| &\leq \big|\gamma(\omega_k) - \gamma_n(\omega_k)\big| + \big|\gamma_n(\omega_k) - \gamma_n(\omega)\big| \\
&\quad + \big|\gamma_n(\omega) - \gamma(\omega)\big| < 2C(\overline{\lambda}/\underline{\lambda})^{-n}
\end{aligned} \tag{16.3}$$

for n sufficiently large and $k \geq m_n$. Given $\varepsilon > 0$, let us choose $n > 0$ such that $2C(\overline{\lambda}/\underline{\lambda})^{-n} < \varepsilon$. Then $|\gamma(\omega_k) - \gamma(\omega)| < \varepsilon$ for $k \geq m_n$. Since F is compact this gives the desired result. ∎

Propositions 16.1 and 16.2 show that the basic sets of a symbolic geometric construction with rectangles cannot be spaced arbitrarily, but are forced to be oriented to approach a continuous vector field.

In [Bar1], Barreira showed that the problem of computing the Hausdorff dimension and box dimension of the limit sets for geometric constructions with rectangles can be reduced to study constructions of a special kind where rectangles are *aligned*, i.e., the sides of length $\overline{\lambda}^n$ for all $n \geq 0$ are parallel (see Figure 10). We present here two of his results in this direction. The proofs are based on Propositions 16.1 and 16.2 and are omitted.

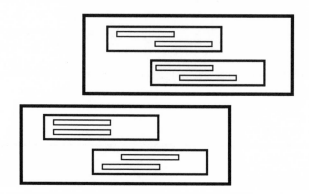

Figure 10. A CONSTRUCTION WITH ALIGNED RECTANGLES.

Theorem 16.2. *Given a geometric construction with rectangles, modeled by a symbolic dynamical system* (Q, σ), *there exists a geometric construction with aligned rectangles, modeled by* (Q, σ), *such that* $\underline{\dim}_B F = \underline{\dim}_B \widetilde{F}$ *and* $\overline{\dim}_B F = \overline{\dim}_B \widetilde{F}$, *where* F *and* $\widetilde{F}$ *are the limit sets of the two constructions.*

We say that a geometric construction has **exponentially large gaps** if there exists $\delta > 0$ such that for each $(i_0 i_1 \dots) \in Q$ and each $n \in \mathbb{N}$ we have that

$\text{dist}(\Delta_{i_0...i_nj}, \Delta_{i_0...i_nk}) \geq \delta \underline{\lambda}^n$ whenever $j \neq k$ and $(i_0...i_nj)$ and $(i_0...i_nk)$ are Q-admissible. With this extra hypothesis we can study the Hausdorff dimension of a geometric construction with rectangles by comparing it with an appropriate geometric construction with aligned rectangles.

Theorem 16.3. *Given a geometric construction with rectangles and exponentially large gaps, modeled by a symbolic dynamical system (Q, σ), there exists a geometric construction with aligned rectangles and exponentially large gaps, modeled by (Q, σ), such that the map $f: F \to \widetilde{F}$ is Lipschitz with Lipschitz inverse and $\dim_H F = \dim_H \widetilde{F}$, where F and $\widetilde{F}$ are the limit sets.*

Non-coincidence of Hausdorff Dimension and Box Dimension

We provide several special examples of geometric constructions with rectangles that illustrate some interesting phenomena and reveal the non-trivial structure of these constructions and the crucial difference between them and Moran-like geometric constructions with stationary ratio coefficients.

We consider a geometric construction with rectangles in $\mathbb{R}^2$ which is generated by finitely many affine maps $f_1, \ldots, f_p: S \to S$, where $S = [0, 1] \times [0, 1]$ is the unit square. Let F be the limit set. Each map f_i can be written as $f_i(x, y) = T_i(x, y) + b_i$, where T_i is a linear contraction and b_i is a two-vector. In [F2] (see also [F4]), Falconer proved that *for almost all $(b_1 \ldots b_p) \in \mathbb{R}^{2p}$* (in the sense of $2p$-dimensional Lebesgue measure), *the Hausdorff dimension and the lower and upper box dimensions of F coincide and the common value is completely determined by $T_1, \ldots, T_p$ provided that* $\|T_i\| < \frac{1}{3}$.

On the other hand, as we mentioned above, the Hausdorff and box dimensions of the limit set for geometric constructions with rectangles may depend on the placement of the basic sets. They may not agree even in the case when the construction is generated by finitely many affine maps. The corresponding example is described by McMullen (see [Mu]; in [LG], Lalley and Gatzouras considered a more general version of his construction).

Given integers $\ell \geq m \geq 2$ choose a set A consisting of pairs of integers (i, j) with $0 \leq i < \ell$ and $0 \leq j < m$. Denote by a the cardinality of A (clearly $a \leq mn$). Let f_k, $k = 1, \ldots, a$ be affine maps given in the following way: if k enumerates the element $(i, j) \in A$ then $f_k(S) = S_{ij}$, where $S = [0, 1] \times [0, 1]$ and $S_{ij} = \left[\frac{i}{\ell}, \frac{i+1}{\ell}\right] \times \left[\frac{j}{m}, \frac{j+1}{m}\right]$. The affine maps f_k generate a simple geometric construction with basic sets $\Delta_{i_0...i_n} = f_{i_0} \circ \cdots \circ f_{i_n}(S)$ being $(\ell^{-n} \times m^{-n})$-rectangles. We allow some of the basic sets $\Delta_1, \ldots, \Delta_a$ on the first step of the construction to intersect each other by either a common vertex or a common edge (see Figure 11 where $\ell = m = 4$ and $a = 6$). The geometric description of this construction is the following: starting from the $(l \times m)$-grid of the unit square S choose rectangles corresponding to $(i, j) \in A$, then repeat this procedure in each chosen rectangle and so on. The limit set F for this construction is

$$F = \left\{ \left(\sum_{n=0}^{\infty} \frac{i_n}{\ell^n}, \sum_{n=0}^{\infty} \frac{j_n}{m^n} \right) : (i_n, j_n) \in A \right\}$$

and is known as a **general Sierpiński carpet** (see [Mu]).

Example 16.1.

(1) $\dim_H F = \log_m \left(\sum_{j=0}^{m-1} t_j^{\log_\ell m} \right) \overset{\text{def}}{=} s$, where t_j is the number of those i for which $(i,j) \in A$.

(2) $\underline{\dim}_B F = \overline{\dim}_B F = \log_m r + \log_\ell(\frac{a}{r})$, where r is the number of those j for which $(i,j) \in A$ for some i.

Remarks.

(1) For the geometric construction shown on Figure 11 we have $t_0 = 1$, $t_1 = 2$, $t_2 = 2$, $t_3 = 1$, and $r = 4$.

(2) The Hausdorff and box dimensions of the limit set F agree if: a) $l = m$; b) the constants t_j take on only one value other then zero.

In the first case the construction is conformal self-similar and the common value for dimensions is $\log_m a$.

Proof. Let $\chi \colon \Sigma_a^+ \to F$ be the coding map given as follows:

$$\chi(k_0 k_1 \dots) = \left(\sum_{n=0}^\infty \frac{i_n}{\ell^n}, \ \sum_{n=0}^\infty \frac{j_n}{m^n} \right),$$

where $k_n = (i_n, j_n)$. This map provides a symbolic representation of the limit set F by the (one-sided) shift on a symbols. It is surjective but may fail to be injective because some points in F may have more than one representation.

Define numbers b_k, $k = 0, \dots, a-1$ in the following way. If $k = (i,j) \in A$ then b_k is the number of i' such that $(i',j) \in A$. Note that by the definition of the number s,

$$m^s = \sum_{k=0}^{a-1} b_k^{\log_\ell m - 1}.$$

We define the Bernoulli measure on Σ_a^+ by assigning probabilities

$$p_k = \frac{b_k^{\log_\ell m - 1}}{m^s}$$

to each symbol $k = 0, \dots, a-1$. In other words, if $C_{k_0 \dots k_n}$ is a cylinder set then $\mu(C_{k_0 \dots k_n}) = \prod_{t=0}^n p_{k_t}$. Note that $\sum_{k=1}^a p_k = 1$. Let λ be the measure on F which is the push forward of the measure μ (i.e., $\lambda(A) = \mu(\chi^{-1}(A))$ for any Borel subset $A \subset F$). We will show that $\dim_H \lambda \geq s$.

Set $q = [n \log_\ell m] \leq n$. Given $(n+1)$-tuple $(k_0 \dots k_n)$ with $k_t = (i_t, j_t) \in A$, consider the set $R_{k_0 \dots k_n} \subset F$ of all points (x,y) for which

$$x = \sum_{t=0}^\infty \frac{i'_t}{\ell^t}, \quad y = \sum_{t=0}^\infty \frac{j'_t}{m^t},$$

where $i'_t = i_t$ for $t = 0, \dots, q$ and $j'_t = j_t$ for $t = 0, \dots, n$. The set $R_{k_0 \dots k_n}$ is "almost" a ball of radius m^{-n}. More precisely, there are constants $C_1 > 0$ and $C_2 > 0$ independent of $(k_0 \dots k_n)$ and a point $(x,y) \in R_{k_0 \dots k_n}$ such that

$$B((x,y), C_1 m^{-n}) \cap F \subset R_{k_0 \dots k_n} \subset B((x,y), C_2 m^{-n}) \cap F. \tag{16.4}$$

We will show that the sets $R_{k_0...k_n}$ comprise an "optimal" cover of J which we use to compute the Hausdorff dimension of the measure λ. Note that each set $R_{k_0...k_n}$ can be decomposed into finitely many basic sets $\Delta_{k_0...k_q k'_{q+1}...k'_n} \cap F$, where $k'_t = (i'_t, j'_t)$ and $j'_t = j_t$ for $t = q+1, ..., n$. Moreover, the number of such sets is equal to $b_{k_{q+1}} b_{k_{q+2}} \cdots b_{k_n}$.

Define now the function ϕ_n on Σ_a^+ by

$$\phi_n(k_0 k_1 \ldots) = \left[\frac{(b_{k_0} b_{k_1} \cdots b_{k_n})^{\log_\ell m}}{(b_{k_0} b_{k_1} \cdots b_{k_q})} \right]^{\frac{1}{n}}.$$

It is easy to see that this function is constant on the cylinder $C_{k_0...k_n}$. We denote the common value by $\phi_{k_0...k_n}$.

Lemma 1. *For any* $(k_0 ... k_n)$ *we have*

$$\lambda(R_{k_0...k_n}) = (\phi_{k_0...k_n} m^{-s})^n.$$

Proof of the lemma. Let us pick an $(n+1)$-tuple $(k_0, k_1, ..., k_n)$ and set

$$(s_0, s_1, ..., s_n) = (b_{k_0}, b_{k_1}, ..., b_{k_n}).$$

Note that this sequence is independent of the choice of $(k_0, k_1, ..., k_n)$. This implies that

$$\lambda(\Delta_{k_0...k_q k'_{q+1}...k'_n} \cap F) = \frac{s_0 s_1 \cdots s_n^{\log_\ell m}}{s_0 s_1 \cdots s_n} m^{-sn}.$$

As we have mentioned the number of basic sets $\Delta_{k_0...k_q k'_{q+1}...k'_n}$ is equal to $b_{k_{q+1}} b_{k_{q+2}} \cdots b_{k_n}$. Therefore,

$$\lambda(R_{k_0...k_n}) = \frac{s_0 s_1 \cdots s_n^{\log_\ell m}}{s_0 s_1 \cdots s_q} m^{-sn} = (\phi_{k_0...k_n} m^{-s})^n.$$

This completes the proof of the lemma. ∎

We now describe the "gaps" between the sets $\Delta_{k_0...k_q k'_{q+1}...k'_n} \cap F$ inside the set $R_{k_0...k_n}$ in terms of the behavior of the functions ϕ_n over n.

Lemma 2.
(1) $\overline{\lim}_{n \to \infty} \phi_n(\omega) \geq 1$ *for all* $\omega \in \Sigma_a^+$.
(2) $\phi_n \to 1$ *as* $n \to \infty$ μ-*almost everywhere.*

Proof of the lemma. Define the functions g_n and h_n on Σ_a^+ by

$$g_n(k_0 k_1 \ldots) = \frac{(b_{k_0} b_{k_1} \cdots b_{k_n})^{\frac{1}{n}}}{(b_{k_0} b_{k_1} \cdots b_{k_q})^{\frac{1}{q}}}, \quad h_n(k_0 k_1 \ldots) = (b_{k_0} b_{k_1} \cdots b_{k_q})^{\frac{1}{q}(\log_\ell m - \frac{q}{n})}.$$

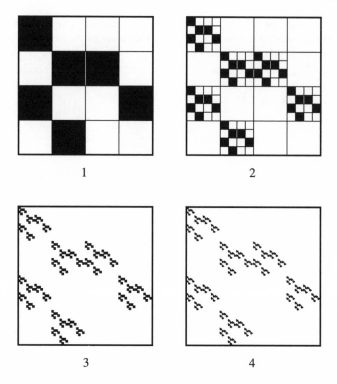

1

2

3

4

Figure 11. A GENERAL SIERPIŃSKI CARPET WITH $l = m = 4$ AND
$A = \{(0,1), (1,0), (1,4), (2,2), (2,3), (3,0)\}.$

Then $\phi_n(\omega) = h_n(\omega) \times g_n(\omega)^{\log_\ell m}$. We claim that the functions g_n and h_n satisfy the following properties which immediately imply the desired result:

1) $h_n(\omega) \to 1$ as $n \to \infty$ for all ω. Indeed, $1 \le b_k \le \ell$ for all k and hence

$$1 \le h_n \le \ell^{\log_\ell m - \frac{q}{n}}.$$

The expression in the right-hand side goes to 1 as $n \to \infty$ since

$$\log_\ell m - \frac{q}{n} = \log_\ell m - \frac{[n \log_\ell m]}{n} \to 0$$

as $n \to \infty$.

2) $\overline{\lim}_{n\to\infty} g_n(\omega) \ge 1$ for all $\omega \in \Sigma_a^+$. Since each $b_k \ge 1$ we have that $s_t \stackrel{\text{def}}{=} (b_{k_0} b_{k_1} \cdots b_{k_t})^{\frac{1}{t}} \ge 1$. Therefore,

$$\overline{\lim}_{n\to\infty} g_n(\omega) = \overline{\lim}_{n\to\infty} \frac{s_n}{s_q} \ge 1$$

since this holds for any positive sequence s_t bounded away from zero.

3) $g_n(\omega) \to 1$ μ-almost everywhere as $n \to \infty$. From the definition of μ it is clear that the functions $(k_0 k_1 \dots) \mapsto p_{k_n}$, $n = 0, 1, 2, \dots$ are independent, identically distributed random variables with respect to μ. Hence, the sequence $(p_{k_0} p_{k_1} \cdots p_{k_n})^{1/n}$, $n = 0, 1, 2, \dots$ converges for almost every $(k_0 k_1 \dots) \in \Sigma_a^+$ by Kolmogorov's strong law of large numbers. Note that by the definition of p_k

$$g_n(k_0 k_1 \dots) = \left[\frac{(p_{k_0} p_{k_1} \cdots p_{k_n})^{\frac{1}{n}}}{(p_{k_0} p_{k_1} \cdots p_{k_q})^{\frac{1}{q}}} \right]^{\frac{1}{\log_\ell m - 1}}$$

and hence $g_n \to 1$ μ-almost everywhere.   ∎

It immediately follows from Lemma 1, the second statement of Lemma 2, and (16.4) that $\underline{d}_\lambda((x,y)) \geq s$ for λ-almost every point $(x,y) \in F$. Thus, $s \leq \dim_H F$. We now show that $\dim_H F \leq s$ by constructing an efficient cover of F. Fix $\varepsilon > 0$ and consider the collection $\mathcal{R}_n$ of those of sets $R_{k_0 \dots k_n}$ for which $\phi_{k_0 \dots k_n} \geq m^{-\varepsilon}$. These sets are disjoint and by Lemma 1 satisfy

$$\lambda(R_{k_0 \dots k_n}) \geq m^{-(s+\varepsilon)n}.$$

Therefore, the number of such sets is bounded by $m^{(s+\varepsilon)n}$ as $\lambda(F) = 1$. Note that any point $(x,y) \in F$ is covered by sets $R_{k_0 \dots k_n} \in \mathcal{R}_n$ for infinitely many n since $\overline{\lim}_{n \to \infty} \phi_n((x,y)) \geq 1 > m^{-\varepsilon}$ (see Statement 1 of Lemma 2). Therefore, $\mathcal{R}(N) = \cup_{n \geq N} \mathcal{R}_n$ is a cover of F for any choice of N. Let us choose N so large that

$$\sum_{n \geq N} m^{-\varepsilon n} < \varepsilon.$$

It follows that

$$\sum_{R_{k_0 \dots k_n} \in \mathcal{R}(N)} (\operatorname{diam} R_{k_0 \dots k_n})^{(s+2\varepsilon)} = \sum_{n \geq N} \operatorname{card} \mathcal{R}_n \, m^{(s+2\varepsilon)n} \leq \sum_{n \geq N} m^{-\varepsilon n} < \varepsilon$$

(here card denotes the cardinality of the corresponding set). This implies that $\dim_H F \leq s + 2\varepsilon$ and the desired result follows since ε can be chosen arbitrarily small.

In order to compute the box dimension of the limit set F let us choose $n > 0$ and set $r_n = m^{-n}$. Consider the finite cover of F by sets $R_{k_0 \dots k_n}$ and let N_n be the number of elements in this cover. It is easy to see that

$$C_1 N_n \geq N(F, r_n) \geq C_2 N_n,$$

where $C_1 > 0$ and $C_2 > 0$ are constants independent of n (recall that $N(F, r)$ is the least number of balls of radius r needed to cover the set F). Note that N_n is precisely the number of ways to choose sequences (i_t), $t = 1, \dots, q$ and (j_t), $t = 1, \dots, n$ (recall that $q = [n \log_\ell m]$) such that
  a) $(i_t, j_t) \in A$ for $t = 1, \dots, q$;
  b) $(\tilde{i}_t, j_t) \in A$ for $t = q + 1, \dots, n$ and some choice of $\tilde{i}_t$.

It follows that $N_n = a^q r^{n-q} = (\frac{a}{r})^q r^n$ (recall that a is the cardinality of A and r is the number of j such that $(i, j) \in A$ for some i). Therefore,

$$\underline{\dim}_B F = \overline{\dim}_B F = \lim_{n \to \infty} \frac{\log N(F, r_n)}{-\log r_n}$$

$$= \lim_{n \to \infty} \frac{\log N_n}{-\log r_n} = \log_m r + \log_m \frac{a}{r} \lim_{n \to \infty} \frac{q}{n} = \log_m r + \log_m \frac{a}{r}.$$

This completes the proof of the statement.   ∎

Following Pesin and Weiss [PW1] we construct a more sophisticated example than in the previous section, which illustrates that all three characteristics — the Hausdorff dimension, the lower and upper box dimensions — may be distinct.

Example 16.2. *There exists a geometric construction with rectangles in the unite square $S \subset \mathbb{R}^2$, modeled by the full shift on two symbols (Σ_2^+, σ), for which $\underline{\lambda}_1 = \underline{\lambda}_2 = \underline{\lambda}$, $\overline{\lambda}_1 = \overline{\lambda}_2 = \overline{\lambda}$, $0 < \underline{\lambda} < \overline{\lambda} < \frac{1}{3}$, and the limit set F satisfies*

$$\dim_H F = \frac{\log 2}{-\log \underline{\lambda}}, \qquad \underline{\dim}_B F = \gamma \frac{\log 2}{-\log \underline{\lambda}}, \qquad \overline{\dim}_B F = \frac{\log 2}{-\log \overline{\lambda}},$$

where $\gamma \in (1, \alpha)$ is an arbitrary number and $\alpha = \log \underline{\lambda} / \log \overline{\lambda}$. Moreover, the induced map G on F is Hölder continuous.

Proof. Let $n_0 = 0$ and for $k = 0, 1, 2, \ldots$, let $n_{k+1} = [\alpha n_k]$ and $\beta_k = 2^{(\gamma - \alpha) n_{3k+1}}$. In order to describe the nth step of the construction we use the basic types of spacings: vertical stacking (A) and horizontal staking (B). See Figure 12.

(1) We start with two horizontally stacked rectangles. During steps $n_{3k} < n \leq n_{3k+1}$ we use (B).

(2) We begin with $2^{n_{3k+1}}$ rectangles. Choose β_k percent of these rectangles arbitrarily and paint them blue; paint the others green. During steps $n_{3k+1} < n \leq n_{3k+2}$, we use (B) in all blue rectangles and use (A) in all green rectangles.

(3) During steps $n_{3k+2} < n \leq n_{3k+3}$, we use (A) in all blue rectangles and use (B) in all green rectangles.

(4) Repaint all $2^{n_{3k+3}}$ rectangles white. Repeat steps 1 through 4.

The collection of rectangles at the nth step of the construction contains 2^n rectangles each with vertical and horizontal sides; the size in the vertical direction is $\underline{\lambda}^n$ and the size in the horizontal direction is $\overline{\lambda}^n$. Any two subrectangles at step $n + 1$ that are contained in the same rectangle at step n are stacked either horizontally or vertically and the distance between them is chosen to be at least $\frac{1}{3} \underline{\lambda}^n$. The projections of any two distinct rectangles at step n onto the two coordinate axes either coincide or are disjoint.

Each point $x \in X$ can be coded by a one-sided infinite sequence of two symbols. The induced map G on the limit set F is easily seen to be Hölder continuous with Hölder exponent $\log \overline{\lambda} / \log \underline{\lambda}$.

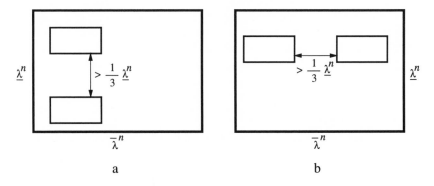

Figure 12. a) Vertical Stacking, b) Horizontal Stacking.

We now compute the Hausdorff dimension and the lower and upper box dimensions of the limit set F.

a) Calculation of Hausdorff Dimension.

Given $\varepsilon > 0$, choose $k > 0$ such that $\underline{\lambda}^{n_{3k+1}} \le \varepsilon$. Consider the cover of F consisting of green rectangles for $n = n_{3k+1}$ and blue rectangles for $n = n_{3k+2}$.

Consider a green rectangle $\Delta_{i_0...i_{n_{3k+1}}}$. By construction the intersection $A = \Delta_{i_0...i_{n_{3k+1}}} \cap F$ is contained in $2^{n_{3k+2}-n_{3k+1}}$ small green rectangles corresponding to $n = n_{3k+2}$. These rectangles are vertically aligned and have size $\underline{\lambda}^{n_{3k+2}} \times \overline{\lambda}^{n_{3k+2}}$. Since $\overline{\lambda}^{n_{3k+2}} = \overline{\lambda}^{[\alpha n_{3k+1}]+2} \le \text{const}\underline{\lambda}^{n_{3k+1}}$ the $(1 - \beta_k)2^{n_{3k+1}}$ green rectangles in the construction of F are each contained in a *green square* of size $\underline{\lambda}^{n_{3k+1}}$.

Now consider a blue rectangle $\Delta_{i_0...i_{n_{3k+1}}}$. By our construction the intersection $B = \Delta_{i_0...i_{n_{3k+2}}} \cap F$ is contained in $2^{n_{3k+3}-n_{3k+2}}$ small blue rectangles corresponding to $n = n_{3k+3}$. They are vertically aligned and have size $\underline{\lambda}^{n_{3k+3}} \times \overline{\lambda}^{n_{3k+3}}$. Since $\overline{\lambda}^{n_{3k+3}} \le \text{const}\underline{\lambda}^{n_{3k+2}}$ the $\beta_k 2^{n_{3k+1}}2^{n_{3k+2}-n_{3k+1}} = \beta_k 2^{n_{3k+2}}$ blue rectangles in the construction of F are each contained in a *blue square* of size $\underline{\lambda}^{n_{3k+2}}$.

The collection of green and blue squares comprises a cover $\mathcal{G} = \{U_i\}$ of F for which

$$\sum_{U_i \in \mathcal{G}} (\text{diam } U_i)^s \le \text{const} \left((1 - \beta_k)2^{n_{3k+1}}(\sqrt{2}\underline{\lambda}^{n_{3k+1}})^s + \beta_k 2^{n_{3k+2}}(\sqrt{2}\underline{\lambda}^{n_{3k+2}})^s \right).$$

The right-hand side of this inequality tends to 0 as $k \to \infty$ if $s > \frac{\log 2}{-\log \underline{\lambda}}$. This implies that $\dim_H F \le \frac{\log 2}{-\log \underline{\lambda}}$. On the other hand, by Theorem 16.1, we know that $\dim_H F \ge \frac{\log 2}{-\log \underline{\lambda}}$. Therefore, we conclude that

$$\dim_H F = \frac{\log 2}{-\log \underline{\lambda}}.$$

We now proceed with the box dimension.

b) Calculation of Lower and Upper Box Dimensions.

Choose $\varepsilon > 0$. We wish to compute explicitly the number $N(F, \varepsilon)$ (the least number of ball of radius ε needed to cover F). There exists a unique integer $n > 0$ such that $\overline{\lambda}^{n+1} < \varepsilon \leq \overline{\lambda}^n$. Denote by

$$A_n = \frac{\log N(F, \varepsilon)}{-\log \overline{\lambda}^n}.$$

We consider the following three cases:

Case 1: $n_{3k} \leq n < n_{3k+1}$. One can easily see that $N(F, \varepsilon) = 2^n$ and hence ,

$$A_n = \frac{\log 2}{-\log \overline{\lambda}}.$$

Case 2: $n_{3k+1} \leq n < n_{3k+2}$. We have $N(F, \varepsilon) = N_{\text{blue}}(F, \varepsilon) + N_{\text{green}}(F, \varepsilon)$, where $N_{\text{blue}}(F, \varepsilon)$ and $N_{\text{green}}(F, \varepsilon)$ are the numbers of ε-balls in the optimal cover that have non-empty intersection with respectively blue and green rectangles at step n. It follows

$$N(F, \varepsilon) = \beta_k 2^n + (1 - \beta_k) 2^{n_{3k+1}}.$$

One can see that for all sufficiently large k (for which $\beta_k \leq \frac{1}{2}$),

$$N(F, \varepsilon) \leq 2 \left(2^{(\gamma-\alpha)n_{3k+1}} 2^{\alpha n_{3k+1} + n - n_{3k+2}} + 2^{n_{3k+1}} \right)$$
$$= 2 \left(2^{\gamma n_{3k+1}} 2^{n - n_{3k+2}} + 2^{n_{3k+1}} \right)$$

and

$$N(F, \varepsilon) \geq \frac{1}{2} \left(2^{(\gamma-\alpha)n_{3k+1}} 2^{\alpha n_{3k+1} + n - n_{3k+2}} + 2^{n_{3k+1}} \right)$$
$$= \frac{1}{2} \left(2^{\gamma n_{3k+1}} 2^{n - n_{3k+2}} + 2^{n_{3k+1}} \right).$$

One can easily check that the following inequality holds

$$2^{\gamma n_{3k+1}} 2^{n - n_{3k+2}} + 2^{n_{3k+1}} \geq 2^{\frac{\gamma n}{\alpha}}$$

provided $n_{3k+1} < n < n_{3k+2}$. This implies that

$$\varliminf_{n \to \infty} A_n \geq \frac{\gamma \log 2}{-\log \underline{\lambda}} = \frac{\gamma \log 2}{-\alpha \log \overline{\lambda}}.$$

Moreover, if $n = n_{3k+1}$ then

$$\lim_{n \to \infty} A_n = \frac{\log 2}{-\log \overline{\lambda}}.$$

Case 3: $n_{3k+2} \leq n < n_{3k+3}$. We have

$$N(F, \varepsilon) = N_{\text{blue}}(F, \varepsilon) + N_{\text{green}}(F, \varepsilon) = \beta_k 2^{n_{3k+2}} + (1 - \beta_k) 2^n.$$

It is easy to see that for sufficiently large k (for which $\beta_k \leq \frac{1}{2}$),

$$N(F,\varepsilon) \leq 2\left(2^{(\gamma-\alpha)n_{3k+1}}2^{\alpha n_{3k+1}} + 2^n\right) = 2\left(2^n + 2^{\gamma n_{3k+1}}\right).$$

and

$$N(F,\varepsilon) \geq \frac{1}{2}\left(2^{(\gamma-\alpha)n_{3k+1}}2^{\alpha n_{3k+1}} + 2^n\right) = \frac{1}{2}\left(2^n + 2^{\gamma n_{3k+1}}\right).$$

Since $2^n + 2^{\gamma n_{3k+1}} \geq 2^n$ we obtain that

$$\varliminf_{n\to\infty} A_n \geq \frac{\log 2}{-\log\underline{\lambda}}.$$

provided $n_{3k+2} < n < n_{3k+3}$. Moreover, if $n = n_{3k+2}$ then

$$\lim_{n\to\infty} A_n = \frac{\log 2}{-\log\overline{\lambda}}.$$

It follows that $\overline{\dim}_B F \geq \frac{\log 2}{-\log\overline{\lambda}}$. Combining this with Theorem 16.1, we conclude that $\overline{\dim}_B F = \frac{\log 2}{-\log\overline{\lambda}}$. It also follows that $\underline{\dim}_B F \geq \gamma\frac{\log 2}{-\log\underline{\lambda}}$. As we have seen above, $\lim_{k\to\infty} A_{n_{3k+1}} = \gamma\frac{\log 2}{-\log\underline{\lambda}}$. Thus, $\underline{\dim}_B F = \gamma\frac{\log 2}{-\log\underline{\lambda}}$. ∎

We consider another example of a simple geometric construction with rectangles in the plane generated by two affine maps which was introduced by Pollicott and Weiss (see [PoW]). It illustrates that the Hausdorff dimension of the limit set may depend on delicate number-theoretic properties of ratio coefficients while the box dimension is much more robust.

Example 16.3 We begin with two disjoint rectangles Δ_1, $\Delta_2 \subset I$ in the unit square $I = [0,1] \times [0,1]$ given by $\Delta_i = [a_i, a_i + \lambda_2] \times J_i$, $i = 1,2$, where $0 < a_1 \leq a_2 < 1$ and J_1, J_2 are disjoint intervals in the vertical axis of the same length λ_1. We assume that $0 < \lambda_1 \leq \lambda_2 < 1$ and $\lambda_1 < \frac{1}{2}$. Consider the two affine maps $h_i \colon I \to \Delta_i$, $i = 1,2$ that contract the unit square by λ_1 in the vertical direction and by λ_2 in the horizontal direction. See Figure 13. These maps generate a simple self-similar geometric construction with rectangles in the plane with basic sets at step n

$$\Delta_{i_0\ldots i_n} = h_{i_0} \circ h_{i_1} \circ \cdots \circ h_{i_n}(I).$$

Let π_k, $k = 1,2$ denote the projections of the unit square I onto the vertical side, for $k = 1$, and onto the horizontal side, for $k = 2$. Obviously, $\pi_1 \circ h_i(x,y) = a_i + \lambda_2 x$ for any $(x,y) \in I$ and $i = 1,2$. Hence, for any basic set $\Delta_{i_0\ldots i_n}$ the left endpoint of the interval $\pi_1(\Delta_{i_0\ldots i_n})$ is given by

$$\pi_1 \circ h_{i_0} \circ h_{i_1} \circ \cdots \circ h_{i_n}(0,0) = \sum_{k=0}^{n-1}(a_1 + i_{k+1}(a_2 - a_1))\lambda_2^k.$$

Taking the limit when $n \to \infty$ yields

$$
\begin{aligned}
\pi_2(F) &= \left\{ \sum_{k=0}^{\infty} (a_1 + i_{k+1}(a_2 - a_1)) \lambda_2^k : \ (i_0, i_1, \ldots) \in \{0,1\}^{\mathbb{N}} \right\} \\
&= \left\{ \frac{a_1}{1 - \lambda_2} + d \sum_{k=0}^{\infty} i_{k+1} \lambda_2^k : \ (i_0, i_1, \ldots) \in \{0,1\}^{\mathbb{N}} \right\} \\
&\subset \left[\frac{a_1}{1 - \lambda_2}, \frac{a_2}{1 - \lambda_2} \right] \equiv J,
\end{aligned}
\tag{16.5}
$$

where $d = a_2 - a_1$.

Note that if $\lambda_1 = \lambda_2$ then the construction is a Moran simple geometric construction (CM1–CM5). It then follows from Theorem 13.1 that

$$
\dim_H F = \underline{\dim}_B F = \overline{\dim}_B F = -\frac{\log 2}{\log \lambda_2}.
\tag{16.6}
$$

There is a very special — *degenerate* — case when $\pi_1(\Delta_1) = \pi_1(\Delta_2)$ (i.e., the rectangle Δ_1 lies directly above the rectangle Δ_2). It is easy to check that in this case (16.6) still holds.

We now compute the Hausdorff dimension and lower and upper box dimensions of the limit set F of the construction assuming that $\lambda_1 < \lambda_2$ and $d = a_2 - a_1 > 0$.

a) Calculation of Box Dimension.

Lemma 1.

(1) $\underline{\dim}_B F = \overline{\dim}_B F \overset{\text{def}}{=} \dim_B F$.

(2)
$$
\dim_B F = \begin{cases} -\dfrac{\log 2}{\log \lambda_2} & \text{if } 0 < \lambda_2 \leq \tfrac{1}{2}, \\[2mm] -\log \dfrac{2\lambda_2}{\lambda_1} \big/ \log \lambda_1 & \text{if } \tfrac{1}{2} \leq \lambda_2 < 1. \end{cases}
$$

Proof. We first consider the case $0 < \lambda_2 \leq \tfrac{1}{2}$. By virtue of (16.5) the set $\pi_2(F)$ is affinely equivalent to the standard Cantor set

$$
\left\{ \sum_{k=0}^{\infty} i_{k+1} \lambda_2^k : \ (i_0 i_1 \ldots) \in \{0,1\}^{\mathbb{N}} \right\}
$$

after scaling by $d = a_2 - a_1 > 0$ and translating by $\frac{a_1}{1-\lambda_2}$. Consider two new disjoint rectangles

$$
\Delta_1^* = \left[\frac{a_1}{1-\lambda_2}, \frac{a_1}{1-\lambda_2} + \frac{d\lambda_2}{1-\lambda_2} \right] \times J_1, \quad \Delta_2^* = \left[\frac{a_2 - d\lambda_2}{1-\lambda_2}, \frac{a_2}{1-\lambda_2} \right] \times J_2.
$$

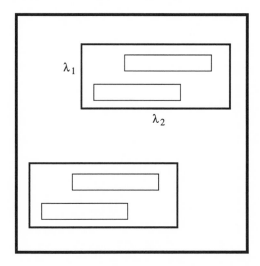

Figure 13. A SELF-SIMILAR CONSTRUCTION WITH RECTANGLES.

It is easy to see that these rectangles have *disjoint* projections into the horizontal line and $\Delta_i^* \subset \Delta_i$, $i = 1, 2$. We consider now the simple self-similar geometric *sub-construction* using the rectangles Δ_1^* and Δ_2^* and the affine maps h_1 and h_2. Obviously, the basic sets of this new construction at step n, i.e., $\Delta_{i_0 \ldots i_n}^*$, satisfy $\Delta_{i_0 \ldots i_n}^* \subset \Delta_{i_0 \ldots i_n}$. Hence, the limit set $F^* \subset F$. This implies that $\underline{\dim}_B F \geq \underline{\dim}_B F^* = -\frac{\log 2}{\log \lambda_2}$. On the other hand, it follows from Theorem 16.1 that $\overline{\dim}_B F \leq -\frac{\log 2}{\log \lambda_2}$.

We turn now to the case $\frac{1}{2} \leq \lambda_2 < 1$. Given n, consider the cover of the limit set F by squares with sides of length λ_1^n such that each rectangle $\Delta_{i_0 \ldots i_n}$ is covered by $\left[\frac{\lambda_2^n}{\lambda_1^n}\right] + 1$ squares aligned in a row. One can see from (16.5) that the projection $\pi_1(F \cap \Delta_{i_0 \ldots i_n})$ contains an interval of length $|J||\pi_1(\Delta_{i_0 \ldots i_n})| = |J|\lambda_2^n = \frac{d\lambda_2^n}{1 - \lambda_2}$. Hence, the proportion of the squares in the cover of $\Delta_{i_0 \ldots i_n}$ required to cover $\Delta_{i_0 \ldots i_n} \cap F$ is at least

$$\frac{|J||\pi_1(\Delta_{i_0 \ldots i_n})|}{\lambda_2^n} = \frac{d}{1 - \lambda_2}.$$

This implies that

$$2^n \frac{\lambda_2^n}{\lambda_1^n} \geq N(F, \lambda_1^n) \geq 2^n \frac{\lambda_2^n}{\lambda_1^n} \frac{1 - \lambda_2}{d}$$

(recall that $N(F, r)$ is the least number of balls of radius r needed to cover F). This implies the desired result. ∎

b) Calculation of Hausdorff Dimension.

Lemma 2. *If* $0 < \lambda_2 < \frac{1}{2}$ *then* $\dim_H F = \dim_B F = -\frac{\log 2}{\log \lambda_2}$.

Proof. First note that $\dim_H(\pi_2(F)) \leq \dim_H F \leq \dim_B F$. As we saw in the proof of Lemma 1 the set $\pi_2(F)$ is the limit set for a simple Moran geometric construction (CM1–CM5) on $[0,1]$ with 2^n basic sets at step n of equal length $\lambda_2^n \frac{d}{1-\lambda_2}$. Hence, $\dim_H(\pi_1(F)) = -\frac{\log 2}{\log \lambda_2}$. The result now follows from Theorem 16.1 and Lemma 1.   ∎

We turn to the case $\frac{1}{2} \leq \lambda_2 < 1$. Following Pollicott and Weiss [PoW], we call a real number $\beta \in [0,1]$ a *GE-number* (after Garsia–Erdős) if there exists a constant $C > 0$ such that for all $x \in [0, +\infty)$

$$\mathrm{card}\left\{ (i_0, \ldots, i_{n-1}) \in \{0,1\}^n : \sum_{r=0}^{n-1} i_r \beta^r \in [x, x + \beta^n) \right\} \leq C \, (2\beta)^n .$$

The following properties of GE-numbers are known (see [PoW]):

(1) no number $0 < \beta < \frac{1}{2}$ is a GE-number;
(2) there exists a *non*-GE-number that is bigger than $\frac{1}{2}$ (for example, the Golden mean, i.e., the positive root of the equation $1 + \frac{1}{\beta} = \frac{1}{\beta^2}$); moreover, the reciprocal of any Pisot–Vijayarghavan number (a root of an algebraic equation whose all conjugates have moduli less than one) is a *non*-GE number [S];
(3) if β is the reciprocal of a root of 2 then it is a GE-number;
(4) almost all numbers on the interval $(\frac{1}{2}, 1)$ are GE-numbers [So];
(5) if $\frac{1}{2} < \beta < 1$, then β is a GE-number if and only if for all sequences $p \in \prod_0^\infty \{0,1\}$ and any $d > 0$ there exists $K > 0$ such that $N_n(p) \leq K(2\beta)^{n-m}$ for any $0 < m < n$, where

$$N_n(p) = \left\{ (i_{m+1} \ldots i_n) \in \{0,1\}^{n-m} : d \left| \sum_{\ell=m+1}^n (p_\ell - i_\ell) \beta^\ell \right| < \beta^n \right\}. \quad (16.7)$$

Lemma 3. *If $\frac{1}{2} < \lambda_2 < 1$ is a GE-number then*

$$\dim_B F = \dim_H F = -\log\left(\frac{2\lambda_2}{\lambda_1}\right) / \log \lambda_1.$$

Proof. Given $r > 0$ choose $n \geq 0$ such that $n = \left[\frac{\log \lambda_2}{\log r}\right] + 1$ and consider a Moran cover $\mathfrak{U}_r = \mathfrak{U}_r(\underline{\lambda})$ constructed in Section 13 with $\underline{\lambda} = (\lambda_2, \lambda_2)$. It is easy to see that this cover consists of all rectangles at step n. Given $x \in F$, we first compute the number $N(x, r)$ of those rectangles at step n that intersect the ball $B(x, r)$. Choose $m = [n \log \lambda_2 / \log \lambda_1]$. Clearly $\lambda_1^m \asymp \lambda_2^n$. Assume that $x \in \Delta_{i_0 \ldots i_n}$ and consider an *asymptotic* square $S(x)$ of dimensions $\lambda_1^m \times \lambda_2^n$ that prolongates the rectangle $\Delta_{i_0 \ldots i_n}$. Let $N_n(x)$ be the number of those rectangles at step n that intersect the square $S(x)$. Obviously, we have $N(x,r) \leq C_1 N_n(x)$, where $C_1 > 0$ is a constant. We now establish an upper estimate for the number $N_n(x)$. First we observe that if a rectangle $\Delta_{j_0 \ldots j_n}$ intersects the square $S(x)$ then $i_0 = j_0, \ldots, i_m = j_m$. We note now that the left endpoint of $\Delta_{i_0 \ldots i_n}$ is $\sum_{\ell=0}^{n}(a_1 + d\,i_\ell)\lambda_2^\ell$ and the left endpoint of $\Delta_{i_0 \ldots i_m j_{m+1} \ldots j_n}$ is $\sum_{\ell=0}^{m}(a_1 + d\,i_\ell)\lambda_2^\ell + \sum_{\ell=m+1}^{n}(a_1 + d\,j_\ell)\lambda_2^\ell$. Hence, for these two rectangles to lie in the same asymptotic square $S(x)$, we should have that

$$d \left| \sum_{\ell=m+1}^{n} (i_\ell - j_\ell)\lambda_2^\ell \right| < \lambda_2{}^n. \tag{16.8}$$

It follows from property (5) of GE-numbers (see (16.7)) that $N_n(x) \le K(2\lambda_2)^{n-m}$. Set $s_\lambda = -\log(\frac{2\lambda_2}{\lambda_1})/\log\lambda_1$. Since the function $s_\lambda \log \lambda_{i_0} = s_\lambda \log \lambda_2$ is constant the corresponding equilibrium measure m_λ is the measure of maximal entropy for the full shift and hence for any basic set $\Delta_{i_0 \ldots i_n}$ we have that $m_\lambda(\Delta_{i_0 \ldots i_n}) = 2^{-n}$. Using (13.5) and following (13.6) we obtain by direct calculation that

$$m_\lambda(B(x,r)) \le K_2 N_n(x) \left(\frac{1}{2}\right)^n \le K_3 r^{s_\lambda},$$

where $K_2 > 0$ and $K_3 > 0$ are constants. The desired result follows now from the uniform mass distribution principle. ∎

c) Non-coincidence of the Hausdorff Dimension and Box Dimension.

We illustrate that the number-theoretic property (16.7) that we have used is not just an artifact of the proof. Consider $\lambda_1 = \frac{1}{2}$. In this very special case the limit set F reduces to the graph of a Weierstrass-like function (modulo a countable set). The dimension of such graphs were studied by several authors. In particular, in [PU], Przytycki and Urbański showed that *if λ_2 is the reciprocal of a PV number then there exist certain configurations (i.e., a choice of numbers a_1 and a_2) such that* $\dim_H F < \dim_B F$. ∎

Chapter 6

Multifractal Formalism

Let f be a dynamical system acting in a domain $U \subset \mathbb{R}^m$ and Z an invariant set. We saw in Chapter 2 that the Hausdorff dimension and box dimension of Z yield information about the geometric (and somehow topological) structure of Z. This information, in fact, may not capture any dynamics. For example, if Z is a periodic orbit then the Hausdorff and box dimensions of Z are zero regardless to whether the orbit is stable, unstable, or neutral.

In order to obtain relevant information about dynamics one should consider not only the geometry of the set Z but also the *distribution* of points on Z under f. In other words one should be interested in how often a given point $x \in Z$ visits a fixed subset $Y \subset Z$ under f. If μ is an f-invariant Borel ergodic measure for which $\mu(Y) > 0$, then for a typical point $x \in Z$ the average number of visits is equal to $\mu(Y)$. Thus, the orbit distribution is completely determined by the measure μ. On the other hand, the measure μ is completely specified by the distribution of a typical orbit.

This fact is widely used in the numerical study of dynamical systems where orbit distributions can easily be generated by a computer. These distributions are, in general, non-uniform and have a clearly visible fine-scaled interwoven structure of *hot* and *cold* spots, i.e., regions where the frequency of visitations is either much greater than average or much less than average respectively (see [GOY] for more details; see Figure 14). For dynamical systems possessing *strange attractors* the computer picture of hot and cold spots reflects the distribution of typical orbits associated with special invariant measures. The latter are naturally interconnected with the geometry of Z and can be used to describe relations between the dynamics on Z and the geometric structure of Z. These measures are often called *natural* measures. If the strange attractor is *hyperbolic*, the corresponding natural measure is the well-known Sinai–Ruelle–Bowen measure. (Definition and properties of these measures can be found in [KH].)

The distribution of hot and cold spots varies with the scale: if a small piece of the invariant set is magnified another picture of hot and cold spots can be seen. In order to obtain a quantitative description of the behavior of hot and cold spots with the scale let us consider a dynamical system f acting on a hypercube K in $\mathbb{R}^m$ and a cover of K by a uniform grid of mesh size r. Let p_i be an average number of visits of a "typical" orbit to a given box B_i of a grid, i.e., $p_i = \mu(B_i)$, where μ is a natural measure. The collection of numbers $\{p_i\}$ determine the distribution of hot and cold spots corresponding to the scale level r. Define **scaling exponents** α_i by $p_i = r^{\alpha_i}$. In the seminal paper [HJKPS], the authors suggested using the limit distribution of numbers α_i when $r \to 0$ as a quantitative

characteristic of the distribution of hot and cold spots. It is intimately related
to the **multifractal structure** of X (see [M2]).

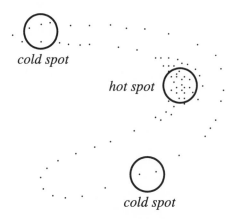

cold spot

hot spot

cold spot

Figure 14. AN ORBIT DISTRIBUTION.

The general concept of multifractality can be formulated as follows (see a
more detailed description in Appendix IV). One can say that the set X has
the multifractal structure if it admits a decomposition — called **multifractal
decomposition** — into subsets that are homogeneous in a sense. Of course, this
is not a rigorous mathematical definition and it depends on how the homogeneity
is interpreted. For example, if $h\colon X \to \mathbb{R}$ is a function then the decomposition of
X into level sets $X_\alpha = \{x \in X : h(x) = \alpha\}$ can be viewed as multifractal with
the sets X_α being homogeneous. There is a multifractal decomposition of X
associated with hot and cold spots of orbit distributions (i.e., the decomposition
that is induced by the invariant measure μ):

$$X = \hat{X} \cup \left(\bigcup_\alpha X_\alpha \right).$$

Here X_α is the set of points where the pointwise dimension takes on the value α
and $\hat{X}$ — the **irregular part** — is the set of points with *no* pointwise dimension. This decomposition can be characterized by the **dimension spectrum
for pointwise dimensions** of the measure μ or $f_\mu(\alpha)$-**spectrum (for dimensions)**, where $f_\mu(\alpha) = \dim_H X_\alpha$. The $f_\mu(\alpha)$-spectrum provides a description of
the fine-scale geometry of the set X (more precisely, the part of it where the
measure μ is concentrated) whose constituent components are the sets X_α (see
Section 18).

The straightforward calculation of the $f_\mu(\alpha)$-spectrum is difficult and one
should relate it to observable properties of the invariant measure μ. For example,
one can use various dimension spectra as most accessible characteristics of orbit

distribution. Among them let us mention the **Rényi spectrum for dimensions** introduced by Tél [Tel]. It is defined as follows:

$$R_q(\mu) = \frac{1}{q-1} \lim_{r \to \infty} \frac{\log \sum_{i=1}^{N} \mu(B_i)^q}{\log r},$$

where $N = N(r)$ is the total number of boxes B_i of the grid with $\mu(B_i) > 0$ (provided the limit exists; see Section 17 and discussion in [V1]).

In [GHP], Grassberger, Hentschel, and Procaccia suggested computing correlations between q-tuples of points in the orbit distribution for $q = 2, 3, \ldots$ (see also [G, GP, HP]). This led to characteristics known as correlation dimensions of order q (see Section 17). These characteristics proved to be experimentally the most accessible and offered a substantial advantage over the other characteristics of dimension type that were used in numerical study of dynamical systems with chaotic behavior. In Section 17 we give a rigorous mathematical substantiation of the fact that the correlation dimension of order q is completely determined by the function $\varphi_q(X, r) = \int_X \mu(B(x, r))^{q-1} d\mu(x)$, $r > 0$ and thus, depends only on the metric on X and the invariant ergodic measure μ, but *not* on the dynamical system itself — the unexpected phenomenon noticed first by Hentschel and Procaccia in [HP]. They also introduced a family of characteristics depending on a *real* parameter $q \geq 0$ (except $q = 1$) of which the correlation dimensions of order q are special cases for integers $q \geq 2$. It is now known as the HP**-spectrum for dimensions** (after Hentschel and Procaccia). The formal definition and a detailed discussion is given in Section 18. For natural measures the HP-spectrum for dimensions can be defined in the following way:

$$HP_q(\mu) = \frac{1}{q-1} \lim_{r \to 0} \frac{\log \inf_{\mathcal{G}} \left\{ \sum_{B(x_i, r) \in \mathcal{G}} \mu(B(x_i, r))^q \right\}}{\log r},$$

where $\mathcal{G}$ is a finite or countable covering of the support of μ by balls of radius ε (provided the limit exists).

A group of physicists in the paper [HJKPS] presented a heuristic argument based on the analogy with statistical mechanics to show that the HP-spectrum for dimensions and the Rényi spectrum for dimensions coincide and that the latter (multiplied by the factor $1 - q$) and the $f_\mu(\alpha)$-spectrum for dimensions form a Legendre transform pair (see Sections 18, 19, and 21; see also Figures 17a and 17b in Chapter 7). Roughly speaking their idea goes as follows (see arguments in [CLP]). Given a grid of size r, consider the partition function

$$Z(q, r) = \sum_{i=1}^{N} \mu(B_i)^q = \sum_{i=1}^{N} e^{-qE_i},$$

where q is the "inverse temperature" and E_i is the "energy" of the element B_i of the grid (the sum is taken over those elements of the grid B_i for which $\mu(B_i)$ is positive). The "free energy" of μ is defined (when it exists) by

$$F(q) = -\lim_{r \to 0} \frac{\log Z(q, r)}{\log r}.$$

The analogy with statistical mechanics is then used to relate the Legendre transform of F (i.e., the function $t \mapsto \inf_q[qt - F(q)]$) to the distribution of the numbers $\mu(B_i)$, i.e., to $f_\mu(\alpha)$.

Once the Legendre transform relation between the two dimension spectra is established, one can compute the delicate and seemingly intractable $f_\mu(\alpha)$-spectrum through the HP-spectrum, since the latter is completely determined by the statistics of a typical trajectory.

In order to convert the above argument into a rigorous mathematical proof one must first establish that the HP-spectrum and the $f_\mu(\alpha)$-spectrum for dimensions are smooth and strictly convex on some intervals (in q and in α respectively). This seems amazing since *a priori* one expects the functions $f_\mu(\alpha)$ and $HP_\mu(q)$ to be only measurable. Furthermore, it is not at all clear whether, even if the measure μ is exact dimensional (i.e., $d_\mu(x)$ exists and is constant, say d, almost everywhere) the pointwise dimension attains *any* values besides d.

In Section 18 we will show that the Hentschel–Procaccia spectrum and the Rényi spectrum for dimensions coincide for any Borel finite measure on $\mathbb{R}^m$. This allows us to set up the concept of a (complete) **multifractal analysis** of dynamical systems as a collection of results which establish smoothness and convexity of these spectra as well as the $f_\mu(\alpha)$-spectrum for dimensions and the Legendre transform relation between them.

Although the multifractal analysis was developed by physicists and applied mathematicians as a tool in numerical study of dynamical systems its significance has not been quite understood. In Section 18 we will demonstrate that dimension spectra — whose study constitutes the multifractal analysis — are Carathéodory dimension characteristics (i.e., they can be introduced within the general Carathéodory construction described in Chapter 1). This alone justifies the study of dimension spectra as part of dimension theory. This also opens new perspectives for the multifractal analysis: one can classify dynamical systems up to an isomorphism that preserves dimension spectra. Such an isomorphism keeps track on stochastic properties of the dynamical system (specified by the invariant measure) as well as dimension properties (of invariant measures or sets) and thus, may be of great importance in describing chaotic dynamical systems (see Appendix IV for more details).

In Chapter 7 we effect a complete multifractal analysis of Gibbs measures for smooth conformal expanding maps and axiom A-diffeomorphisms on surfaces. As a part of this analysis we show that any equilibrium measure on a conformal repeller corresponding to a Hölder continuous function is diametrically regular (see Proposition 21.4). The same result holds for equilibrium measures on conformal locally maximal hyperbolic sets (see Proposition 24.1).

In Section 19 of this chapter we consider another class of examples which includes Gibbs measures on limit sets for a large class of geometric constructions of type (CG1–CG3). Our analysis is based on the dynamical properties of the induced map on the limit set generated by the shift map on the symbolic space. Our main assumption is that this map, being continuous, is expanding and conformal in a weak sense. Whether the induced map satisfies these properties strongly depends on the symbolic dynamics and *its embedding* into Euclidean space, i.e., the gaps between the basic sets. Examples include self-similar geometric constructions and geometric constructions associated with some (classical) Schottky groups (see Section 13), as well as geometric constructions effected by a sequence of similarity maps whose ratio coefficients admit some asymptotic estimates. The multifractal analysis of these "expanding and conformal" geometric constructions is intimately related to the analysis for smooth conformal expanding maps (see Section 21).

17. Correlation Dimension

One can obtain information about a dynamical system that is related to invariant ergodic measures by observing individual trajectories. In [GHP], Grassberger, Hentschel, and Procaccia introduced the notion of correlation dimension in an attempt to produce a characteristic of dynamics that captures information on the global behavior of "typical" trajectories by observing a single one (see also [G]). This trajectory should be typical with respect to an invariant measure.

The formal definition of the correlation dimension is as follows. Let (X, ρ) be a complete separable metric space with metric ρ and $f \colon X \to X$ a continuous map. Given $x \in X$, $n > 0$, and $r > 0$, we define the **correlation sum** as

$$C(x, n, r) = \frac{1}{n^2} \text{ card } \{(i, j) : \rho(f^i(x), f^j(x)) \leq r \text{ and } 0 \leq i, j \leq n\}, \qquad (17.1)$$

where card A denotes the cardinality of the set A. Given a point $x \in X$, we call the quantities

$$\underline{\alpha}(x) = \varliminf_{r \to 0} \lim_{n \to \infty} \frac{\log C(x, n, r)}{\log(1/r)}, \quad \overline{\alpha}(x) = \varlimsup_{r \to 0} \lim_{n \to \infty} \frac{\log C(x, n, r)}{\log(1/r)} \qquad (17.2)$$

the **lower** and **upper correlation dimensions at the point** x respectively. We shall first discuss the existence of the limit as $n \to \infty$. Let μ be an f-invariant Borel probability measure on X. Consider the function

$$\varphi^+(r) = \int_X \mu(D(x, r)) \, d\mu(x),$$

where $D(x, r)$ is the *closed* ball of radius r centered at x. It is clearly nondecreasing in r and hence may have only a finite or countable set of discontinuity points. Thus, it is right-continuous; it is continuous at r if and only if $\mu(S(x, r)) = 0$ for μ-almost every $x \in X$ ($S(x, r)$ is the sphere of radius r centered at x).

Theorem 17.1. [P3, PT] *Assume that μ is ergodic. Then there exists a set $Z \subset X$ of full measure such that for any $\varepsilon > 0$, $R > 0$, and any $x \in Z$ one can find a positive integer $N = N(x, \varepsilon, R)$ for which the inequality*

$$|C(x, n, r) - \varphi^+(r)| \leq \varepsilon$$

holds for every $n \geq N$ *and* $0 < r \leq R$. *In other words,* $C(x, n, r)$ *tends to* $\varphi^+(r)$ *when* $n \to \infty$ *for* μ-*almost every* $x \in X$ *uniformly over* r *for* $0 < r \leq R$.

Proof. For simplicity we provide the proof under the additional assumptions that the function $\varphi^+(r)$ is continuous and the map f is invertible. The general case is considered in [PT]. We first show that given a number $r > 0$, there exists a set $X_r \subset X$ of full measure such that for any $x \in X_r$,

$$\lim_{n \to \infty} C(x, n, r) = \varphi^+(r). \tag{17.3}$$

Given a point $x \in X$, a measurable set $A \subset X$, and integers $n \geq 0$, $m \geq 0$ denote by $N(x, A, n, m)$ the number of points $f^i(x)$, $-m \leq i \leq n$ for which $f^i(x) \in A$. Let us fix a countable collection of closed balls $D_k = D(y_k, r_k)$, $k \geq 0$ forming a basis of the topology in X. We can assume that $\mu(\partial D_k) = 0$ for all $k \geq 0$. Since μ is ergodic there exists a set $Y \subset X$ of full measure such that for any $x \in Y$ and $k \geq 0$,

$$\lim_{n \to \infty} \frac{N(x, D_k, n, m)}{n + m} = \mu(D_k).$$

It follows that given $x \in Y$, $\varepsilon \geq 0$, and $k \geq 0$, there exists $n(x, \varepsilon, k)$ such that for any $n \geq 0$, $m \geq 0$, and $n + m \geq n(x, \varepsilon, k)$,

$$\mu(D_k) - \varepsilon \leq \frac{N(x, D_k, n, m)}{n + m} \leq \mu(D_k) + \varepsilon. \tag{17.4}$$

Let us fix $r > 0$, $\varepsilon > 0$, and $x \in Y$ and choose sequences $x_{k,1}$, $r_{k,1}$ and $x_{k,2}$, $r_{k,2}$ such that $x_{k,1} \to x$, $x_{k,2} \to x$, $r_{k,1} \to r$, $r_{k,2} \to r$, and

$$D(x_{k,1}, r_{k,1}) \subset D(x, r) \subset D(x_{k,2}, r_{k,2}).$$

This implies that

$$\mu(D(x_{k,1}, r_{k,1})) \leq \mu(D(x, r)) \leq \mu(D(x_{k,2}, r_{k,2})). \tag{17.5}$$

Moreover, since $\mu(\partial D_k) = 0$ for all k, one can find $K = K(\varepsilon)$ such that for any $k \geq K(\varepsilon)$ and $i = 1, 2$,

$$|\mu(D(x_{k,i}, r_{k,i})) - \mu(D(x, r))| \leq \varepsilon. \tag{17.6}$$

It follows from (17.4), (17.5), and (17.6) that for any $k \geq K(\varepsilon)$ and $n \geq 0$, $m \geq 0$ with $n + m \geq n(x, \varepsilon, k)$,

$$\frac{N(x, D(x, r), n, m)}{n + m} \leq \frac{N(x, D(x_{k,2}, r_{k,2}), n, m)}{n + m}$$
$$\leq \mu(D(x_{k,2}, r_{k,2})) + \varepsilon \leq \mu(D(x, r)) + 2\varepsilon,$$
$$\frac{N(x, D(x, r), n, m)}{n + m} \geq \frac{N(x, D(x_{k,1}, r_{k,2}), n, m)}{n + m}$$
$$\geq \mu(D(x_{k,1}, r_{k,1})) - \varepsilon \geq \mu(D(x, r)) - 2\varepsilon.$$

Thus, for any $x \in Y$ and $\varepsilon > 0$ there exists $n_1 = n_1(x, \varepsilon) \geq 0$ such that for any $n \geq 0$, $m \geq 0$ with $n + m \geq n_1$,

$$\left| \frac{N(x, D(x, r), n, m)}{n + m} - \mu(D(x, r)) \right| \leq 2\varepsilon. \tag{17.7}$$

Since μ is ergodic and $\mu(D(x, r))$ is a bounded Borel function over $x \in X$ we have by virtue of the Birkhoff ergodic theorem that for μ-almost every $x \in X$,

$$\lim_{n \to \infty} \frac{1}{n} \sum_{i=0}^{n-1} \mu(D(f^i(x), r)) = \int_X \mu(D(y, r)) d\mu(y) = \varphi^+(r).$$

This implies that for any $\varepsilon > 0$ there exist a measurable set $X_{r,\varepsilon}^{(1)} \subset Y$ with $\mu(X_{r,\varepsilon}^{(1)}) \geq 1 - \varepsilon$ and a number $n_2 = n_2(\varepsilon)$ such that for any $x \in X_{r,\varepsilon}^{(1)}$ and $n \geq n_2$,

$$\left| \frac{1}{n} \sum_{i=0}^{n-1} \mu(D(f^i(x), r)) - \varphi^+(r) \right| \leq \varepsilon. \tag{17.8}$$

Moreover, for μ-almost every $x \in X_{r,\varepsilon}^{(1)}$,

$$\lim_{n \to \infty} \frac{N(x, X_{r,\varepsilon}^{(1)}, n, 0)}{n} = \mu(X_{r,\varepsilon}^{(1)}).$$

Therefore, there exists a set $X_{r,\varepsilon}^{(2)} \subset X_{r,\varepsilon}^{(1)}$ with $\mu(X_{r,\varepsilon}^{(2)}) \geq \mu(X_{r,\varepsilon}^{(1)}) - \varepsilon \geq 1 - 2\varepsilon$ and a number $n_3 = n_3(\varepsilon)$ such that for any $x \in X_{r,\varepsilon}^{(2)}$ and $n \geq n_3$,

$$\left| \frac{1}{n} N(x, X_{r,\varepsilon}^{(2)}, n, 0) - 1 \right| \leq 2\varepsilon. \tag{17.9}$$

Let us fix $x \in X_{r,\varepsilon}^{(2)}$ and $n \geq n(\varepsilon) = \max\{n_1, n_2, n_3\}$. Write $n = tn(\varepsilon) + \tilde{n}_\varepsilon$ with $0 \leq \tilde{n}_\varepsilon < n(\varepsilon)$. Denote

$$N_i = N(f^i(x), D(f^i(x), r), n - i, i), \quad i \geq 0.$$

One can rewrite the expression for $C(x, n, r)$ (see (17.1)) in the following form:

$$C(x, n, r) = \frac{1}{n} \sum_{i=0}^{n} \frac{N_i}{n} = \frac{1}{n} \left(\sum_1 \frac{N_i}{n} + \sum_2 \frac{N_i}{n} + \sum_3 \frac{N_i}{n} \right),$$

where the first sum is taken over all i, $0 \leq i \leq tn(\varepsilon)$ with $f^i(x) \in X_{r,\varepsilon}^{(2)}$, the second sum over all i such that $tn(\varepsilon) + 1 \leq i \leq n$, and the third sum over all other i (i.e., those i, $0 \leq i \leq tn(\varepsilon)$ for which $f^i(x) \notin X_{r,\varepsilon}^{(2)}$). Since $N_i \leq n + 1$ this implies that

$$\frac{1}{n} \sum_2 \frac{N_i}{n} \leq \frac{\tilde{n}_\varepsilon}{n} \leq \varepsilon \tag{17.10}$$

if n is sufficiently large. It follows from (17.9) that

$$\frac{1}{n}\sum_3 \frac{N_i}{n} \leq \frac{1}{n}(n - N(x, X_{r,\varepsilon}^{(2)}, n, 0)) \leq 2\varepsilon. \tag{17.11}$$

Let us now estimate the first sum. It follows from (17.8) that

$$\left|\frac{1}{n}\sum_1 \frac{N_i}{r_i} - \varphi^+(r)\right| \leq \left|\frac{1}{n}\sum_1 \frac{N_i}{n} - \frac{1}{n}\sum_{i=0}^{n-1}\mu(D(f^i(x), r))\right| + \varepsilon$$

$$\leq \frac{1}{n}\sum_1 \left|\frac{N_i}{n} - \mu(D(f^i(x), r))\right| \tag{17.12}$$

$$+ \frac{1}{n}\sum_2 \mu(D(f^i(x), r)) + \frac{1}{n}\sum_3 \mu(D(f^i(x), r)) + \varepsilon,$$

where the first, second, and third sums are taken over i as above. Therefore, by (17.7) and (17.9)–(17.11), we find

$$\left|\frac{1}{n}\sum_1 \frac{N_i}{n} - \varphi^+(r)\right| \leq 2\varepsilon + \frac{\tilde{n}_\varepsilon}{n} + 2\varepsilon + \varepsilon \leq 6\varepsilon.$$

This inequality along with (17.10), (17.11), and (17.12), imply (17.3) for all $x \in X_{r,\varepsilon}^{(2)}$, and hence for any $x \in X_r = \cup_{\varepsilon>0}X_{r,\varepsilon}^{(2)}$. Obviously, X_r is a set of full measure.

We now show the uniform convergence of $C(x, n, r)$ to $\varphi^+(r)$ over r on an interval $0 < r \leq R$. Let us fix $\varepsilon > 0$. Since $\varphi^+(r)$ is continuous there exists $\delta > 0$ such that $|\varphi^+(r_1) - \varphi^+(r_2)| \leq \varepsilon$ for any $r_1, r_2 \in (0, R]$ with $|r_1 - r_2| \leq \delta$. Let us fix a countable everywhere dense set $T \subset \mathbb{R}$ and put $Y = \cap_{r\in T}X_r$, where X_r is the set defined in (17.3). Obviously, $\mu(Y) = 1$. Given $r \in \mathbb{R}$, there exist $r_1, r_2 \in T$ satisfying $r_1 < r < r_2$ and $r_2 - r_1 \leq \delta$. It is easy to see that for any $n > 0$,

$$C(x, n, r_1) \leq C(x, n, r) \leq C(x, n, r_2).$$

If n is sufficiently large we have also that

$$C(x, n, r_1) \geq \varphi^+(r_1) - \varepsilon \geq \varphi^+(r) - 2\varepsilon,$$

$$C(x, n, r_2) \leq \varphi^+(r_2) + \varepsilon \leq \varphi^+(r) + 2\varepsilon.$$

These inequalities imply that $|C(x, n, r) - \varphi^+(r)| \leq 2\varepsilon$ for all $0 < r \leq R$ if n is sufficiently large. ∎

We now extend the notion of the correlation dimension by considering correlations of higher order. Namely, given $x \in X$, $n > 0$, $r > 0$, and an integer $q \geq 2$, we define the **correlation sum of order** q (specified by the points $\{f^i(x)\}$, $i = 0, 1, \ldots, n$) by

$$C_q(x, n, r) = \frac{1}{n^q} \text{ card } \{(i_1 \ldots i_q) \in \{0, 1, \ldots, n\}^q :$$

$$\rho(f^{i_j}(x), f^{i_k}(x)) \leq r \text{ for any } 0 \leq j, k \leq q\}.$$

The quantities

$$\underline{\alpha}_q(x) = \frac{1}{q-1}\varliminf_{r \to 0}\lim_{n \to \infty}\frac{\log C_q(x,n,r)}{\log(1/r)},$$

$$\overline{\alpha}_q(x) = \frac{1}{q-1}\varlimsup_{r \to 0}\lim_{n \to \infty}\frac{\log C_q(x,n,r)}{\log(1/r)}$$

are called respectively the **lower** and **upper correlation dimensions of order q at the point** x or **lower** and **upper q-correlation dimensions at** x.

The existence of the limit as $n \to \infty$ is guaranteed by a more general version of Theorem 17.1 which we shall state now. For any $r > 0$ and any integer $q \geq 2$ consider the function

$$\varphi_q^+(r) = \int_X \mu(D(x,r))^{q-1}\, d\mu(x),$$

where μ is an f-invariant Borel probability measure on X (recall that $D(x,r)$ is the *closed* ball of radius r centered at x). The function $\varphi_q^+(r)$ is non-decreasing and hence may have only a finite or countable set of discontinuity points. It is clearly right-continuous.

The following statement is proved by Pesin and Tempelman in [PT] and is an extension of Theorem 17.1 to correlation dimensions of higher order.

Theorem 17.2. *If μ is an ergodic measure then for any integer $q \geq 2$ there exists a set $Z \subset X$ of full measure such that for any $\varepsilon > 0$, $R > 0$, and any $x \in Z$ one can find $N = N(x, \varepsilon, R)$ for which*

$$|C_q(x,n,r) - \varphi_q^+(r)| \leq \varepsilon$$

with arbitrary $n \geq N$ and $0 < r \leq R$.

Remarks.

(1) We stress that the lower and upper q-correlation dimensions do not depend either on the dynamical system f or on the point x for μ-almost every x (provided μ is ergodic). Instead, they are completely specified by the measure μ. This allows us to introduce the notion of q-correlation dimension for any finite Borel measure μ on a complete separable metric space X (see [PT]). Let $Z \subset X$ be a Borel subset of positive measure. Define the **lower** and **upper q-correlation dimensions of the measure** μ **on** Z for $q = 2, 3, \ldots$ by

$$\underline{\mathrm{Cor}}_q(\mu, Z) = \frac{1}{q-1}\underline{\dim}_q Z, \quad \overline{\mathrm{Cor}}_q(\mu, Z) = \frac{1}{q-1}\overline{\dim}_q Z.$$

For $q = 2$ these formulae define the **lower** and **upper correlation dimensions of the measure** μ **on** Z. We also define the **(q, H)-correlation dimension of the measure** μ **on** Z for $q = 2, 3, \ldots$ by

$$\mathrm{Cor}_{q,H}(\mu, Z) = \frac{1}{q-1}\dim_q Z.$$

Consider the function $\varphi_q(r) = \varphi_q(X, r)$ defined by (8.7). We stress that in (8.7) we use *open* balls of radius r centered at points in X. A simple argument (which we leave to the reader) shows that

$$\varliminf_{r \to 0} \frac{\log \varphi_q^+(r)}{\log(1/r)} = \varliminf_{r \to 0} \frac{\log \varphi_q(r)}{\log(1/r)}, \quad \varlimsup_{r \to 0} \frac{\log \varphi_q^+(r)}{\log(1/r)} = \varlimsup_{r \to 0} \frac{\log \varphi_q(r)}{\log(1/r)}.$$

If μ is an invariant ergodic measure for a dynamical system f acting on X it follows that for μ-almost every $x \in X$,

$$\underline{\alpha}_q(x) = \underline{\mathrm{Cor}}_q(\mu, X), \quad \overline{\alpha}_q(x) = \overline{\mathrm{Cor}}_q(\mu, X)$$

(X can be also replaced by any set of full measure). These relations expose the "dimensional" nature of the notions of lower and upper q-correlation dimensions: *they coincide respectively* (up to the constant $\frac{1}{q-1}$) *with the upper and lower q-box dimensions of the space X for $q = 2, 3, \ldots$* (or any other set of full measure).

(2) Example 8.1 shows that the lower and upper q-correlation dimensions of μ may not coincide: namely, for any integer $q \geq 2$ there exists a Borel finite measure μ on $I = [0, 1]$ for which

$$\underline{\mathrm{Cor}}_q(\mu, I) < \overline{\mathrm{Cor}}_q(\mu, I)$$

(and μ is absolutely continuous with respect to the Lebesgue measure on I). One can check that μ is an invariant measure for a continuous map f on I, where $f(x) = \mu([0, x))$. Thus, with respect to this map

$$\underline{\alpha}_q(x) = \underline{\mathrm{Cor}}_q(\mu, I) < \overline{\mathrm{Cor}}_q(\mu, I) = \overline{\alpha}_q(x) \tag{17.13}$$

for μ-almost every $x \in I$. Furthermore, using methods in [K] one can construct a diffeomorphism of a compact surface preserving an absolutely continuous Bernoulli measure with non-zero Lyapunov exponents for which (17.13) holds.

(3) We describe a more general setup for introducing the lower and upper q-correlation dimensions. Namely, given $x, y \in X$, $n > 0$, $r > 0$, and an integer $q \geq 2$, we define the correlation sum of order q (specified by the points $\{f^i(x)\}$ and $\{f^i(y)\}$, $i = 0, 1, \ldots, n$) by

$$C_q(x, y, n, r) = \frac{1}{n^q} \, \mathrm{card} \, \{(i_1 \ldots i_q) \in \{0, 1, \ldots, n\}^q :$$
$$\rho(f^{i_j}(x), f^{i_k}(y)) \leq r \text{ for any } 0 \leq j, k \leq q\}.$$

Consider the direct product space $Y = X \times X$. We call the quantities

$$\underline{\alpha}_q(x, y) = \varliminf_{r \to 0} \lim_{n \to \infty} \frac{\log C_q(x, y, n, r)}{\log(1/r)}, \quad \overline{\alpha}_q(x, y) = \varlimsup_{r \to 0} \lim_{n \to \infty} \frac{\log C_q(x, y, n, r)}{\log(1/r)}$$

the **lower** and **upper** q-**correlation dimensions at the point** $(x, y) \in Y$. One can prove the following statement: *let μ be an f-invariant ergodic Borel* *probability measure and* $\nu = \mu \times \mu$; *then for any integer* $q \geq 2$ *and* ν-*almost every* $(x, y) \in Y$ *the limit*

$$\lim_{n \to \infty} C_q(x, y, n, r) = \varphi_q^+(X, r).$$

exists.

(4) The functions $\varphi_q(X, r)$, $q = 2, 3, \ldots$ admit the following interpretation. Consider the direct product space (Y_q, ρ_q, ν_q), where

$$Y_q = \underbrace{X \times \cdots \times X}_{q \text{ times}}, \quad \nu_q = \underbrace{\mu \times \cdots \times \mu}_{q \text{ times}}$$

and $\rho_q(\bar{x}, \bar{y}) = \sum_{k=1}^{q} \rho(x_k, y_k)$ for $\bar{x} = (x_1, \ldots, x_q)$, $\bar{y} = (y_1, \ldots, y_q)$. Let $\Lambda = \{(x, \ldots, x) \in Y_q : x \in X\}$ be the diagonal. Then for any $r > 0$,

$$\varphi_q(X, r) = \nu_q(U(\Lambda, r)),$$

where $U(\Lambda, r)$ is the r-neighborhood of Λ.

(5) Let μ be a Borel finite measure on $\mathbb{R}^m$ with bounded support. Following Sauer and Yorke [SY] we describe another approach to the notion of correlation dimension of μ based on the potential theoretic method (see Section 7). Define the quantity

$$D(\mu) = \sup\{s : I_s(\mu) < \infty\}, \tag{17.14}$$

where $I_s(\mu)$ is the s-energy of μ (see (7.5)). The "correlation dimension" $D(\mu)$ is interpreted as the supremum of those s for which the measure μ has finite s-energy.

Given a subset $Z \subset \mathbb{R}^m$, one can compute the Hausdorff dimension of Z via the quantity $D(\mu)$. Namely, by the potential principle (see (7.6)),

$$\dim_H Z = \sup_{\mu}\{D(\mu) : \mu(Z) = 1\}. \tag{17.15}$$

A slight modification of the argument by Sauer and Yorke in [SY] shows that

$$\overline{\text{Cor}}_2(\mu, X) = -D(\mu),$$

where X is the support of μ. Indeed, set $D = \underline{\text{Cor}}_2(\mu, X)$. By the definition of the lower correlation dimension for any $\varepsilon > 0$ there exists $r_0 > 0$ such that for any $0 < r \leq r_0$,

$$\varphi_2(X, r) \leq r^{D-\varepsilon}.$$

Given $0 \leq s < D$, set $\varepsilon = (D - s)/2 > 0$. We have that

$$\int_0^\infty \frac{d\varphi_2(X, r)}{r^s} = \int_0^{r_0} \frac{d\varphi_2(X, r)}{r^s} + \int_{r_0}^\infty \frac{d\varphi_2(X, r)}{r^s}$$

$$\leq D\frac{r_0^\varepsilon}{\varepsilon} + \text{constant} < \infty.$$

This implies that $I_s(\mu)$ is finite. On the other hand, let $0 < D < s$. Set $\varepsilon = (s - D)/2 > 0$. By the definition of the lower correlation dimension there exists a sequence of numbers $r_n > 0$ such that

$$\varphi_2(X, r_n) \geq r_n^{D+\varepsilon}.$$

Given $n > 0$, one can find a number $m = m(n) > n$ such that $\varphi_2(X, r_m) < \varphi_2(X, r_n)/2$. We have that

$$\int_{r_m}^{r_n} \frac{d\varphi_2(X, r)}{r^s} \geq r_n^{-s}[\varphi_2(X, r_n) - \varphi_2(X, r_m)]$$

$$\geq \frac{1}{2} r_n^{-s} \varphi_2(X, r_n) \geq \frac{1}{2} r_n^{-s} r_n^{D+\varepsilon} = \frac{1}{2} r_n^{-\varepsilon}.$$

Since r_n can be chosen arbitrarily small there is no upper bound for $I_s(\mu)$.

Sauer and Yorke also studied the problem whether the quantity $D(\mu)$ is preserved under a smooth map. In particular, they proved the following statement: *assume that $D(\mu) \leq p$; then there exists a residual subset $\mathcal{A} \subset C^1(\mathbb{R}^m, \mathbb{R}^p)$* (the space of smooth maps from $\mathbb{R}^m$ to $\mathbb{R}^p$) *such that the equality $D(g_*\mu) = D(\mu)$ holds for every $g \in \mathcal{A}$.* In view of (17.15) a similar result holds for the Hausdorff dimension of a subset $Z \subset \mathbb{R}^m$. On the other hand, it fails in the case of box dimension: there exists a compact subset $Z \subset \mathbb{R}^m$ with $\underline{\dim}_B Z = \overline{\dim}_B Z \stackrel{\text{def}}{=} D \leq p$ such that $\overline{\dim}_B g(Z) < D$ for *every $g \in C^1(\mathbb{R}^m, \mathbb{R}^p)$.*

(6) Let (Y, ν) be a Lebesgue space (ν is a probability measure) and $f: Y \to Y$ a measurable transformation. Let also (X, ρ) be a metric space and $h: Y \to X$ a Borel function called *observable*. In practice, f-orbits may not be accessible but the values $h_n = h(f^n(y))$ for some $y \in Y$ can often be computed. One can analyze the data $\{h_n, \ n = 1, 2, \dots\}$ in the following way. For $q = 2, 3, \dots$ we define the **h-correlation sum of order** q (specified by the data $\{h_n, \ n = 1, 2, \dots\}$) by

$$C_q(\{h_n\}, r) = \frac{1}{n^q} \text{ card } \{(i_1 \dots i_q) \in \{0, 1, \dots, n\}^q :$$

$$\rho(h_{i_j}, h_{i_k}) \leq r \text{ for any } 0 \leq j, k \leq q\}.$$

The quantities

$$\underline{\alpha}_q(\{h_n\}) = \varliminf_{r \to 0} \lim_{n \to \infty} \frac{\log C_q(\{h_n\}, r)}{\log(1/r)}, \quad \overline{\alpha}_q(\{h_n\}) = \varlimsup_{r \to 0} \lim_{n \to \infty} \frac{\log C_q(\{h_n\}, r)}{\log(1/r)}$$

are called respectively the **lower** and **upper q-correlation dimensions specified by the data** $\{h_n, \ n = 0, 1, 2, \dots\}$. One can prove the following statement: *For any $q = 2, 3, \dots$ and ν-almost every $y \in Y$ the limit*

$$\lim_{n \to \infty} C_q(\{h_n\}, r) = \int_X \mu(D(x, r)) d\mu(x)$$

exists where $\mu = h_*\nu$.

We consider the case when $Y \subset \mathbb{R}^m$ is a compact subset, f is a C^1-diffeomorphism of an open domain $U \subset \mathbb{R}^m$ into $\mathbb{R}^m$ for which Y is an invariant set, and $h: U \to \mathbb{R}$ is a smooth function. In [T1], Takens developed a method of reconstructing the Hausdorff dimension and box dimensions of Y using the data $h_n = h(f^n(y))$ for some $y \in Y$. Given an integer $p \geq 1$, define the *delay coordinate map* $G_{f,h,p}: Y \to \mathbb{R}^p$ by

$$G_{f,h,p}(x) = \big(h(x), h(f(x)), \ldots, h(f^{p-1}(x))\big).$$

Takens proved that *for a residual set of C^1-diffeomorphisms of Y and a residual set of functions h the delay coordinate map $G_{f,h,p}$ with $p = 2m$ is a C^1-embedding.* In particular, the Hausdorff dimension and lower and upper box dimensions of Y are preserved under $G_{f,h,p}$. Sauer and Yorke [SY] proved a similar result involving the quantity $D(\mu)$ (see (17.14)): *assume that $D(\mu) < p$ (p is an integer) and that f has at most countably many periodic points; then $D((G_{f,h,p})_*\mu) = D(\mu)$ for a residual set of functions h.*

18. Dimension Spectra: Hentschel–Procaccia, Rényi, and $f(\alpha)$-Spectra; Information Dimension

Hentschel–Procaccia Spectrum for Dimensions

Let μ be a Borel finite measure on $\mathbb{R}^m$ and X the support of μ. We introduce the *HP-spectrum for dimensions* (after Hentschel and Procaccia; see [HP]), specified by the measure μ, as the one-parameter family of pairs of quantities $\big(\underline{HP}_q(\mu), \overline{HP}_q(\mu)\big)$, $q > 1$, where

$$\underline{HP}_q(\mu) = \frac{1}{q-1} \varliminf_{r \to 0} \frac{\log \int_X \mu(B(x,r))^{q-1}\, d\mu(x)}{\log(1/r)},$$

$$\overline{HP}_q(\mu) = \frac{1}{q-1} \varlimsup_{r \to 0} \frac{\log \int_X \mu(B(x,r))^{q-1}\, d\mu(x)}{\log(1/r)}.$$

It follows from (8.10) that

$$\underline{HP}_q(\mu) = \frac{1}{q-1}\underline{\dim}_q X, \qquad \overline{HP}_q(\mu) = \frac{1}{q-1}\overline{\dim}_q X. \tag{18.1}$$

Thus, the HP-spectrum for dimensions is a one-parameter family of characteristics of dimension type: *for every $q > 1$ the quantities $\underline{HP}_q(\mu)$ and $\overline{HP}_q(\mu)$ coincide (up to a normalizing factor $\frac{1}{q-1}$) with the lower and upper q-box dimensions of the set X respectively.*

Equalities (18.1) allow us to rewrite the definition of the HP-spectrum for dimensions in the following way: using (8.5) we obtain that

$$\underline{HP}_q(\mu) = \frac{1}{q-1} \varliminf_{r \to 0} \frac{\log \Delta_{q,\gamma}(X,r)}{\log(1/r)},$$

$$\overline{HP}_q(\mu) = \frac{1}{q-1} \varlimsup_{r \to 0} \frac{\log \Delta_{q,\gamma}(X,r)}{\log(1/r)},$$

where $\gamma > 1$ is an arbitrary number and $\Delta_{q,\gamma}(X, r)$ is defined by (8.6).

As Remark 1 in Section 17 shows for $q = 2, 3, \ldots$ the values $\underline{HP}_q(\mu)$ and $\overline{HP}_q(\mu)$ coincide with the lower and upper correlation dimensions of X, i.e., $\underline{\mathrm{Cor}}_q(\mu, X)$ and $\overline{\mathrm{Cor}}_q(\mu, X)$.

Following Pesin and Tempelman [PT] we introduce the **modified HP-spectrum for dimensions** specified by the measure μ as a one-parameter family of pairs of quantities $\left(\underline{HPM}_q(\mu), \overline{HPM}_q(\mu)\right)$, $q > 1$, where

$$\underline{HPM}_q(\mu) = \frac{1}{q-1} \lim_{\delta \to 0} \sup_Z \varliminf_{r \to 0} \frac{\log \int_Z \mu(B(x,r))^{q-1} d\mu(x)}{\log(1/r)},$$

$$\overline{HPM}_q(\mu) = \frac{1}{q-1} \lim_{\delta \to 0} \sup_Z \varlimsup_{r \to 0} \frac{\log \int_Z \mu(B(x,r))^{q-1} d\mu(x)}{\log(1/r)}$$

and the supremum is taken over all sets $Z \subset X$ with $\mu(Z) \geq 1 - \delta$. One can derive from Theorem 8.4 that for any $q > 1$,

$$\underline{HPM}_q(\mu) = \frac{1}{q-1} \underline{\dim}_q \mu, \quad \overline{HPM}_q(\mu) = \frac{1}{q-1} \overline{\dim}_q \mu.$$

Thus, the modified HP-spectrum for dimensions is a one-parameter family of Carathéodory dimension characteristics specified by the measure μ. It follows from Theorem 9.2 that for any $q > 1$,

$$-\operatorname*{ess\,inf}_{x \in X} \overline{d}_\mu(x) \leq \underline{HPM}_q(\mu) \leq \overline{HPM}_q(\mu) \leq -\operatorname*{ess\,inf}_{x \in X} \underline{d}_\mu(x).$$

In particular, if the measure μ satisfies $\underline{d}_\mu(x) = \overline{d}_\mu(x) = d_\mu(x)$ then

$$\underline{HPM}_q(\mu) = \overline{HPM}_q(\mu) = -\operatorname*{ess\,inf}_{x \in X} d_\mu(x)$$

(compare to (9.4)).

The modified HP-spectrum for dimensions is completely specified by the equivalence class of μ. This was shown in [PT]. We present the corresponding result omitting the proof.

Theorem 18.1. *Let μ_1 and μ_2 be Borel measures on X. If these measures are equivalent then*

$$\underline{HPM}_q(\mu_1) = \underline{HPM}_q(\mu_2), \quad \overline{HPM}_q(\mu_1) = \overline{HPM}_q(\mu_2).$$

It follows from the definitions that for $q > 1$

$$\underline{HPM}_q(\mu) \geq \underline{HP}_q(\mu), \quad \overline{HPM}_q(\mu) \geq \overline{HP}_q(\mu).$$

As Example 8.1 shows there exists a Borel finite measure μ on $[0, p]$ for some $p > 0$ such that

$$\underline{HP}_q(\mu) < \overline{HP}_q(\mu) < \underline{HPM}_q(\mu) = \overline{HPM}_q(\mu) = -1.$$

for all $1 < q \leq Q$, where $Q > 0$ is a given number (the reader can easily check that this holds provided $\beta > \alpha^{\frac{q-1}{q}}$).

Rényi Spectrum for Dimensions

Let $U \subset \mathbb{R}^m$ be an open domain. A finite or countable partition $\xi = \{C_i, \ i \geq 1\}$ of U is called a (β, r)-**grid** for some $0 < \beta < 1$, $r > 0$ if for any $i \geq 0$ one can find a point $x_i \in U$ such that

$$B(x_i, \beta r) \subset C_i \subset B(x_i, r).$$

The Euclidean space $\mathbb{R}^m$ admits (β, r)-grids for every $0 < \beta < 1$ and $r > 0$.

Let μ be a Borel finite measure on $\mathbb{R}^m$ and X the support of μ. We assume that X is contained in an open bounded domain $U \subset \mathbb{R}^m$. Given numbers $q \geq 0$ and $\beta > 0$, we set

$$\underline{\operatorname{Dim}}_q X = \varliminf_{r \to 0} \frac{\log \tilde{\Lambda}_q(X, r)}{\log(1/r)}, \quad \overline{\operatorname{Dim}}_q X = \varlimsup_{r \to 0} \frac{\log \tilde{\Lambda}_q(X, r)}{\log(1/r)},$$

where

$$\tilde{\Lambda}_q(X, r) = \inf_\xi \left\{ \sum_{C_i \in \xi} \mu(C_i)^q \right\}$$

and the infimum is taken over all (β, r)-grids in U (compare to (8.6)). We wish to compare the quantities $\underline{\operatorname{Dim}}_q X, \overline{\operatorname{Dim}}_q X$ with the quantities $\underline{\dim}_q X, \overline{\dim}_q X$. The following result obtained by Guysinsky and Yaskolko [GY] establishes the coincidence of these quantities for $q > 1$ and demonstrates that one can use grids instead of covers in the definition of q-dimension.

Theorem 18.2. *If μ is a Borel finite measure on $\mathbb{R}^m$ with a compact support X then for any $q > 1$,*

$$\underline{\operatorname{Dim}}_q X = \underline{\dim}_q X, \quad \overline{\operatorname{Dim}}_q X = \overline{\dim}_q X.$$

Proof. Let $\xi = \{C_i, \ i \geq 1\}$ be a (β, r)-grid. There are points $\{x_i\}$ such that $B(x_i, \beta r) \subset C_i \subset B(x_i, r)$. Therefore,

$$\sum_{i \geq 1} \mu(C_i)^q = \sum_{i \geq 1} \int_{C_i} \mu(C_i)^{q-1} \, d\mu$$

$$\leq \sum_{i \geq 1} \int_{C_i} \mu(B(x, 2r))^{q-1} \, d\mu = \int_X \mu(B(x, 2r))^{q-1} \, d\mu = \varphi_q(2r).$$

This implies that $\tilde{\Lambda}_q(X, r) \leq \varphi_q(2r)$.

We now prove the opposite inequality. Consider again a (β, r)-grid $\{C_i, i \geq 1\}$ and choose points $\{x_i\}$ such that $B(x_i, \beta r) \subset C_i \subset B(x_i, r)$. We have

$$\varphi_q(r) = \int_X \mu(B(x,r))^{q-1} d\mu = \sum_{i \geq 1} \int_{C_i} \mu(B(x,r))^{q-1} d\mu$$

$$\leq \sum_{i \geq 1} \int_{C_i} \mu(B(x_i, 2r))^{q-1} d\mu \leq \sum_{i \geq 1} \mu(B(x_i, 2r))^q.$$

Therefore,

$$\varphi_q(r) \leq \sum_{i \geq 1} \mu(B(x_i, 2r))^q.$$

Let us fix a ball $B(x_i, 2r)$ and consider the set $D_i = \{x_j : C_j \cap B(x_i, 2r) \neq \varnothing\}$. Note that $D_i \subset B(x_i, 3r)$ and that any two points $x_{j_1}, x_{j_2} \in D_i$ are at least $2\beta r$ apart (recall that balls $B(x_i, \beta r)$ are disjoint).

This implies that there exists $N = N(\beta)$ such that $\operatorname{card} D_i \leq N$ (we stress that N does not depend on r). This implies that any ball $B(x_i, 2r)$ can be covered by at most N elements C_{j_k} of the grid.

We will exploit the following well-known inequality: for any $N > 0$ and $q > 1$ there exists $K = K(N, q) > 0$ such that for any collection of numbers $\{a_i, 0 \leq a_i \leq 1, i = 1, \ldots, N\}$,

$$\left(\sum_{i=1}^N a_i\right)^q \leq K \sum_{i=1}^N a_i^q.$$

We have that for any $i \geq 1$,

$$\mu(B(x_i, 2r))^q \leq K \sum_{k=1}^N \mu(C_{j_k})^q.$$

Now let us fix an element C_i of the grid and consider the set $A_i = \{x_j : B(x_j, 2r) \cap C_i \neq \varnothing\}$. Again, one can see that $A_i \subset B(x_i, 3r)$ and that any two points $x_{j_1}, x_{j_2} \in A_i$ are at least $2\beta r$ apart. Therefore, we conclude that $\operatorname{card} A_i \leq N$ and hence

$$\varphi_q(r) \leq NK \sum_{i \geq 1} \mu(C_i)^q$$

for any (β, r)-grid. This implies that $NK\tilde{\Lambda}_q(X, r) \geq \varphi_q(r)$ and completes the proof of the theorem. ∎

We introduce the **Rényi spectrum for dimensions** (see [Tel]), specified by the measure μ, as a one-parameter family of pairs of quantities $\left(\underline{R}_q(\mu), \overline{R}_q(\mu)\right)$, for $q \geq 0$, $q \neq 1$, where

$$\underline{R}_q(\mu) = \frac{1}{q-1} \underline{\operatorname{Dim}}_q X = \frac{1}{q-1} \varliminf_{r \to 0} \frac{\log \tilde{\Lambda}_q(X, r)}{\log(1/r)},$$

$$\overline{R}_q(\mu) = \frac{1}{q-1} \overline{\operatorname{Dim}}_q X = \frac{1}{q-1} \varlimsup_{r \to 0} \frac{\log \tilde{\Lambda}_q(X, r)}{\log(1/r)}.$$

It follows from Theorem 18.2 that for any $q > 1$,

$$\underline{R}_q(\mu) = \underline{HP}_q(\mu), \quad \overline{R}_q(\mu) = \overline{HP}_q(\mu). \tag{18.3}$$

Information Dimension

Let ξ be a finite partition of X. We define the **entropy of the partition** ξ with respect to μ by

$$H_\mu(\xi) = -\sum \mu(C_\xi) \log \mu(C_\xi),$$

where C_ξ is an element of the partition ξ. Given a number $r > 0$, we set

$$H_\mu(r) = \inf_\xi \left\{ H_\mu(\xi) \colon \operatorname{diam} \xi \leq r \right\},$$

where $\operatorname{diam} \xi = \max \operatorname{diam} C_\xi$.

We define the **lower** and **upper information dimensions** of μ by

$$\underline{I}(\mu) = \varliminf_{r \to 0} \frac{H_\mu(r)}{\log(1/r)}, \quad \overline{I}(\mu) = \varlimsup_{r \to 0} \frac{H_\mu(r)}{\log(1/r)}. \tag{18.4}$$

Obviously, $\underline{I}(\mu) \leq \overline{I}(\mu)$.

Proposition 18.1. [Y2]

(1) $\overline{I}(\mu) \leq \overline{\dim}_B \mu$.

(2) If $\underline{d}_\mu(x) \geq d$ for μ-almost every x then $\underline{I}(\mu) \geq d$.

Proof. Fix $\delta > 0$, $r > 0$ and let $B_1, \ldots, B_N$, $N = N(\delta, r)$ be r-balls that cover a set of measure $\geq 1 - \delta$. Set $\tilde{B}_1 = B_1$ and for $k = 2, 3, \ldots, N$,

$$\tilde{B}_k = B_k \setminus \bigcup_{i=1}^{k-1} \tilde{B}_i.$$

The sets $\tilde{B}_k$ are disjoint and have diameter $\leq r$. Therefore, one can find measurable sets $U_j, j = 1, \ldots, M(r)$ such that together with B_k they comprise a finite measurable partition ξ of X of diameter $\leq r$.

We recall the following inequality: if numbers $0 < t \leq 1$ and $0 < p_k \leq 1$, $k = 1, \ldots, s$ are such that $\sum_{k=1}^{s} p_k = t$ then

$$-\sum_{k=1}^{s} p_k \log p_k \leq -t \log t + t \log s.$$

It follows that

$$H_\mu(r) \leq H_\mu(\xi) \leq \log N(\delta, r) - \delta \log \delta + \delta \log M(r)$$

Therefore,

$$\overline{I}(\mu) \leq \varlimsup_{r \to 0} \frac{\log N(\delta, r)}{\log(1/r)} + \varlimsup_{r \to 0} \frac{-\delta \log \delta + \delta \log M(r)}{\log(1/r)} \leq \overline{\dim}_B \mu + \delta \overline{\dim}_B X.$$

Since $\overline{\dim}_B X$ is finite the first statement follows by letting $\delta \to 0$.

In order to prove the second statement choose $\alpha < d$ and $\delta > 0$. There exists a set $Z \subset X$ with $\mu(Z) \geq 1 - \delta$ and a number $r_0 > 0$ such that $\mu(B(x, r)) \leq r^\alpha$ for any $0 < r \leq r_0$ and $x \in Z$. Fix $r \leq r_0$ and consider a finite partition $\xi = \{C_1, \ldots, C_n\}$ of diameter $\leq r$. Let $\mathcal{B}_1$ be the collection of sets $C_i \in \xi$ that have non-empty intersection with Z and $\mathcal{B}_2$ the collection of other sets $C_i \in \xi$. We have that $\mu(\cup_{C_i \in \mathcal{B}_2} C_i) \leq \delta$. Each $C_i \in \mathcal{B}_1$ is contained in a ball $B(x_i, r)$ centered at some point $x_i \in C_i$ and therefore, has μ-measure $\leq r^\alpha$. Thus,

$$H_\mu(\xi) \geq \sum_{C_i \in \mathcal{B}_1} \mu(C_i) \log \mu(C_i) \geq \frac{1 - 2\delta}{r^\alpha}(-r^\alpha \log r^\alpha) = (1 - 2\delta)\alpha \log(1/r).$$

It follows that

$$\frac{H_\mu(\xi)}{\log(1/r)} \geq (1 - 2\delta)\alpha.$$

This implies the second statement.   ∎

As an immediate consequence of Proposition 18.1 and Theorem 7.1 we obtain the following claim.

Proposition 18.2. *Let μ be a finite Borel measure on $\mathbb{R}^m$. Assume that μ is exact dimensional (see Section 7). Then $\underline{I}(\mu) = \overline{I}(\mu) = \dim_H \mu$.*

It is conjectured that for "good" measures

$$\underline{I}(\mu) = \lim_{q \to 1, q > 1} \underline{R}_q(\mu) = \lim_{q \to 1, q > 1} \underline{HP}_q(\mu),$$

$$\overline{I}(\mu) = \lim_{q \to 1, q > 1} \overline{R}_q(\mu) = \lim_{q \to 1, q > 1} \overline{HP}_q(\mu).$$

Below we will show that this conjecture holds for equilibrium measures corresponding to Hölder continuous functions on conformal repellers for smooth expanding maps (see Section 21) and on basic sets for "conformal" axiom A diffeomorphisms (see Section 24).

$f(\alpha)$-Spectrum for Dimensions

We now introduce another dimension spectrum in order to describe the distribution of values of pointwise dimension of a measure. Let μ be a Borel finite measure on $\mathbb{R}^m$ and X the support of μ. There is the multifractal decomposition of X associated with the pointwise dimension of μ:

$$X = \hat{X} \cup \left(\bigcup_\alpha X_\alpha\right),$$

where
$$\hat{X} = \{x \in X : \underline{d}_\mu(x) < \overline{d}_\mu(x)\}$$
is the **irregular part** and the level sets X_α are defined by

$$X_\alpha = \{x \in X : \underline{d}_\mu(x) = \overline{d}_\mu(x) \overset{\text{def}}{=} d_\mu(x) = \alpha\}. \tag{18.5}$$

In order to characterize this multifractal decomposition quantitatively we introduce the **dimension spectrum for pointwise dimensions** of the measure μ or $f_\mu(\alpha)$-**spectrum (for dimensions)** by

$$f_\mu(\alpha) = \dim_H X_\alpha.$$

We first consider those values α for which $\mu(X_\alpha) > 0$.

Theorem 18.3. *If $\mu(X_\alpha) > 0$ then $\dim_H X_\alpha = \alpha$.*

Proof. The statement immediately follows from Theorems 7.1 and 7.2. ∎

If $\mu(X_\alpha) = 0$ the set X_α may still have positive Hausdorff dimension and hence be "observable" from a physical point of view. Thus, the $f_\mu(\alpha)$-spectrum can be used to describe the fine-scale structure of the measure μ provided the function $d_\mu(x)$ exists on an "observable" set of points in X.

An invariant ergodic measure μ supported on a repeller of a smooth expanding map (see definition in Section 20) or on a hyperbolic set of an axiom A diffeomorphism (see definition in Section 22) can be shown to be exact dimensional (see Theorems 21.2 and 24.2). This means that the function $d_\mu(x)$ exists and is constant on a set of full measure. Therefore, $\mu(X_\alpha) = 0$ for all values α but one. *A priori*, it is not clear at all why the sets X_α are not empty for all values α except this special one. Even if these sets are not empty, it seems unclear from a point of view of the classical measure theory that the function $f_\mu(\alpha)$ would behave "nicely" and provide meaningful information about the measure μ. Furthermore, Barreira and Schmeling [BS] showed that for any equilibrium measure (corresponding to a Hölder continuous function) for a smooth expanding map or for an axiom A diffeomorphism the set $\hat{X}$ is *observable*; moreover, it has *full* Hausdorff dimension (see Appendix IV for details; some very special cases should be excluded).

Despite this it was conjectured in [HJKPS] that for "good" measures, which are invariant under dynamical systems, the function $f_\mu(\alpha)$ is correctly defined on some interval $[\alpha_1, \alpha_2]$, real analytic and convex. This was supported by a computer simulation of some dynamical systems: a typical "computer made" graph of the function $f_\mu(\alpha)$ is drawn on Figure 17b in Chapter 7.

Furthermore, a strong connection between the HP-spectrum and $f_\mu(\alpha)$-spectrum was discovered to support an important role that these spectra play in numerical study of dynamical systems: heuristic arguments (based on an analogy with statistical mechanics; see introduction to this chapter) show that the HP-spectrum for dimensions (multiplied by $q-1$) and the $f_\mu(\alpha)$-spectrum for dimensions form a Legendre transform pair.

The rigorous mathematical study of the above multifractal decomposition and $f_\mu(\alpha)$-spectrum is based upon constructing a one-parameter family of equilibrium measures ν_α, $\alpha \in [\alpha_1, \alpha_2]$ which are supported on X_α (i.e., $\nu_\alpha(X_\alpha) = 1$) and have full dimension (i.e., $\dim_H \nu_\alpha = \dim_H X_\alpha$). We will construct such families of measures for multifractal decomposition generated by equilibrium measures invariant under conformal dynamical systems of hyperbolic type (see Chapter 7).

We will also reveal extremely complicated "bizarre" structure of this multifractal decomposition: each set X_α is everywhere dense in X and so is the set $\hat{X}$. This also shows that the Hausdorff dimension in the definition of the $f_\mu(\alpha)$-spectrum cannot be replaced by the box dimension.

Finally, we will observe that the $f_\mu(\alpha)$-spectrum is **complete**, i.e., for any α outside the interval $[\alpha_1, \alpha_2]$, the set X_α is empty.

19. Multifractal Analysis of Gibbs Measures on Limit Sets of Geometric Constructions

In this section we will show how to compute the Hentschel–Procaccia spectrum, Rényi spectrum, and $f(\alpha)$-spectrum for dimensions of equilibrium measures supported on the limit sets of some geometric constructions (CG1–CG3) in $\mathbb{R}^m$ (see Section 13). We will also undertake a complete multifractal analysis of these measures.

Continuous Expanding Maps

Let $X \subset \mathbb{R}^m$ be a compact set. We say that a continuous map $f: X \to X$ is **expanding** if f is a local homeomorphism and f satisfies Condition (15.9), i.e., there exist constants $b \geq a > 1$ and $r_0 > 0$ such that

$$B(f(x), ar) \subset f(B(x, r)) \subset B(f(x), br) \qquad (19.1)$$

for every $x \in X$ and $0 < r < r_0$.

Without loss of generality we may assume that for any $x \in X$ the map f restricted to the ball $B(x, r_0)$ is a homeomorphism. Note that a continuous expanding expanding map is (locally) bi-Lipschitz.

We recall that a **Markov partition** for an expanding map f is a finite cover $\mathcal{R} = \{R_1, \ldots, R_p\}$ of X by elements (called **rectangles**) such that:

(a) each rectangle R_i is the closure of its interior $\operatorname{int} R_i$;
(b) $\operatorname{int} R_i \cap \operatorname{int} R_j = \varnothing$ unless $i = j$;
(c) each $f(R_i)$ is a union of rectangles R_j.

An expanding map has Markov partitions of arbitrary small diameters. Let us fix a Markov partition $\mathcal{R}$. It generates a symbolic model of the map f by a subshift of finite type (Σ_A^+, σ), where $A = (a_{ij})$ is the transfer matrix of the Markov partition, namely, $a_{ij} = 1$ if $\operatorname{int} R_i \cap f^{-1}(\operatorname{int} R_j) \neq \varnothing$ and $a_{ij} = 0$ otherwise. This gives the coding map $\chi: \Sigma_A^+ \to X$ such that

$$\chi(\omega) = \bigcap_{n \geq 0} h(R_{i_n}), \text{ for } \omega = (i_0 i_1 \ldots) \qquad (19.2)$$

(where h is an appropriate branch of f^{-n}) and the following diagram

$$
\begin{array}{ccc}
\Sigma_A^+ & \xrightarrow{\ \sigma\ } & \Sigma_A^+ \\
\chi \downarrow & & \downarrow \chi \\
X & \xrightarrow{\ f\ } & X
\end{array}
$$

is commutative. Under the coding map the cylinder sets $C_{i_0 \ldots i_n} \subset \Sigma_A^+$ correspond to the **basic sets** in X generated by the Markov partition

$$
R_{i_0 \ldots i_n} = R_{i_0} \cap f^{-1}(R_{i_1} \cap f^{-1}(\cdots \cap f^{-1}(R_{i_n}) \ldots)). \tag{19.3}
$$

The map χ is Hölder continuous and injective on the set of points whose trajectories never hit the boundary of any element of the Markov partition. The pullback by χ of any Hölder continuous function on X is a Hölder continuous function on Σ_A^+.

Let f be a continuous expanding map of a compact set $X \subset \mathbb{R}^m$. We obtain effective estimates for the Hausdorff dimension and box dimension of X following the approach suggested by Barreira in [Bar2]. Let $\mathcal{R} = \{R_1, \ldots, R_p\}$ be a Markov partition of X of a small diameter ε (which should be less than the number r_0 in (19.1)) and (Σ_A^+, σ) the symbolic representation of X by a subshift of finite type generated by $\mathcal{R}$. The Markov partition allows one to view X as the basic set for a geometric construction (modeled by the subshift of finite type) whose basic sets are defined by (19.3) and whose induced map is f. We will extend the approach, developed in Section 15 for expanding induced maps, to arbitrary continuous expanding maps, and we will apply the non-additive version of thermodynamic formalism to compute the Hausdorff dimension and box dimension of X. Fix a number $k > 0$. For each $\omega = (i_0 i_1 \ldots) \in \Sigma_A^+$ and $n \geq 1$ define numbers

$$
\begin{aligned}
\underline{\lambda}_k(\omega, n) &= \inf\left\{ \frac{\|f^n(x) - f^n(y)\|}{\|x - y\|} \right\}, \\
\overline{\lambda}_k(\omega, n) &= \sup\left\{ \frac{\|f^n(x) - f^n(y)\|}{\|x - y\|} \right\},
\end{aligned} \tag{19.4}
$$

where the infimum and the supremum are taken over all distinct $x, y \in X \cap R_{i_0 \ldots i_{n+k}}$ (compare to (15.7)). Note that since f is expanding Condition (15.8) holds. Define two sequences of functions on Σ_A^+

$$
\underline{\varphi}^{(k)} = \{\underline{\varphi}_n^{(k)}(\omega) = -\log \overline{\lambda}_k(\omega, n)\}, \quad \overline{\varphi}^{(k)} = \{\overline{\varphi}_n^{(k)}(\omega) = -\log \underline{\lambda}_k(\omega, n)\}. \tag{19.5}
$$

Consider Bowen's equations

$$
\overline{CP}_{\Sigma_A^+}(s\underline{\varphi}^{(k)}) = 0, \quad P_{\Sigma_A^+}(s\overline{\varphi}^{(k)}) = 0. \tag{19.6}
$$

One can show that for any sufficiently large k, any $\omega \in \Sigma_A^+$, and $n \geq 1$,

$$
a^n \leq \underline{\lambda}_k(\omega, n) \leq \overline{\lambda}_k(\omega, n) \leq b^n
$$

(see the proof of Theorem 15.5). Thus, Condition (A2.23) (see Appendix II) holds. By Theorem A2.6 this ensures the existence of unique roots of Bowen's equations (19.6). We denote these roots by $\underline{s}^{(k)}$ and $\overline{s}^{(k)}$ respectively. Clearly, $\underline{s}^{(k)} \leq \overline{s}^{(k)}$. The following result establishes dimension estimates for X.

Proposition 19.1. [Bar2] *Assume that f is topologically mixing. Then*

$$\sup_{k \geq 0} \underline{s}^{(k)} \leq \dim_H X \leq \underline{\dim}_B X \leq \overline{\dim}_B X \leq \inf_{k \geq 0} \overline{s}^{(k)}.$$

We say that a continuous expanding map f is **quasi-conformal** if there exist numbers $C > 0$ and $k > 0$ such that for each $\omega \in \Sigma_A^+$ and $n \geq 0$ Condition (15.11) holds. The following result is similar to Theorem 15.5.

Proposition 19.2. [Bar2] *Assume that f is topologically mixing. Then for any open set $U \subset X$ and all sufficiently large k,*

$$\dim_H(U \cap X) = \underline{\dim}_B(U \cap X) = \overline{\dim}_B(U \cap X) = \underline{s}^{(k)} = \overline{s}^{(k)} \overset{\text{def}}{=} s$$

and s is a unique number satisfying

$$\lim_{n \to \infty} \frac{1}{n} \log \sum_{\substack{(i_0 \ldots i_n) \\ \Sigma_A^+\text{-admissible}}} \left(diam\, (X \cap R_{i_0 \ldots i_n}) \right)^s = 0.$$

Weakly-conformal Maps

We say that a map f of a compact set X is **weakly-conformal** if there exist a Hölder continuous function $a(x)$ with $|a(x)| > 1$ on X and positive constants C_1, C_2, and r_0 such that for any two points $x, y \in X$ and any integer $n \geq 0$ we have: if $\rho(f^k(x), f^k(y)) \leq r_0$ for all $k = 0, 1, \ldots n$ then

$$C_1 \rho(x, y) \prod_{k=0}^{n} |a(f^k(x))|^{-1} \leq \rho(f^n(x), f^n(y))$$

$$\leq C_2 \rho(x, y) \prod_{k=0}^{n} |a(f^k(x))|^{-1}. \tag{19.7}$$

Obviously, a weakly-conformal map is continuous, expanding, and quasi-conformal and Proposition 19.2 applies.

We now discuss the diametrically regular property of equilibrium measures corresponding to Hölder continuous functions for weakly-conformal maps (see Condition (8.15)). If f were a smooth map this property would follow from Proposition 21.4. The proof in the general case is a modification of arguments in the proof of this proposition and is omitted.

Proposition 19.3. *Let f be a weakly-conformal map of a compact set X and φ a Hölder continuous function on X. Then any equilibrium measure corresponding to φ is diametrically regular.*

Te diametrically regular property of an equilibrium measure demonstrates a close connection between the structure of X induced by this measure and metric structure of X induced by the metric ρ. This property plays an important role in effecting a multifractal analysis of the measure.

Multifractal Analysis of Equilibrium Measures for Weakly-conformal Maps

From now on we assume that the map f is topologically mixing. Let φ be a Hölder continuous function on X and $\nu = \nu_\varphi$ an equilibrium measure corresponding to φ. Note that since f is topologically mixing the measure ν_φ is unique and is ergodic (in fact it is a Bernoulli measure). Define the function ψ such that $\log \psi = \varphi - P_X(\varphi)$. Clearly ψ is a Hölder continuous function on X such that $P_X(\log \psi) = 0$ and ν is a unique equilibrium measure for $\log \psi$.

We denote by m a unique equilibrium measure corresponding to the function $x \mapsto -s \log |a(x)|$ on X, where s is the unique root of Bowen's equation

$$P_X(-s \log |a(x)|) = 0.$$

Define the one-parameter family of functions φ_q, $q \in (-\infty, \infty)$ on X by

$$\varphi_q(x) = -T(q) \log |a(x)| + q \log \psi(x),$$

where $T(q)$ is chosen such that $P_X(\varphi_q) = 0$. One can show that for every $q \in \mathbb{R}$ there exists only one number $T(q)$ with the above property. It is obvious that the functions φ_q are Hölder continuous on X.

The following statement effects the complete multifractal analysis of equilibrium measures corresponding to Hölder continuous functions for weakly-conformal maps. Its proof uses the diametrically regular property of these measures (see Proposition 19.3) and is a slight modification of arguments in the proof of Theorem 21.1 where we consider the case of smooth expanding conformal maps.

Theorem 19.1. [PW2] *Let f be a topologically mixing weakly-conformal map on X. Then for any Hölder continuous function φ on X we have*

(1) *the pointwise dimension $d_\nu(x)$ exists for ν-almost every $x \in X$ and*

$$d_\nu(x) = -\frac{\int_X \log \psi(x)\, d\nu(x)}{\int_X \log |a(x)|\, d\nu(x)};$$

(2) *the function $T(q)$ is real analytic for all $q \in \mathbb{R}$; $T(0) = \dim_H F$ and $T(1) = 0$; $T'(q) \leq 0$ and $T''(q) \geq 0$ (see Figure 17a in Chapter 7);*

(3) *the function $\alpha(q) = -T'(q)$ takes on values in an interval $[\alpha_1, \alpha_2]$, where $0 \leq \alpha_1 = \alpha(\infty) \leq \alpha_2 = \alpha(-\infty) < \infty$; moreover, $f_\nu(\alpha(q)) = T(q) + q\alpha(q)$ (see Figure 17b in Chapter 7);*

(4) *for any $q \in \mathbb{R}$ there exists a unique equilibrium measure ν_q supported on the set $X_{\alpha(q)}$, i.e., $\nu_q(X_{\alpha(q)}) = 1$ (the sets X_α are defined by (18.5) with respect to the measure ν) and $\underline{d}_{\nu_q}(x) = \overline{d}_{\nu_q}(x) = T(q) + q\alpha(q)$ for ν_q-almost every $x \in X_{\alpha(q)}$;*

(5) *if $\nu \neq m$ then the functions $f_\nu(\alpha)$ and $T(q)$ are strictly convex and form a Legendre transform pair (see Appendix V);*

(6) *the ν-measure of any open ball centered at points in X is positive and for any $q \in \mathbb{R}$ we have*

$$T(q) = -\lim_{r \to 0} \frac{\log \inf_{\mathcal{G}_r} \sum_{B \in \mathcal{G}_r} \nu(B)^q}{\log r},$$

where the infimum is taken over all finite covers $\mathcal{G}_r$ of X by open balls of radius r. In particular, for $q > 1$,

$$\frac{T(q)}{1-q} = \underline{HP}_q(\nu) = \overline{HP}_q(\nu) = \underline{R}_q(\nu) = \overline{R}_q(\nu).$$

The following statement is an immediate corollary of Theorem 19.1 (see Statement 1).

Proposition 19.4. *Any equilibrium measure corresponding to a Hölder continuous function for a topologically mixing weakly-conformal map f on X is exact dimensional.*

In fact, this result holds for an arbitrary (not necessarily equilibrium) ergodic measure for a weakly-conformal map. The proof is quite similar to the proof of Theorem 21.3 (which deals with the smooth case).

Induced Maps for Geometric Constructions (CG1–CG3): Multifractal Analysis of Equilibrium Measures

Consider a geometric construction (CG1–CG3) in $\mathbb{R}^m$ (see Section 13) and assume that it is modeled by a transitive subshift of finite type (Σ_A^+, σ). Let F be the limit set.

Since we require the separation condition (CG3) the coding map χ is a homeomorphism and we can consider the induced map $G = \chi \circ \sigma \circ \chi^{-1}$ on the limit set F. It is a local homeomorphism since the geometric constructions we consider are modeled by a subshift of finite type (Σ_A^+, σ). Moreover, if one builds a geometric construction modeled by an arbitrary symbolic system (Q, σ) with the expanding induced map on the limit set then $\sigma|Q$ must be topologically conjugate to a subshift of finite type. This follows from a result of Parry [Pa]. Therefore, the induced map G is expanding if and only if it satisfies Condition (15.9).

Note that the placement of basic sets of a geometric construction, whose induced map is expanding, cannot be arbitrary and must satisfy the following special property: there are constants $C_1 > 0$ and $C_2 > 0$ such that for every $x \in F$ and any $r > 0$ there exists $n > 0$ and basic sets $\Delta^{(1)}, \ldots, \Delta^{(m)}$ (with $m = m(x, r)$) for which

$$B(x, C_1 r) \cap F \subset \bigcup_{1 \leq k \leq m} \Delta^{(k)} \cap F \subset B(x, C_2 r) \cap F.$$

We assume that the induced map is weakly-conformal. Note that in this case it is also quasi-conformal (see Condition (15.11)). *We conjecture that one can*

*build a geometric construction modeled by the full shift with disjoint basic sets
such that the induced map is quasi-conformal but not weakly-conformal.*

Theorem 19.1 can be used to effect the complete multifractal analysis of
equilibrium measures, corresponding to Hölder continuous functions, supported
on the limit sets of geometric constructions (CG1–CG3) with weakly-conformal
induced map. We apply this result to a special class of self-similar geometric
constructions (see Section 13). Recall that this means that the basic sets $\Delta_{i_0 \ldots i_n}$
are given by

$$\Delta_{i_0 \ldots i_n} = h_{i_0} \circ \cdots \circ h_{i_n}(D),$$

where $h_1, \ldots, h_p \colon D \to D$ are conformal affine maps, i.e., $\|h_i(x) - h_i(y)\| =
\lambda_i \|x - y\|$ with $0 < \lambda_i < 1$ and $x, y \in D$ (the unit ball in $\mathbb{R}^m$). Assuming that the
basic sets Δ_i, $i = 1, \ldots, p$ are disjoint, one can easily see that the induced map G
on the limit set F is weakly-conformal (with $a(x) = \lambda_{i_0}^{-1}$, where $\chi(x) = (i_0 i_1 \ldots)$).
Thus, Theorem 19.1 applies. For Bernoulli measures this result was obtained by
several authors who used various methods (see, for example, [CM], [EM], [F1],
[O], [Ri]).

We describe a more general class of geometric constructions to which Theorem 19.1 applies. Namely, consider geometric constructions build up by p sequences of bi-Lipschitz contraction maps $h_i^{(n)} \colon D \to D$ such that

$$\Delta_{i_0 \ldots i_n} = h_{i_0}^{(0)} \circ h_{i_1}^{(1)} \circ \cdots \circ h_{i_n}^{(n)}(D)$$

and for any $x, y \in D$,

$$\underline{\lambda}_i^{(n)} \operatorname{dist}(x, y) \leq \operatorname{dist}(h_i^{(n)}(x), h_i^{(n)}(y)) \leq \overline{\lambda}_i^{(n)} \operatorname{dist}(x, y),$$

where $0 < \underline{\lambda}_i^{(n)} \leq \overline{\lambda}_i^{(n)} < 1$ (see [PW2]). We assume that the following *asymptotic
estimates* hold: there exist $0 < \lambda_i < 1$ such that

$$\left| \frac{\underline{\lambda}_i^{(n)}}{\lambda_i} - 1 \right| \leq e^{-n}, \quad \left| \frac{\overline{\lambda}_i^{(n)}}{\lambda_i} - 1 \right| \leq e^{-n}. \tag{19.8}$$

One can check that the induced map G is weakly-conformal (with $a(x) = \lambda_{i_0}^{-1}$,
where $\chi(x) = (i_0 i_1 \ldots)$) and Theorem 19.1 applies.

Example 25.2 shows that there exists a geometric construction produced by
three sequences of bi-Lipschitz contraction maps which do *not* satisfy the asymptotic estimates (19.8). Although the basic sets at each step of the construction
are disjoint it does *not* admit the multifractal analysis described by Theorem
19.1 (otherwise, by Proposition 19.3 *any* equilibrium measure corresponding to
a Hölder continuous function on the limit set F of this construction would be
exact dimensional which is false for the geometric construction described in this
example).

It is still an open problem in dimension theory whether one can effect the
complete multifractal analysis of Gibbs measures supported on the limit set of
a Moran geometric construction (CM1–CM5) modeled by a transitive subshift

of finite type (for some results in this direction see [PW2, LN]). Notice that by Theorem 15.4, the pointwise dimension of such a measure exists (and is constant) almost everywhere.

Remarks.

(1)   One can show that if $\nu = m_\lambda$ then $T(q) = (1 - q)s$ (thus, $T(q)$ is a *linear* function) and $f_\nu(\alpha)$ is the δ-function (i.e., $f_\nu(s) = s$ and $f_\nu(\alpha) = 0$ for all $\alpha \neq s$; see Remark 1 in Section 21; this case was studied by Lopes [Lo]).

(2)   The graphs of functions $T(q)$ and $f(\alpha)$ are shown on Figures 17a and 17b in Chapter 7. Note that the function $f(\alpha)$ is defined on the interval $[\alpha_1, \alpha_2]$, where

$$\alpha_1 = - \lim_{q \to +\infty} T'(q), \quad \alpha_2 = - \lim_{q \to -\infty} T'(q).$$

It attains its maximal value s at $\alpha = \alpha(0)$. Furthermore, $f(\alpha(1)) = \alpha(1)$ is the common value of the lower and upper information dimensions of ν (see Section 21) and $f'(\alpha(1)) = 1$. We also note that the $f(\alpha)$-spectrum is **complete**, i.e., for any α outside the interval $[\alpha_1, \alpha_2]$ the corresponding set X_α is empty (see Section 21).

(3)   Consider again a self-similar geometric construction modeled by the full shift σ and assume that ν is the Bernoulli measure defined by the vector $(p_1, \ldots, p_r)$, where $0 < p_k < 1$ and $\sum_{k=1}^{r} p_k = 1$. It is easily seen from Theorem A2.8 (see Appendix II) that

$$P_{\Sigma_p^+}\left(\log(\lambda_{i_0}^{T(q)} p_{i_0}^q)\right) = 0$$

is equivalent to

$$\sum_{k=1}^{r} \lambda_k^{T(q)} p_k^q = 1.$$

Chapter 7

Dimension of Sets and Measures Invariant under Hyperbolic Systems

In this Chapter we study the Hausdorff dimension and box dimension of sets invariant under smooth dynamical systems of hyperbolic type. This includes repellers for expanding maps and basic sets for Axiom A diffeomorphisms. The reader who is not quite familiar with these notions, can find all the necessary definitions and brief description of basic results relevant to our study in this chapter. For more complete information we refer the reader to [KH].

We recover two major results in the area: Ruelle's formula for the Hausdorff dimension of conformal repellers (see Theorem 20.1) and Manning and Mc-Cluskey's formula for the Hausdorff dimension of two-dimensional locally maximal hyperbolic sets (see Theorem 22.2). Our approach differs from the original ones and is a manifestation of our general Carathéodory construction (see Chapter 1) and the dimension interpretation of the thermodynamic formalism: it systematically exploits the "dimension" definition of the topological pressure described in Chapter 4 and Appendix II. This unifies and simplifies proofs and reveals a non-trivial relation between the topological pressure of some special functions on the invariant set and the Hausdorff dimension of this set.

Furthermore, we use Markov partitions to lay down a deep analogy between conformal repellers (as well as two-dimensional locally maximal hyperbolic sets) and the limit sets for Moran-like geometric constructions. We then apply methods developed in Chapter 5 (see Theorem 13.1) to study dimension of these invariant sets. In particular, this allows us to strengthen the results of Ruelle and of Manning and McCluskey by including the box dimension of repellers and hyperbolic sets into consideration. We also cover the case of multidimensional *conformal* hyperbolic sets.

The crucial feature of the dynamics, to which our methods can be applied, is its conformality. In the non-conformal case (multidimensional non-conformal repellers and hyperbolic sets) the approach, based upon the non-additive version of the thermodynamic formalism, allows us to obtain sharp dimension estimates. We stress that in this case the Hausdorff dimension and box dimension may not agree.

We describe the most famous examples of repellers (including hyperbolic Julia sets, repellers for one-dimensional Markov maps, and limit sets for reflection groups; see Section 20) and hyperbolic sets (including Smale horseshoes and Smale–Williams solenoids; see Section 23). We also provide a brief exposition of recent results on the Hausdorff dimension of a class of three-dimensional solenoids by Bothe and on the Hausdorff dimension and box dimension of attractors for generalized baker's transformations by Alexander and Yorke, by Falconer, and by Simon (see Section 23). These results illustrate some new methods of study that have been recently developed, as well as reveal some obstructions in studying the Hausdorff dimension in non-conformal and multidimensional cases.

A significant part of the chapter is devoted to the recent innovation in the dimension theory of dynamical systems — the multifractal analysis of equilibrium measures (corresponding to Hölder continuous functions) supported on conformal repellers (see Theorem 21.1) or two-dimensional locally maximal hyperbolic sets (see Theorem 24.1; in fact, we cover more general multidimensional conformal hyperbolic sets). The first rigorous multifractal analysis of measures invariant under smooth dynamical systems with hyperbolic behavior was carried out by Collet, Lebowitz, and Porzio in [CLP] for a special class of measures invariant under some one-dimensional Markov maps. Lopes [Lo] studied the measure of maximal entropy for a hyperbolic Julia set. Pesin and Weiss [PW2] effected a complete multifractal analysis of equilibrium measures for conformal repellers and conformal Axiom A diffeomorphisms. In this chapter we follow their approach. Simpelaere [Si] used another approach, which is based on large deviation theory, to effect a multifractal analysis of equilibrium measures for Axiom A surface diffeomorphisms.

There are two main by-products of our multifractal analysis of equilibrium measures on conformal repellers and hyperbolic sets. The first one is that these measures are exact dimensional (see Theorems 21.3 and 24.2; in Chapter 8 we extend these results to arbitrary hyperbolic measures). The second one is the complete description of the dimension spectrum for Lyapunov exponents for expanding maps on conformal repellers (see Theorem 21.4) and diffeomorphisms on locally maximal conformal hyperbolic sets (see Theorem 24.3 and also Appendix IV). This spectrum provides important additional information on the deviation of Lyapunov exponents from the mean value given by the Multiplicative Ergodic Theorem.

20. Hausdorff Dimension and Box Dimension of Conformal Repellers for Smooth Expanding Maps

Repellers for Smooth Expanding Maps

Let $\mathcal{M}$ be a smooth Riemannian manifold and $f\colon \mathcal{M} \to \mathcal{M}$ a $C^{1+\alpha}$-map. Let J be a compact subset of $\mathcal{M}$ such that $f(J) = J$. We say that f is **expanding** on J and J is a **repeller** if

(a) there exist $C > 0$ and $\lambda > 1$ such that $\|df_x^n v\| \geq C\lambda^n \|v\|$ for all $x \in J$, $v \in T_x\mathcal{M}$, and $n \geq 1$ (with respect to a Riemannian metric on $\mathcal{M}$);

(b) there exists an open neighborhood V of J (called a *basin*) such that $J = \{x \in V : f^n(x) \in V \text{ for all } n \geq 0\}$

Obviously, f is a local homeomorphism, i.e., there exists $r_0 > 0$ such that for every $x \in J$ the map $f|B(x, r_0)$ is a homeomorphism onto its image. Thus, f is expanding as a continuous map (see Section 19 and Condition (19.1)).

We recall some facts about expanding maps. A point $x \in \mathcal{M}$ is called *non-wandering* if for each neighborhood U of x there exists $n \geq 1$ such that $f^n(U) \cap U \neq \varnothing$. We denote by $\Omega(f)$ the set of all non-wandering points of f. It is a closed f-invariant set.

The **Spectral Decomposition Theorem** claims that the set $\Omega(f)$ can be decomposed into finitely many disjoint closed f-invariant subsets, $\Omega(f) = J_1 \cup \cdots \cup J_m$, such that $f \mid J_i$ is topologically transitive. Moreover, for each i there exist a number n_i and a set $A_i \subset J_i$ such that the sets $f^k(A_i)$ are disjoint for $0 \leq k < n_i$, their union is the set J_i, $f^{n_i}(A_i) = A_i$, and the map $f^{n_i} \mid A_i$ is topologically mixing (see [KH] for more details).

From now on we will assume that the map f is topologically mixing. This is just a technical assumption that will allow us to simplify proofs. In view of the Spectral Decomposition Theorem our results can be easily extended to the general case (with some obvious modifications).

An expanding map f has Markov partitions of arbitrarily small diameter (see definition of the Markov partition in Section 19; see also [R2]). Let $\mathcal{R} = \{R_1, \ldots, R_p\}$ be a Markov partition for f. It generates a symbolic model of the repeller by a subshift of finite type (Σ_A^+, σ). Namely, consider the basic sets $R_{i_0 \ldots i_n}$ defined by (19.3) and the coding map $\chi \colon \Sigma_A^+ \to J$ defined by (19.2). Then the following diagram

$$
\begin{array}{ccc}
\Sigma_A^+ & \xrightarrow{\ \sigma\ } & \Sigma_A^+ \\
\chi \downarrow & & \downarrow \chi \\
J & \xrightarrow{\ f\ } & J
\end{array}
$$

is commutative. We remind the reader that the map χ is Hölder continuous and injective on the set of points whose trajectories never hit the boundary of any element of the Markov partition.

We need the following well-known estimates of the Jacobian of $C^{1+\alpha}$ expanding maps.

Proposition 20.1.

(1) *Let ψ be a Hölder continuous function on J such that $\psi(x) \geq c > 0$. There exist positive constants $L_1 = L_1(\psi)$ and $L_2 = L_2(\psi)$ such that for any $(n+1)$-tuple $(i_0 \ldots i_n)$ and any $x, y \in R_{i_0 \ldots i_n}$,*

$$
L_1 \leq \prod_{k=0}^{n} \frac{\psi(f^k(x))}{\psi(f^k(y))} \leq L_2.
$$

(2) *Let ψ be a Hölder continuous function on J such that $\psi(x) \geq c > 0$. There exist positive constants $L_1 = L_1(\psi)$ and $L_2 = L_2(\psi)$ such that for any $n > 0$, any branch h of f^{-n}, and any points $x \in J$, $y \in h(B(x, r_0))$ Statement 1 holds.*

(3) *There exist positive constants L_1 and L_2 such that for any $(n+1)$-tuple $(i_0 \ldots i_n)$ and any $x, y \in R_{i_0 \ldots i_n}$,*

$$
L_1 \leq \frac{|\mathrm{Jac}\, f^n(x)|}{|\mathrm{Jac}\, f^n(y)|} \leq L_2,
$$

where $\mathrm{Jac}\, f^n$ denotes the Jacobian of f^n.

Proof. Let $\beta > 0$ be the Hölder exponent and $C_1 > 0$ the Hölder constant of ψ. Then

$$\prod_{k=0}^{n} \frac{\psi(f^k(x))}{\psi(f^k(y))} \leq \prod_{k=0}^{n} (1 + \frac{C_1}{C} |f^k(x) - f^k(y)|^\beta).$$

By the expanding property we find that

$$|f^k(x) - f^k(y)| \leq C_2 \lambda^{k-n} |f^n(x) - f^n(y)|,$$

where $C_2 > 0$ is a constant. Therefore,

$$\prod_{k=0}^{n} \frac{\psi(f^k(x))}{\psi(f^k(y))} \leq \prod_{k=0}^{n-1} \left(1 + C_1 C_2^\alpha (\lambda^\alpha)^k\right).$$

This implies the upper bound. The lower bound follows by interchanging x and y. This completes the proof of the first statement. The second statement can be proved in a similar fashion. The third statement follows by applying the first one to the function $\psi(x) = \operatorname{Jac} f(x)$ which is Hölder continuous since f is of class $C^{1+\alpha}$. ∎

A Markov partition $\mathcal{R} = \{R_1, \ldots, R_p\}$ allows one to set up a complete analogy between limit sets of geometric constructions (CG1–CG2) and repellers of expanding maps by considering the sets $R_{i_0 \ldots i_n}$ as basic sets. Namely,

$$J = \bigcap_{n \geq 0} \bigcup_{(i_0 \ldots i_n)} R_{i_0 \ldots i_n},$$

where the union is taken over all admissible $(n + 1)$-tuples $(i_0 \ldots i_n)$. By the Markov property every basic set $R_{i_0 \ldots i_n} = h(R_{i_n}) \cap R_{i_0}$ for some branch h of f^{-n+1}.

A smooth map $f : \mathcal{M} \to \mathcal{M}$ is called **conformal** if for each $x \in X$ we have $df_x = a(x) \operatorname{Isom}_x$, where Isom_x denotes an isometry of $T_x \mathcal{M}$ and $a(x)$ is a scalar. A smooth conformal map f is expanding if $|a(x)| > 1$ for every point $x \in \mathcal{M}$. The repeller J for a conformal expanding map is called a **conformal repeller**. Note that a smooth conformal expanding map is weakly conformal (as a continuous expanding map; see Condition 19.7). The converse is not true in general: there exists a C^∞-map which is *not* conformal in the above sense but is weakly-conformal (see [Bar2]).

Conformal repellers can be viewed as limit sets for Moran-like geometric constructions with non-stationary ratio coefficients since the basic sets $R_{i_0 \ldots i_n}$ satisfy Condition (B2).

Proposition 20.2.

(1) *Each basic set $R_{i_0 \ldots i_n}$ contains a ball of radius $\underline{r}_{i_0 \ldots i_n}$ and is contained in a ball of radius $\overline{r}_{i_0 \ldots i_n}$.*

(2) *There exist positive constants K_1 and K_2 such that for every basic set $R_{i_0 \ldots i_n}$ and every $x \in R_{i_0 \ldots i_n}$,*

$$K_1 \prod_{j=0}^{n} |a(f^j(x))|^{-1} \leq \underline{r}_{i_0 \ldots i_n} \leq \overline{r}_{i_0 \ldots i_n} \leq K_2 \prod_{j=0}^{n} |a(f^j(x))|^{-1}. \tag{20.1}$$

Proof. Since f is conformal and expanding on J we have for every $x \in J$,

$$\|df_x^n\| = \prod_{j=0}^{n} |a(f^j(x))| = |\operatorname{Jac} f^n(x)|.$$

This fact and the third statement of Proposition 20.1 imply that for every $x \in R_{i_0 \ldots i_n}$,

$$\operatorname{diam} R_{i_0 \ldots i_n} \leq \operatorname{diam} R_{i_n} \times \max_{y \in R_{i_n}} \|dh_y\| = \operatorname{diam} R_{i_n} \times \max_{y \in R_{i_n}} |\operatorname{Jac} h(y)|$$

$$= \operatorname{diam} R_{i_n} \left(\frac{\max_{y \in R_{i_n}} |\operatorname{Jac} h(y)|}{|\operatorname{Jac} h(f^n(x))|} \right) |\operatorname{Jac} h(f^n(x))| \leq C_1 \prod_{j=0}^{n} |a(f^j(x))|^{-1},$$

where h is some branch of f^{-n} and $C_1 > 0$ is a constant. Since each R_j is the closure of its interior we have

$$\operatorname{diam} R_{i_0 \ldots i_n}(x) \geq \operatorname{diam} R_{i_n} \times \min_{y \in R_{i_n}} \|dh_y\| = \operatorname{diam} R_{i_n} \times \min_{y \in R_{i_n}} |\operatorname{Jac} h(y)|$$

$$= \operatorname{diam} R_{i_n} \left(\frac{\min_{y \in R_{i_n}} |\operatorname{Jac} h(y)|}{|\operatorname{Jac} h(f^n(x))|} \right) |\operatorname{Jac} h(f^n(x))| \geq C_2 \prod_{j=0}^{n} |a(f^j(x))|^{-1},$$

where $C_2 > 0$ is a constant. This completes the proof of Proposition 20.2. ∎

We use the analogy with geometric constructions to define a **Moran cover** of the conformal repeller. It allows us to build up an "optimal" cover for computing the Hausdorff dimension and lower and upper box dimensions of the repeller. Given $r > 0$ and a point $\omega \in \Sigma_A^+$, let $n(\omega)$ denote the unique positive integer such that

$$\prod_{k=0}^{n(\omega)} |a(\chi(\sigma^k(\omega)))|^{-1} > r, \quad \prod_{k=0}^{n(\omega)+1} |a(\chi(\sigma^k(\omega)))|^{-1} \leq r. \tag{20.2}$$

It is easy to see that $n(\omega) \to \infty$ as $r \to 0$ uniformly in ω. Fix $\omega \in \Sigma_A^+$ and consider the cylinder set $C_{i_0 \ldots i_{n(\omega)}} \subset \Sigma_A^+$. We have that $\omega \in C_{i_0 \ldots i_{n(\omega)}}$. Furthermore, if $\omega' \in C_{i_0 \ldots i_{n(\omega)}}$ and $n(\omega') \leq n(\omega)$ then

$$C_{i_0 \ldots i_{n(\omega)}} \subset C_{i_0 \ldots i_{n(\omega')}}.$$

Let $C(\omega)$ be the largest cylinder set containing ω with the property that $C(\omega) = C_{i_0 \ldots i_{n(\omega'')}}$ for some $\omega'' \in C(\omega)$ and $C_{i_0 \ldots i_{n(\omega')}} \subset C(\omega)$ for any $\omega' \in C(\omega)$. The sets $C(\omega)$ corresponding to different $\omega \in \Sigma_A^+$ either coincide or are disjoint. We denote these sets by $C^{(j)}$, $j = 1, \ldots, N_r$. There exist points $\omega_j \in \Sigma_A^+$ such that $C^{(j)} = C_{i_0 \ldots i_{n(\omega_j)}}$. These sets comprise a disjoint cover of Σ_A^+ which we denote by $\mathfrak{U}_r$ and call a Moran cover. The sets $R^{(j)} = \chi(C^{(j)})$, $j = 1, \ldots, N_r$ may overlap along their boundaries. They comprise a cover of J (which we will denote by the same symbol $\mathfrak{U}_r$ if it does not cause any confusion). We have that $R^{(j)} = R_{i_0 \ldots i_{n(x_j)}}$ for some $x_j \in J$.

Let $Q \subset \Sigma_A^+$ be a (not necessarily invariant) subset. One can repeat the above arguments to construct a Moran cover of the set Q. It consists of cylinder sets $C^{(j)}$, $j = 1, \ldots, N_r$ for which there exist points $\omega_j \in Q$ such that $C^{(j)} = C_{i_0 \ldots i_{n(\omega_j)}}$ and the intersection $C^{(j)} \cap C^{(i)} \cap Q$ is empty if $j \neq i$ (while the intersection $C^{(j)} \cap C^{(i)}$ may *not* be empty). We denote this cover by $\mathfrak{U}_{r,Q}$.

Moran covers have a property that plays a crucial role in studying the Hausdorff dimension and box dimension of the conformal repeller J. Namely, given a point $x \in J$ and a sufficiently small $r > 0$, the number of basic sets $R^{(j)}$ in a Moran cover $\mathfrak{U}_r$ that have non-empty intersection with the ball $B(x, r)$ is bounded from above by a number M, which is independent of x and r. Analogously to Moran-like geometric constructions (CPW1–CPW4) we call this number a **Moran multiplicity factor**.

In order to explain this property of a Moran cover let $\hat{r} = \max\{\operatorname{diam} R_i : i = 1, \ldots, p\}$. Since the sets R_i are the closure of their interiors there exists a number $0 < r_1 \leq \hat{r}$ such that each R_i contains a ball of radius r_1. By Proposition 20.2 each basic set $R^{(j)}$ in the Moran cover contains a ball of radius Cr, where $C > 0$ is a constant independent of r and j. This implies the desired property of the Moran cover.

Examples of Conformal Expanding Maps

(1) **Rational Maps** [CG]. Let $R: \widehat{\mathbb{C}} \to \widehat{\mathbb{C}}$ be a rational map of degree ≥ 2, where $\widehat{\mathbb{C}}$ denotes the Riemann sphere. The map R, being holomorphic, is clearly conformal. The **Julia set** J of R is the closure of the set of repelling periodic points of R (recall that a periodic point p of period m is repelling if $|(R^m)'(p)| > 1$). One says that R is a hyperbolic rational map (and that J is a hyperbolic Julia set) if the map R is expanding on J (i.e., it satisfies Conditions (1)–(2) in the definition of smooth expanding map with respect to the spherical metric on $\widehat{\mathbb{C}}$). It is known that the map $z \mapsto z^2 + c$ is hyperbolic provided $|c| < \frac{1}{4}$ (see Figure 15). It is conjectured that there is a dense set of hyperbolic quadratic maps.

(2) **One-Dimensional Markov Maps** [Ra]. Assume that there exists a finite family of disjoint closed intervals $I_1, I_2, \ldots I_p \subset I$ and a map $f : \bigcup_j I_j \to I$ such that

(a) for every j, there is a subset $P = P(j)$ of indices with $f(I_j) = \bigcup_{k \in P} I_k$ (mod 0);

(b) for every $x \in \cup_j \operatorname{int} I_j$, the derivative of f exists and satisfies $|f'(x)| \geq \alpha$ for some fixed $\alpha > 0$;

(c) there exists $\lambda > 1$ and $n_0 > 0$ such that if $f^m(x) \in \bigcup_j \operatorname{int} I_j$, for all $0 \leq m \leq n_0 - 1$ then $|(f^{n_0})'(x)| \geq \lambda$.

Let $J = \{x \in I : f^n(x) \in \bigcup_{j=1}^p I_j \text{ for all } n \in \mathbb{N}\}$. The set J is a repeller for the map f. It is conformal because the domain of f is one-dimensional (see Figure 16).

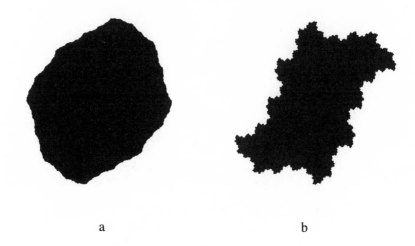

a

b

Figure 15. THE BOUNDARIES OF THESE "BLACK SPOTS" ARE
JULIA SETS FOR THE POLYNOMIAL $z^2 + c$ WITH
a) $c = -\frac{1}{10} + \frac{1}{5}i$ and b) $c = \frac{1}{10} - \frac{1}{2}i$.

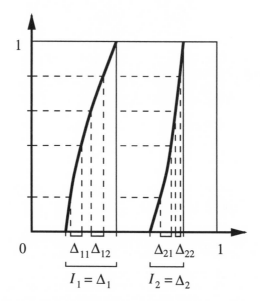

Figure 16. A ONE-DIMENSIONAL MARKOV MAP.

(3) **Conformal Toral Endomorphisms.** This is a map of multidimensional
torus defined by a diagonal matrix $(k, \ldots, k)$, where k is an integer and $|k| > 1$.

(4) **Induced Maps.** Let $h_1, \ldots, h_p \colon D \to D$ be conformal affine maps of the unit ball D in $\mathbb{R}^m$. Assume that the sets $h_i(D)$ are disjoint. Consider the self-similar construction generated by these maps and modeled by the full shift (or a subshift of finite type; see Section 13). Define the map G on $\bigcup_i \mathrm{int} h_i(D)$ by $G(x) = h_i^{-1}(x)$ if $x \in \mathrm{int} h_i(D)$. Clearly, G is a smooth conformal expanding map and the repeller for G is the limit set of the geometric construction (i.e., G is a *smooth extension* of the induced map).

Similar result holds for the induced map on the limit set of the geometric construction generated by reflection groups (see Section 15).

Hausdorff Dimension and Box Dimension of Conformal Repellers

We compute the Hausdorff dimension and box dimension of a conformal repeller. Let $f \colon \mathcal{M} \to \mathcal{M}$ be a $C^{1+\alpha}$-conformal expanding map with a conformal repeller J. Let m be a unique equilibrium measure corresponding to the Hölder continuous function $-s \log|a(x)|$ on $\mathcal{M}$, where s is the unique root of Bowen's equation

$$P_J(-s \log|a|) = 0$$

(see Appendix II).

Theorem 20.1.

(1) $\dim_H J = \underline{\dim}_B J = \overline{\dim}_B J = s$; *moreover*

$$s = \frac{h_m(f)}{\int_J \log|a(x)| \, dm(x)}.$$

(2) *The s-Hausdorff measure of J is positive and finite; moreover, it is equivalent to the measure m.*

(3) $s = \dim_H m$, *in other words, the measure m is an invariant measure of full Carathéodory dimension (see Section 5).*

In [R2], Ruelle proved that the Hausdorff dimension of the conformal repeller J of a $C^{1+\alpha}$-map is given by the root of Bowen's equation $P_J(-s \log|a|) = 0$. He also showed that the s-Hausdorff measure of J is positive and finite. In [F4], Falconer showed that the Hausdorff and box dimensions of J coincide. One can derive Theorem 20.1 from a more general result for continuous weakly-conformal expanding maps (see [Bar2] and Section 19). In particular, this shows that Theorem 20.1 holds for C^1-conformal expanding maps. This result was also established by Gatzouras and Peres [GaP]; see also Takens [T2] for a particular case. We provide here an independent and straightforward proof of Theorem 20.1 which is similar to the proof of Theorem 13.1 (where we dealt with limit sets of geometric constructions (CPW1–CPW4)). ∎

Proof of the theorem. Set $d = \dim_H J$. We first show that $s \leq d$. Fix $\varepsilon > 0$. By the definition of the Hausdorff dimension there exists a number $r > 0$ and a cover of J by balls B_ℓ, $\ell = 1, 2, \ldots$ of radius $r_\ell \leq r$ such that

$$\sum_\ell r_\ell^{d+\varepsilon} \leq 1. \tag{20.3}$$

Let $\mathcal{R} = \{R_1, \ldots, R_p\}$ be a Markov partition for f. For every $\ell > 0$ consider a Moran cover $\mathfrak{U}_{r_\ell}$ of J and choose those basic sets from the cover that intersect B_ℓ. Denote them by $R_\ell^{(1)}, \ldots, R_\ell^{(m(\ell))}$. Note that $R_\ell^{(j)} = R_{i_0 \ldots i_{n(\ell,j)}}$ for some $(i_0 \ldots i_{n(\ell,j)})$. By Proposition 20.2 and (20.2) it follows that for every $x \in R_\ell^{(j)}$,

$$K_1 \prod_{k=0}^{n(\ell,j)} |a(f^k(x))|^{-1} \leq \underline{r}_{i_0 \ldots i_{n(l,j)}} \leq \overline{r}_{i_0 \ldots i_{n(l,j)}} \leq C_1 r_\ell, \qquad (20.4)$$

where $C_1 > 0$ is a constant independent of ℓ and j. By the property of the Moran cover we conclude that $m(\ell) \leq M$, where $M > 0$ is a Moran multiplicity factor (which is independent of ℓ). The sets $\{R_\ell^{(j)}, \ j = 1, \ldots, m(\ell), \ \ell = 1, 2, \ldots\}$ comprise a cover $\mathcal{G}$ of J, and the corresponding cylinder sets $C_\ell^{(j)} = C_{i_0 \ldots i_{n(l,j)}}$ comprise a cover of Σ_A^+. By (20.3) and (20.4) we have

$$\sum_{R_\ell^{(j)} \in \mathcal{G}} \sup_{x \in R_\ell^{(j)}} \prod_{k=0}^{n(\ell,j)} |a(f^k(x))|^{-d-\varepsilon} = \sum_\ell \sum_{j=1}^{m(\ell)} \sup_{x \in R_\ell^{(j)}} \prod_{k=0}^{n(\ell,j)} |a(f^k(x))|^{-d-\varepsilon}$$

$$\leq M \left(\frac{C_1}{K_1}\right)^{d+\varepsilon} \sum_\ell r_\ell^{d+\varepsilon} \leq M \left(\frac{C_1}{K_1}\right)^{d+\varepsilon}.$$

Given a number $N > 0$, choose r so small that $n(\ell, j) \geq N$ for all ℓ and j. We now have that for any $n > 0$ and $N > n$,

$$M(\Sigma_A^+, 0, \varphi, \mathcal{U}_n, N) \leq \sum_{R_\ell^{(j)} \in \mathcal{G}} \exp \left(\sup_{\omega \in C_\ell^{(j)}} \sum_{k=0}^{n(\ell,j)} \varphi(\sigma^k(\omega)) \right)$$

$$= \sum_{R_\ell^{(j)} \in \mathcal{G}} \sup_{x \in R_\ell^{(j)}} \prod_{k=0}^{n(\ell,j)} |a(f^k(x))|^{-d-\varepsilon} \leq M \left(\frac{C_1}{K_1}\right)^{d+\varepsilon},$$

where $M(\Sigma_A^+, 0, \varphi, \mathcal{U}_n, N)$ is defined by (A2.19) (see Appendix II) with $\alpha = 0$ and

$$\varphi(\omega) = -(d + \varepsilon) \log |a(\chi(\omega))|.$$

This implies that

$$P_{\Sigma_A^+} \left(-(d + \varepsilon) \log |a \circ \chi| \right) \leq 0,$$

where $P_{\Sigma_A^+}$ is the topological pressures on Σ_A^+. Hence, by Theorem A2.5 (see Appendix II), $s \leq d + \varepsilon$. Since this inequality holds for all ε we conclude that $s \leq d$.

Denote $\bar{d} = \overline{\dim}_B J$. We now proceed with the upper estimate and show that $\bar{d} \leq s$. Fix $\varepsilon > 0$. By the definition of the upper box dimension (see Section 6) there exists a number $r = r(\varepsilon) > 0$ such that $N(J, r) \geq r^{\varepsilon - \bar{d}}$ (recall that $N(J, r)$ is the least number of balls of radius r needed to cover the set J).

Let $\mathcal{R} = \{R_1, \ldots, R_r\}$ be a Markov partition for f and $\mathfrak{U}_r$ a Moran cover of J by basic sets $R^{(j)} = R_{i_0 \ldots i_{n(x_j)}}$, $j = 1, \ldots, N_r$. The sets $C^{(j)} = \chi^{-1}(R^{(j)}) = C_{i_0 \ldots i_{n(\omega_j)}}$ (where $\omega_j = \chi^{-1}(x_j)$) comprise a Moran cover of Σ_A^+ (recall that χ is the coding map generated by the Markov partition). Repeating arguments in the proof of Theorem 13.1 one can show that there exist $A > 0$ and a positive integer N such that for any sufficiently small r,

$$\text{card } \{j : n(\omega_j) = N\} \geq r^{2\varepsilon - \bar{d}}.$$

Consider an arbitrary cover $\mathcal{G}$ of Σ_A^+ by cylinder sets $C_{i_0 \ldots i_N}$. It follows that

$$\sum_{C_{i_0 \ldots i_N} \in \mathcal{G}} \sup_{x \in R_\ell^{(j)}} \prod_{k=0}^{N} |a(f^k(x))|^{-\bar{d}+2\varepsilon} \geq \sum_{j : n(j) = N} \sup_{x \in R_\ell^{(j)}} \prod_{k=0}^{n(x_j)} |a(f^k(x))|^{-\bar{d}+2\varepsilon}$$

$$\geq \sum_{j : n(j) = N} \left(\frac{r}{A}\right)^{\bar{d} - 2\varepsilon} \geq A^{2\varepsilon - \bar{d}} r^{\bar{d} - 2\varepsilon} r^{2\varepsilon - \bar{d}} \geq C_2,$$

where $C_2 > 0$ is a constant. We now have that for any $n > 0$ and $N > n$,

$$R(\Sigma_A^+, 0, \varphi, \mathcal{U}_n, N) = \sum_{C_{i_0 \ldots i_N} \in \mathcal{G}} \exp \left(\sup_{\omega \in C_{i_0 \ldots i_N}} \sum_{k=0}^{N} \varphi(\sigma^k(\omega)) \right)$$

$$= \sum_{C_{i_0 \ldots i_N} \in \mathcal{G}} \sup_{x \in R_\ell^{(j)}} \prod_{k=0}^{N} |a(f^k(x))|^{-\bar{d}+2\varepsilon} \geq C_2,$$

where $R(\Sigma_A^+, 0, \varphi, \mathcal{U}_n, N)$ is defined by (A2.19′) (see Appendix II) with $\alpha = 0$ and

$$\varphi(\omega) = -(\bar{d} - 2\varepsilon) \log |a(\chi(\omega))|.$$

By Theorem 11.5 this implies that

$$\overline{CP}_J\big(-(\bar{d} - 2\varepsilon) \log |a|\big) = P_J\big(-(\bar{d} - 2\varepsilon) \log |a|\big) = P_{\Sigma_A^+}\big(-(\bar{d} - 2\varepsilon) \log |a \circ \chi|\big) \geq 0$$

and hence $\bar{d} - 2\varepsilon \leq s$. Since this inequality holds for all ε we conclude that $\bar{d} \leq s$. The last equality in Statement 1 follows from the variational principle (see Theorem A2.1 in Appendix II). This completes the proof of the first statement.

We now prove the other two statements. Note that

$$P_J(-s \log |a|) = h_m(f) - s \int_J \log |a(x)| \, dm(x) = 0,$$

where $h_m(f)$ is the measure-theoretic entropy of m. One can use formulae (21.20) and (21.21) below to conclude that $d_m(x) = s$ for m-almost every $x \in J$. The third statement follows now from Theorem 7.1.

However, we present here a simple and straightforward proof of the third statement. Having in mind that the measure m is an equilibrium measure and

$P_J(-s \log |a|) = 0$, there exists a constant $C_3 > 0$ such that for any $x \in \mathcal{M}$ and any basic set $R_{i_0 \ldots i_n}(x)$

$$m(R_{i_0 \ldots i_n}(x)) \leq C_3 \prod_{k=0}^{n} |a(f^k x)|^{-s} \tag{20.5}$$

(see Condition (A2.20) in Appendix II). Given $r > 0$, consider a Moran cover $\mathfrak{U}_r = \{R^{(j)}\}$ of J constructed from a Markov partition $\mathcal{R}$ for f. It follows from the property of the Moran cover and (20.5) that

$$m(B(x,r)) \leq \sum_{R^{(j)} \in \mathfrak{U}_r} m(R_r^{(j)}) \leq M C_3 r^s.$$

Thus, m satisfies the uniform mass distribution principle (see Section 7) and hence $\dim_H m \geq s$.

The above arguments also imply the second statement. ∎

Remark.

By slight modification of arguments in the proof of Theorem 20.1 one can strengthen the first statement of this theorem and prove the following result (see [Bar2]; compare to Statement 4 of Theorem 13.4): *given an open set $U \subset \mathcal{M}$ such that $U \cap J \neq \varnothing$ we have*

$$\dim_H(U \cap J) = \underline{\dim}_B(U \cap J) = \overline{\dim}_B(U \cap J) = s$$

(recall that s is the unique root of Bowen's equation $P_J(-s \log |a|) = 0$). Analysing the proof of Theorem 20.1 one can also obtain a lower bound for the Hausdorff dimension as well as an upper bound of the upper box dimension of *any* set $Z \subset J$ (which need not be invariant or compact). Namely,

$$s \leq \dim_H Z, \quad \overline{\dim}_B Z \leq \overline{s},$$

where s and $\overline{s}$ are unique roots of Bowen's equations $P_Z(-s \log |a|) = 0$ and $\overline{CP}_Z(-\overline{s} \log |a|) = 0$ (existence and uniqueness of these roots are guaranteed by Theorem A2.5 in Appendix II).

One can apply Theorem 20.1 to conformal expanding maps described in Examples 1–4 above. We leave it as an exercise to the reader to show that:

1) if f is a *linear* one-dimensional Markov map with the repeller J then

$$\dim_H J = \underline{\dim}_B J = \overline{\dim}_B J = s,$$

where s is the unique root of the equation

$$c_1^{-s} + \cdots + c_p^{-s} = 1$$

(here $c_j > 1$ is the slope of f on the interval I_j, $j = 1, \ldots, p$).

2) if f is the induced map generated by conformal affine maps $h_1, \ldots, h_p \colon D \to D$ (D is the unit ball in $\mathbb{R}^m$), $h_j(x) = \lambda_j x + a_j$, then the number s (that is the common value of the Hausdorff dimension and lower and upper box dimensions) is the unique root of the equation

$$\lambda_1{}^s + \cdots + \lambda_p{}^s = 1.$$

One can also use Theorem 20.1 to compute dimension of hyperbolic Julia sets of rational maps. We present two additional results (without proofs) that provide more detailed information on the Hausdorff dimension of Julia sets. The first result is due to Ruelle [R2] and deals with the two-parameter family of rational maps $z \mapsto z^q - p$. Let $J_{q,p}$ be the corresponding Julia set.

Proposition 20.3. *If $J_{q,p}$ is hyperbolic then*

$$\dim_H J_{q,p} = 1 + \frac{|p|^2}{4 \log q} + (\textit{terms of order} > 2 \textit{ in } p).$$

Consider a family $\{R_\lambda : \lambda \in \Lambda\}$ of rational maps. Given λ, let J_λ be the corresponding Julia set. The family R_λ is said to be J-stable at $\lambda_0 \in \Lambda$ if there exists a continuous map $h \colon \Lambda' \times J_{\lambda_0} \to \overline{\mathbb{C}}$ such that Λ' is a neighborhood of λ_0 in Λ and $h(\lambda, \cdot)$ is a conjugacy from $(J_{\lambda_0}, R_{\lambda_0})$ to (J_λ, R_λ) satisfying $h(\lambda_0, \cdot) = \mathrm{id}|J_{\lambda_0}$.

The second result was obtained by Shishikura [Shi]. Consider the one-parameter family of complex quadratic polynomials $R_\lambda(z) = z^2 + \lambda$. The set $\partial \mathcal{M} = \{\lambda \in \mathbb{C} : R_\lambda \text{ is } not \ J\text{-stable}\}$ is known to be the boundary of a set $\mathcal{M} \subset \mathbb{C}$ called the **Mandelbrot set**.

Proposition 20.4. *There exists a residual subset $\mathcal{A} \subset \partial \mathcal{M}$ such that $\dim_H J_\lambda = 2$ for any $\lambda \in \mathcal{A}$.*

Estimates of Hausdorff Dimension and Box Dimension of Repellers: Non-conformal Case

Let J be a repeller for an expanding $C^{1+\alpha}$-map f. If f is not conformal the Hausdorff dimension and the lower and upper box dimension of J may not coincide. An example is given by the induced map on the limit set of the self-similar geometric construction, described in Section 16.1 (or Section 16.3), which is smooth expanding but is not conformal. Nevertheless, a slight modification of the above approach, which is relied upon the thermodynamic formalism (and its non-additive version), can be still used to establish effective dimension estimates (i.e., estimates that can not be improved). We follow Barreira [Bar2].

Consider two Hölder continuous functions φ and $\overline{\varphi}$ on J

$$\underline{\varphi}(x) = -\log \|d_x f\|, \quad \overline{\varphi}(x) = \log \|(d_x f)^{-1}\|. \tag{20.6}$$

Let $\underline{t}$ and $\overline{t}$ be the unique roots of Bowen's equations

$$P_J(t\,\underline{\varphi}) = 0, \quad P_J(t\,\overline{\varphi}) = 0.$$

Let $\mathcal{R} = \{R_1, \ldots, R_p\}$ be a Markov partition of J of a small diameter and (Σ_A^+, σ) the symbolic representation of J by a subshift of finite type. Since f is a continuous expanding map we can consider two sequences of functions $\underline{\varphi}^{(k)}$ and $\overline{\varphi}^{(k)}$ on Σ_A^+ defined by (19.5). One can verify that given $\varepsilon > 0$, there exists $k \geq 0$ such that for every $\omega \in \Sigma_A^+$ and $n \geq 1$ we have

$$-n\varepsilon + \sum_{j=0}^{n} \underline{\varphi}(f^j(x)) \leq \underline{\varphi}_n^{(k)}(\omega) \leq \overline{\varphi}_n^{(k)}(\omega) \leq n\varepsilon + \sum_{j=0}^{n} \overline{\varphi}(f^j(x)),$$

where $x = \chi(\omega) \in J$. This implies that $\underline{t} \leq \underline{s}^{(k)} \leq \overline{s}^{(k)} \leq \overline{t}$ (we remind the reader that $\underline{s}^{(k)}$ and $\overline{s}^{(k)}$ are unique roots of Bowen's equations (19.6)). By Proposition 19.3 we obtain the following dimension estimates of the repeller J:

$$\underline{t} \leq \dim_H J \leq \underline{\dim}_B J \leq \overline{\dim}_B J \leq \overline{t}. \tag{20.7}$$

We describe another method which allows one to obtain even sharper dimension estimates. Consider two sequences of functions on J defined as follows:

$$\underline{\varphi} = \{\underline{\varphi}_n(x) = -\log \|d_x f^n\|\}, \quad \overline{\varphi} = \{\overline{\varphi}_n(x) = \log \|(d_x f^n)^{-1}\|\}. \tag{20.8}$$

Note that there exists a constant $C > 1$ such that

$$C^n \leq \|(d_x f^n)^{-1}\|^{-1} \leq \|d_x f^n\| \leq K^n$$

for any $x \in J$ and $n \geq 1$, where $K = \max\{\|d_x f\| : x \in J\}$.

We wish to apply the non-additive version of thermodynamic formalism and find roots of Bowen's equations

$$\overline{CP}_J(s\,\underline{\varphi}) = 0, \quad P_J(s\,\overline{\varphi}) = 0. \tag{20.9}$$

In order to do this we ought to establish Property (12.1). Given $\delta \in (0, 1]$, we call the map f δ-bunched if for every $x \in J$ we have

$$\|(d_x f)^{-1}\|^{1+\alpha} \|d_x f\| < 1. \tag{20.10}$$

Proposition 20.5. [Bar2] *If f is a $C^{1+\alpha}$-expanding δ-bunched map (for some $\delta > 0$) then the sequences of functions $\underline{\varphi}$ and $\overline{\varphi}$ satisfy Condition (12.1). Moreover, there exists $\varepsilon \geq 0$ such that for every $\omega \in \Sigma_A^+$,*

$$-\varepsilon + \underline{\varphi}_n(x) \leq \underline{\varphi}_n^{(k)}(\omega) \leq \overline{\varphi}_n^{(k)}(\omega) \leq \overline{\varphi}_n(x) + \varepsilon$$

for all sufficiently large n and some $k \geq 1$ (where $x = \chi(\omega)$).

In view of Proposition 19.1 this implies dimension estimates for the repeller J,

$$\underline{s} \leq \dim_H J \leq \underline{\dim}_B J \leq \overline{\dim}_B J \leq \overline{s}, \tag{20.11}$$

where $\underline{s}$ and $\overline{s}$ are unique roots of Bowen's equations (20.9) (one can show that under the assumptions of Proposition 20.5 Bowen's equations (20.9) have unique roots).

The lower and upper estimates in (20.7) and (20.11) cannot be improved. Note that

$$\sum_{j=0}^{n} \underline{\varphi}(f^j(x)) \leq \underline{\varphi}_n(x) \leq \overline{\varphi}_n(x) \leq \sum_{j=0}^{n} \overline{\varphi}(f^j(x))$$

for every $x = \chi(\omega) \in J$ and $n \geq 1$. These inequalities imply that if f is an δ-bunched $C^{1+\alpha}$-expanding map then $\underline{t} \leq \underline{s} \leq \overline{s} \leq \overline{t}$. If f is conformal it is easy-to see that $\underline{t} = \underline{s} = \overline{s} = \overline{t}$. If f is not conformal the numbers $\underline{s}$ and $\overline{s}$ may provide sharper estimates then the numbers $\underline{t}$ and $\overline{t}$. Indeed, in [Bar2], Barreira constructed an example of a 1-bunched C^∞-expanding map of a compact manifold for which $\underline{t} < \underline{s} = \overline{s} < \overline{t}$. This map is *not* conformal but can be shown to be weakly-conformal (as a continuous expanding map; see (19.7)).

21. Multifractal Analysis of Gibbs Measures for Smooth Conformal Expanding Maps

We undertake the complete multifractal analysis of Gibbs measures for smooth conformal expanding maps. Let J be a conformal repeller for a $C^{1+\alpha}$-conformal expanding map $f: \mathcal{M} \to \mathcal{M}$ of a compact smooth Riemannian manifold $\mathcal{M}$. We assume that f is topologically mixing. The general case can be reduced to this one (with obvious modifications) using the Spectral Decomposition Theorem.

Thermodynamic Description of the Dimension Spectrum

Let $\mathcal{R} = \{R_1, \dots, R_p\}$ be a Markov partition for f and χ the corresponding coding map from Σ_A^+ to J (see Section 20). Consider a Hölder continuous function φ on J. The pull back by χ of φ is a Hölder continuous function $\tilde{\varphi}$ on Σ_A^+, i.e., $\tilde{\varphi}(\omega) = \varphi(\chi(\omega))$ for $\omega \in \Sigma_A^+$. Since f is topologically mixing (and so is the shift σ) an equilibrium measure corresponding to this function, $\mu = \mu_{\tilde{\varphi}}$, is the unique Gibbs measure for σ (see Appendix II). Its push forward is a measure on J which is a unique equilibrium measure corresponding to φ. We denote it by $\nu = \nu_\varphi$.

Define the function ψ on J such that $\log \psi = \varphi - P_J(\varphi)$. Clearly ψ is a Hölder continuous function such that $P_J(\log \psi) = 0$ and ν is a unique equilibrium measure for $\log \psi$. By the variational principle (see Theorem A2.1 in Appendix II) we obtain that

$$\int_J \log \psi(x) \, d\mu(x) = h_\nu(f) = h_\mu(\sigma).$$

Define the one parameter family of functions φ_q, $q \in (-\infty, \infty)$ on J by

$$\varphi_q(x) = -T(q) \log |a(x)| + q \log \psi(x),$$

where $T(q)$ is chosen in such a way that $P_J(\varphi_q) = 0$. It is obvious that the functions φ_q are Hölder continuous. Clearly, $T(0) = \dim_H J = s$.

The function $T(q)$ can also be described in terms of symbolic representation of the repeller by a subshift of finite type (Σ_A^+, σ). Namely, let $\tilde{a}$, $\tilde{\psi}$, and $\tilde{\varphi}_q$ be the pull back by the coding map χ of the functions a, ψ, and φ_q respectively. Note that $\log \tilde{\psi} = \tilde{\varphi} - P_{\Sigma_A^+}(\tilde{\varphi})$. Therefore, the function $T(q)$ satisfies

$$\tilde{\varphi}_q(\omega) = -T(q) \log |\tilde{a}(\omega)| + q \log \tilde{\psi}(\omega)$$

with $P_{\Sigma_A^+}(\tilde{\varphi}_q) = 0$. This simple observation allows us to work with the function $T(q)$ using either the dynamical system (J, f) or the underlying symbolic dynamical system (Σ_A^+, σ).

We first study some basic properties of the function $T(q)$.

Proposition 21.1. *The function $T(q)$ is real analytic for all $q \in \mathbb{R}$.*

Proof. Consider the function $c \colon \mathbb{R}^2 \to C^\alpha(\Sigma_A^+, \mathbb{R})$ defined by $c(r, q) = -r \log |\tilde{a}| + q \log \tilde{\psi}$. This function is clearly real analytic. It is known that the pressure $P = P_{\Sigma_A^+}$ is a real analytic function on the space of Hölder continuous functions (see for example, [R1]), i.e. the function $P(r, q) = P_{\Sigma_A^+}(c(r, q))$ is real analytic with respect to r and q. The desired result follows immediately from the Implicit Function Theorem once we verify the non-degeneracy hypothesis. The latter is

$$\frac{\partial P(c(r, q))}{\partial r}\bigg|_{(r_0, q_0)} \neq 0.$$

In order to compute the partial derivative we use the well-known formula for the derivative of the pressure (see [R1]): given two Hölder continuous functions h_1 and h_2 on Σ_A^+, we have

$$\frac{d}{d\varepsilon}\bigg|_{\varepsilon=0} P(h_1 + \varepsilon h_2) = \int_{\Sigma_A^+} h_2 \, d\mu_{h_1}, \tag{21.1}$$

where μ_{h_1} denotes the Gibbs measure for the function h_1. Applying this formula with $h_1 = -r_0 \log |\tilde{a}| + q_0 \log \tilde{\psi}$, $h_2 = \log |\tilde{a}|$, and $\varepsilon = r_0 - r$ we obtain that

$$\frac{\partial P(c(r, q))}{\partial r}\bigg|_{(r_0, q_0)} = -\int_{\Sigma_A^+} \log |\tilde{a}| \, d\mu_{r_0, q_0} \neq 0, \tag{21.2}$$

where μ_{r_0, q_0} is the Gibbs measure for $c(r_0, q_0)$. ∎

We show that the function $T(q)$ is strictly decreasing by computing its first derivative.

Fix $q \in \mathbb{R}$ and let μ_q denote the Gibbs measure corresponding to the function $\tilde{\varphi}_q$. We also denote the push forward of μ_q to J by ν_q. The measure ν_q is a unique equilibrium measure corresponding to the function φ_q. Given $q \in \mathbb{R}$, let us set

$$\alpha(q) = \frac{\int_{\Sigma_A^+} \log(\tilde{\psi}(\omega)) \, d\mu_q}{\int_{\Sigma_A^+} \log |\tilde{a}(\omega)|^{-1} \, d\mu_q} = \frac{\int_J \log(\psi(x)) \, d\nu_q}{\int_J \log |a(x)|^{-1} \, d\nu_q} > 0. \tag{21.3}$$

The following statement establishes monotonicity of the function $T(q)$.

Proposition 21.2. *For all q we have $\alpha(q) = -T'(q)$. In particular, $T'(q) < 0$ for all q.*

Proof. Since $P(\varphi_q) = 0$ for all q we have

$$\frac{d}{dq}P(\varphi_q) = \frac{\partial P(c(q,r))}{\partial q} + \frac{\partial P(c(q,r))}{\partial r}\bigg|_{r=T(q)} T'(q) = 0.$$

Hence,

$$T'(q) = -\left.\frac{\partial P(c(q,r))}{\partial q}\right|_{r=T(q)} \left(\left.\frac{\partial P(c(q,r))}{\partial r}\right|_{r=T(q)}\right)^{-1}.$$

In order to compute the partial derivative $\frac{\partial P(c(q,r))}{\partial q}$ we apply (21.1) with $h_1 = -r_0\log|\tilde{a}| + q_0\log\tilde{\psi}$, $h_2 = \log\tilde{\psi}$, and $\varepsilon = q_0 - q$. This results in

$$\left.\frac{\partial P(c(r,q))}{\partial q}\right|_{(r_0,q_0)} = -\int_{\Sigma_A^+}\log\tilde{\psi}\,d\mu_{r_0,q_0} \neq 0.$$

Using (21.2) and assuming that $r_0 = T(q_0)$ we conclude that

$$T'(q_0) = -\frac{\int_{\Sigma_A^+}\log(\tilde{\psi}(\omega))\,d\mu_{q_0}}{\int_{\Sigma_A^+}\log|\tilde{a}(\omega)|\,d\mu_{q_0}} = -\alpha(q_0).$$

The lemma follows. ∎

We show now that the function $T(q)$ is convex by computing its second derivative. We recall that by m we denote the measure of full dimension (see Theorem 20.1), i.e., a unique equilibrium measure corresponding to the Hölder continuous function $-s\log|a(x)|$ on J, where s is the unique root of Bowen's equation $P_J(-s\log|a|) = 0$.

Proposition 21.3. *The function $T(q)$ is convex, i.e., $T''(q) \geq 0$. It is strictly convex if and only if $\nu \neq m$ (see Figure 17b).*

Proof. Differentiating twice the equation $P(\varphi_q) = 0$ we obtain that

$$T''(q) = -\frac{T'(q)^2\left(\frac{\partial^2 P(c(q,r))}{\partial r^2}\right) + 2T'(q)\left(\frac{\partial^2 P(c(q,r))}{\partial q\,\partial r}\right) + \left(\frac{\partial^2 P(c(q,r))}{\partial q^2}\right)}{\left(\frac{\partial P(c_{q,r})}{\partial r}\right)},$$

where the partial derivatives are evaluated at $(q,r) = (q, T(q))$.

We use the explicit formula for the second derivative of pressure for the shift map on Σ_A^+ obtained by Ruelle in [R1]. Namely,

$$\left.\frac{\partial^2 P(h + \epsilon_1 h_1 + \epsilon_2 h_2)}{\partial\epsilon_1\partial\epsilon_2}\right|_{\epsilon_1=\epsilon_2=0} = Q_h(f_1, f_2),$$

where Q_h is the bilinear form defined for $h_1, h_2 \in C^\alpha(\Sigma_A^+, \mathbb{R})$ by

$$Q_h(h_1, h_2) = \sum_{k=0}^\infty \left(\int_{\Sigma_A^+} h_1(h_2 \circ \sigma^k) \, d\mu_h - \int_{\Sigma_A^+} h_1 \, d\mu_h \int_{\Sigma_A^+} h_2 \, d\mu_h \right)$$

and μ_h is the Gibbs measure for the potential h. Ruelle also showed that $Q_h(g, g) \geq 0$ for all functions $g \in C^\alpha(\Sigma_A^+, \mathbb{R})$ and that $Q_h(g, g) > 0$ if and only if g is not cohomologous to a constant function (see definition of cohomologous functions in Appendix V).

Applying the second derivative formula we obtain that

$$\frac{\partial^2}{\partial r^2} P(-r \log |\tilde{a}| + q \log \tilde{\psi}) = \frac{\partial^2}{\partial r^2} P(-(r_0 + \varepsilon_1 + \varepsilon_2) \log |\tilde{a}| + q \log \tilde{\psi}),$$
$$= Q_h(\log |\tilde{a}|, \, \log |\tilde{a}|).$$

Arguing similarly we find that

$$\frac{\partial^2}{\partial q^2} P(-r \log |\tilde{a}| + q \log \tilde{\psi}) = Q_h(\log \tilde{\psi}, \, \log \tilde{\psi}),$$

$$\frac{\partial^2}{\partial q \, \partial r} P(-r \log |\tilde{a}| + q \log \tilde{\psi}) = Q_h(\log \tilde{\psi}, \, \log |\tilde{a}|).$$

This implies that

$$T''(q) = \frac{Q_q\big(\log \tilde{\psi} - T'(q) \log |\tilde{a}|, \, \log \tilde{\psi} - T'(q) \log |\tilde{a}| \big)}{\int_{\Sigma_A^+} \log |\tilde{a}| d\mu_q}.$$

It follows that $T''(q) \geq 0$ for any q. Moreover, $T''(q) > 0$ for some q if the function $\log \tilde{\psi}(\omega) - T'(q) \log |\tilde{a}(\omega)|$ is not cohomologous to a constant function. The latter can be assured provided that the functions $\log \tilde{\psi}(\omega)$ and $-T'(q) \log |\tilde{a}(\omega)|$ are not cohomologous. On the other hand, if they are cohomologous for some q then

$$P_{\Sigma_A^+}(-T'(q) \log |\tilde{a}|) = P_{\Sigma_A^+}(\log \tilde{\psi}) = 0.$$

Hence, $-T'(q) = s$. This implies that $\nu = m$. ∎

Diametrical Regularity of Equilibrium Measures

An important ingredient of the multifractal analysis of equilibrium measures is the remarkable fact that these measures are diametrically regular (see Condition (8.5)) as the following statement shows.

Proposition 21.4. *Let φ be a Hölder continuous function on a conformal repeller J. Then any equilibrium measure for φ with respect to f is diametrically regular.*

Proof. Let $\nu = \nu_\varphi$ be an equilibrium measure for φ. Choose a Markov partition $\mathcal{R}$ of J. Given a number $r > 0$, consider a Moran cover $\mathfrak{U}_r$ of J. Fix a point $x \in J$

and choose those elements $R^{(1)}, \ldots, R^{(m)}$ from the Moran cover that intersect the ball $B(x, 2r)$. We recall the following properties of the Moran cover: for every $j = 1, \ldots, m$,

   (1) $R^{(j)} = R_{i_1 \ldots i_{n(x_j)}}$, where $x_j \in J$ is a point;
   (2) diam $R^{(j)} \leq r$;
   (3) $m \leq K$, where K is a constant independent of x and r.

There is an element $R^{(t)}$ of the Moran cover that contains x. We have that

$$R^{(t)} \subset B(x, r) \subset B(x, 2r) \subset \bigcup_{j=1}^{m} R^{(j)}.$$

Define the function ψ such that $\log \psi = \varphi - P(\varphi)$. Clearly, ψ is a Hölder continuous function on X such that $P(\log \psi) = 0$ and ν is an equilibrium measure for $\log \psi$. By (A2.20) (see Appendix II) we also have that for any $j = 1, \ldots, m$,

$$D_1 \leq \frac{\nu(R^{(j)})}{\prod_{k=0}^{n(x_j)-1} \psi(f^k(x_j))} \leq D_2.$$

In view of the second statement of Proposition 20.1, if the diameter of the Moran cover does not exceed r_0, then for any $j = 1, \ldots, m$,

$$C_1 \leq \frac{\prod_{k=0}^{n(x_t)-1} \psi(f^k(x_t))}{\prod_{k=0}^{n(x_j)-1} \psi(f^k(x_j))} \leq C_2.$$

It now follows that

$$\nu(B(x, 2r)) \leq \sum_{j=1}^{m} \nu(R^{(j)}) \leq D_2 \sum_{j=1}^{m} \prod_{k=0}^{n(x_j)-1} \psi(f^k(x_j))$$

$$\leq KC_2 D_2 \prod_{k=0}^{n(x_t)-1} \psi(f^k(x_t)) \leq KC_2 \frac{D_2}{D_1} \nu(R^{(t)}) \leq KC_2 \frac{D_2}{D_1} \nu(B(x, r)).$$

This completes the proof.   ■

Multifractal Analysis of Equilibrium Measures

   We state the result that establishes the multifractal analysis for equilibrium measures (corresponding to Hölder continuous functions) supported on repellers of smooth conformal expanding maps. One can apply this result to conformal expanding maps listed in the previous section (i.e., hyperbolic rational maps, one-dimensional Markov maps, etc.).

   For $\alpha \geq 0$ consider the sets

$$J_\alpha = \{x \in J : d_\nu(x) = \alpha\}$$

(compare to (18.5)) and the $f_\nu(\alpha)$-spectrum for dimensions $f_\nu(\alpha) = \dim_H J_\alpha$.

Theorem 21.1. [PW2]

(1) *The pointwise dimension $d_\nu(x)$ exists for ν-almost every $x \in J$ and*

$$d_\nu(x) = \frac{\int_{\Sigma_A^+} \log \tilde{\psi}(\omega)\, d\mu(\omega)}{-\int_{\Sigma_A^+} \log |\tilde{a}(\omega)|\, d\mu(\omega)} = \frac{h_\mu(f)}{-\int_J \log |a(x)|\, d\nu(x)}.$$

(2) *The function $f_\nu(\alpha)$ is defined on the interval $[\alpha_1, \alpha_2]$ which is the range of the function $\alpha(q)$ (i.e., $0 \le \alpha_1 \le \alpha_2 < \infty$, $\alpha_1 = \alpha(\infty)$ and $\alpha_2 = \alpha(-\infty)$); this function is real analytic and $f_\nu(\alpha(q)) = T(q) + q\alpha(q)$ (see Figure 17b).*

(3) *For any $q \in \mathbb{R}$ we have that $\nu_q(J_{\alpha(q)}) = 1$ and $\underline{d}_{\nu_q}(x) = \overline{d}_{\nu_q}(x) = T(q) + q\alpha(q)$ for ν_q-almost every $x \in J_{\alpha(q)}$.*

(4) *If $\nu \ne m$ (i.e., ν is not the measure of full dimension) then the functions $f_\nu(\alpha)$ and $T(q)$ are strictly convex and form a Legendre transform pair (see Appendix V).*

(5) *The ν-measure of any open ball centered at points in J is positive and for any $q \in \mathbb{R}$ we have*

$$T(q) = -\lim_{r \to 0} \frac{\log \inf_{\mathcal{G}_r} \sum_{B \in \mathcal{G}_r} \nu(B)^q}{\log r},$$

where the infimum is taken over all finite covers $\mathcal{G}_r$ of J by open balls of radius r. In particular, for every $q > 1$,

$$\frac{T(q)}{1-q} = \underline{HP}_q(\nu) = \overline{HP}_q(\nu) = \underline{R}_q(\nu) = \overline{R}_q(\nu).$$

Proof. We begin with the following lemma.

Lemma 1. *There exist constants $C_1 > 0$ and $C_2 > 0$ such that for every $q \in \mathbb{R}$ and for all basic sets $R_{i_0 \ldots i_n}$,*

$$C_1 \le \frac{\nu_q(R_{i_0 \ldots i_n})}{m(R_{i_0 \ldots i_n})^{T(q)/s} \nu(R_{i_0 \ldots i_n})^q} \le C_2. \tag{21.4}$$

Proof of the lemma. Note that μ and μ_q are Gibbs measures corresponding to the Hölder continuous functions $\log \tilde{\psi}$ and $\tilde{\varphi}_q = -T(q) \log |\tilde{a}| + q \log \tilde{\psi}$ whose topological pressure is zero. Note also that m is a unique equilibrium measure corresponding to the Hölder continuous function $-s \log |a|$ whose topological pressure is zero. It now follows from (A2.20) (see Appendix II) that the ratios

$$\frac{\nu(R_{i_0 \ldots i_n})}{\prod_{k=0}^n \psi(f^k(x))}, \quad \frac{\nu_q(R_{i_0 \ldots i_n})}{\prod_{k=0}^n |a(f^k(x))|^{-T(q)} \psi(f^k(x))^q}, \quad \frac{m(R_{i_0 \ldots i_n})}{\prod_{k=0}^n |a(f^k(x))|^{-s}}$$

(where $x \in R_{i_0 \ldots i_n}$) are bounded from below and above by constants independent of n. The desired result follows. ∎

Given $0 < r < 1$, consider a Moran cover $\mathfrak{U}_r$ of the repeller J by basic sets $R_r^{(j)} = R_{i_0 \ldots i_{n(x_j)}}$ with radius approximately equal to r (see Section 20). Let $N(x, r)$ denote the number of sets $R_r^{(j)}$ that have non-empty intersection with a given ball $B(x, r)$ centered at x of radius r. By the property of the Moran cover $N(x, r) \leq M$ uniformly in x and r, where M is a Moran multiplicity factor (recall that M does not depend on x and r).

It follows from (20.1) and (20.2) that there exist positive numbers C_3 and C_4 such that for every $R_r^{(j)} \in \mathfrak{U}_r$,

$$C_3 r^s \leq m(R_r^{(j)}) \leq C_4 r^s. \tag{21.5}$$

Since $\mathfrak{U}_r$ is a disjoint cover of J we have

$$\sum_{R_r^{(j)} \in \mathfrak{U}_r} \nu_q(R_r^{(j)}) = 1.$$

Hence, summing (21.4) over the elements of the cover $\mathfrak{U}_r$, we obtain that there exist positive constants C_5 and C_6 such that

$$C_5 \leq r^{T(q)} \sum_{R_r^{(j)} \in \mathfrak{U}_r} \nu(R_r^{(j)})^q \leq C_6.$$

Taking logarithm and dividing by $\log r$ yields for all $q \in \mathbb{R}$,

$$-\lim_{r \to 0} \frac{\log \sum_{R_r^{(j)} \in \mathfrak{U}_r} \nu(R_r^{(j)})^q}{\log r} = T(q). \tag{21.6}$$

Given $0 < r < 1$ and a point $\omega \in \Sigma_A^+$, consider the unique positive integer $n(\omega) = n(\omega, r)$ defined by (20.2). For a number $\alpha \geq 0$ denote by $\tilde{J}_\alpha$ the "symbolic" level set, i.e., the set of points $\omega \in \Sigma_A^+$ for which the limit

$$\lim_{r \to 0} \frac{\sum_{k=0}^{n(\omega,r)-1} \log \tilde{\psi}(\sigma^k(\omega))}{\sum_{k=0}^{n(\omega,r)-1} \log |\tilde{a}(\sigma^k(\omega))|^{-1}} \tag{21.7}$$

exists and is equal to α. Define the "symbolic" spectrum for dimensions by

$$\tilde{f}_\nu(\alpha) = \dim_H \tilde{J}_\alpha. \tag{21.8}$$

Lemma 2. *For every $q \in \mathbb{R}$ we have*

(1) $\nu_q(\chi(\tilde{J}_{\alpha(q)})) = 1$;

(2) $d_{\nu_q}(x) = T(q) + q\alpha(q)$ *for ν_q-almost all $x \in \chi(\tilde{J}_{\alpha(q)})$ and $\overline{d}_{\nu_q}(x) \leq T(q) + q\alpha(q)$ for all $x \in \chi(\tilde{J}_{\alpha(q)})$;*

(3) $\dim_H \chi(\tilde{J}_{\alpha(q)}) = T(q) + q\alpha(q)$.

Proof of the lemma. Consider the functions $\omega \mapsto \log |\tilde{a}(\omega)|$ and $\omega \mapsto \log \tilde{\psi}(\omega)$. Since μ_q is ergodic the Birkhoff ergodic theorem yields that

$$\lim_{n \to \infty} \frac{\sum\limits_{k=0}^{n} \log \tilde{\psi}(\sigma^k(\omega))}{\sum\limits_{k=0}^{n} \log |\tilde{a}(\sigma^k(\omega))|^{-1}} = \alpha(q)$$

for μ_q-almost every $\omega \in \Sigma_A^+$. This implies the first statement.

It follows that for any $\varepsilon > 0$ and every $\omega \in \tilde{J}_{\alpha(q)}$ there exists $r(\omega)$ such that for any $r \leq r(\omega)$,

$$\alpha(q) - \varepsilon \leq \frac{\sum\limits_{k=0}^{n(\omega,r)-1} \log \tilde{\psi}(\sigma^k(\omega))}{\sum\limits_{k=0}^{n(\omega,r)-1} \log |\tilde{a}(\sigma^k(\omega))|^{-1}} \leq \alpha(q) + \varepsilon. \tag{21.9}$$

Given $\ell > 0$, denote by $Q_\ell = \{\omega \in \tilde{J}_{\alpha(q)} : r(\omega) \leq \frac{1}{\ell}\}$. It is easy to see that $Q_\ell \subset Q_{\ell+1}$ and $\tilde{J}_{\alpha(q)} = \bigcup_{\ell=1}^{\infty} Q_\ell$. Thus, there exists $\ell_0 > 0$ such that $\eta_q(Q_\ell) > 0$ if $\ell \geq \ell_0$. Let us choose $\ell \geq \ell_0$.

Fix r, $0 < r < 1$ and consider a Moran cover $\mathfrak{U}_{r,Q_\ell}$ of the set Q_ℓ (see Section 20). It consists of cylinder sets $C_\ell^{(j)}$, $j = 1, \ldots, N_{r,\ell}$ for which there exist points $\omega_j \in Q_\ell$ such that $C_\ell^{(j)} = C_{i_0 \ldots i_{n(\omega_j)}}$. If r is sufficiently small we have $n(\omega_j) \geq \ell$ for all j.

Since μ_q is a Gibbs measure we obtain by (A2.20) (see Appendix II) that for every $\omega = (i_0 i_1 \ldots) \in \Sigma_A^+$ and $n > 0$,

$$C_7 \leq \frac{\mu_q(C_{i_0 \ldots i_n})}{\prod_{k=0}^{n} |\tilde{a}(\sigma^k(\omega))|^{-T(q)} \tilde{\psi}(\sigma^k(\omega))^q} \leq C_8, \tag{21.10}$$

where C_7 and C_8 are positive constants. It follows from (21.9), (21.10), and (20.2) that for all $n \geq \ell$ and any $x \in \chi(Q_\ell)$,

$$\nu_q(B(x,r) \cap \chi(Q_\ell)) \leq \sum_{j=1}^{M} \mu_q(C_\ell^{(j)})$$

$$\leq C_8 \sum_{j=1}^{M} \prod_{k=0}^{n(\omega_j)-1} |\tilde{a}(\sigma^k(\omega))|^{-T(q)} \tilde{\psi}(\sigma^k(\omega))^q$$

$$\leq C_8 \sum_{j=1}^{M} \prod_{k=0}^{n(\omega_j)-1} |\tilde{a}(\sigma^k(\omega))|^{-T(q)-q(\alpha(q)-\varepsilon)} \leq C_9 r^{T(q)+q(\alpha(q)-\varepsilon)},$$

where $C_9 > 0$ is a constant and M is a Moran multiplicity factor (we stress that both C_9 and M do not depend on l and j). Since $\nu_q(\chi(Q_\ell)) > 0$, by the Borel

Density Lemma (see Appendix V), for ν_q-almost every $x \in \chi(Q_\ell)$ there exists a number $r_0 = r_0(x)$ such that for every $0 < r \leq r_0$ we have

$$\nu_q(B(x,r)) \leq 2\nu_q(B(x,r) \cap \chi(Q_\ell)).$$

This implies that for any $\ell > \ell_0$ and almost every $x \in \chi(Q_\ell)$,

$$\underline{d}_{\nu_q}(x) = \lim_{r \to 0} \frac{\log \nu_q(B(x,r))}{\log r} \geq \lim_{r \to 0} \frac{\log \nu_q(B(x,r) \cap \chi(Q_\ell))}{\log r}$$
$$\geq T(q) + q(\alpha(q) - \varepsilon).$$

Since sets Q_ℓ are nested and exhaust the set Q we obtain that $\underline{d}_{\nu_q}(x) \geq T(q) + q(\alpha(q) - \varepsilon)$ for ν_q-almost every $x \in \tilde{J}_{\alpha(q)}$. Since ε is arbitrary this implies that $\underline{d}_{\nu_q}(x) \geq T(q) + q\alpha(q)$ for ν_q-almost every $x \in \chi(\tilde{J}_{\alpha(q)})$. By Theorem 7.1 we obtain that $\dim_H \chi(\tilde{J}_{\alpha(q)}) \geq T(q) + q\alpha(q)$.

Fix $0 < r < 1$. For each $\omega = (i_0 i_1 \ldots) \in Q_\ell$ choose $n(\omega) = n(\omega, r)$ according to (20.2). It follows that $R_{i_0 \ldots i_{n(\omega)}} \subset B(x,r)$, where $x = \chi(\omega)$. By virtue of (21.9) and (21.10) for all $\omega \in Q_\ell$,

$$\nu_q(B(x, Kr)) \geq \nu_q(R_{i_0 \ldots i_{n(\omega)}}) \geq C_7 \prod_{k=0}^{n(\omega)} |\tilde{a}(\sigma^k(\omega))|^{-T(q)} \tilde{\psi}(\sigma^k(\omega))^q$$

$$\geq C_7 \prod_{k=0}^{n(\omega)} |\tilde{a}(\sigma^k(\omega)))|^{-T(q)-q(\alpha(q)+\varepsilon)} \geq C_7 r^{T(q)+q(\alpha(q)+\varepsilon)}.$$

It follows that for all $x \in \chi(Q_\ell)$,

$$\overline{d}_{\nu_q}(x) = \overline{\lim_{r \to 0}} \frac{\log \nu_q(B(x,r))}{\log r} \leq T(q) + q(\alpha(q) + \varepsilon).$$

Since ε is arbitrary this implies that $\overline{d}_{\nu_q}(x) \leq T(q) + q\alpha(q)$ for every $x \in \chi(\tilde{J}_{\alpha(q)})$. Therefore, $\underline{d}_{\nu_q}(x) = \overline{d}_{\nu_q}(x) = T(q) + q\alpha(q)$ for ν_q-almost every $x \in \chi(\tilde{J}_{\alpha(q)})$ and the second statement of the lemma follows. Moreover, by Theorem 7.2, we obtain that $\dim_H \chi(\tilde{J}_{\alpha(q)}) \leq T(q) + q\alpha(q)$. This implies the last statement and completes the proof of the lemma. ∎

It immediately follows from (21.6) that $T(1) = 0$ and thus, $\mu = \mu_1$. The first statement of the theorem now follows from Lemma 2 and the following lemma.

Lemma 3. $J_\alpha = \chi(\tilde{J}_\alpha)$.

Proof of the lemma. Fix $0 < r < 1$. For each $\omega = (i_0 i_1 \ldots) \in Q_\ell$ choose $n = n(\omega, r)$ according to (20.2). It follows that

$$B(y, C_{10}r) \subset R_{i_0 \ldots i_n} \subset B(x,r),$$

where $x = \chi(\omega)$, $y \in R_{i_0 \ldots i_n}$ is a point, and $C_{10} > 0$ is a constant. Since the measure ν is diametrically regular (see Proposition 21.4) we obtain

$$\nu(R_{i_0 \ldots i_n}) \leq \nu(B(x, r)) \leq \nu(B(y, 2r))$$
$$\leq C_{11} \nu(B(y, C_{10} r)) \leq \nu(R_{i_0 \ldots i_n}),$$

where $C_{11} > 0$ is a constant. Note that μ is a Gibbs measure corresponding to the Hölder continuous function $\log \tilde{\psi}$ whose topological pressure is zero. Note also that m is a unique equilibrium measure corresponding to the Hölder continuous function $-s \log |a|$ whose topological pressure is zero. It now follows from (A2.20) (see Appendix II) that the ratios

$$\frac{\nu(R_{i_0 \ldots i_n})}{\prod_{k=0}^{n} \psi(f^k(x))}, \quad \frac{m(R_{i_0 \ldots i_n})}{\prod_{k=0}^{n} |a(f^k(x))|^{-s}}$$

(where $x \in R_{i_0 \ldots i_n}$) are bounded from below and above by constants independent of n. This implies that the two limits

$$\lim_{r \to 0} \frac{\log \nu(B(x, r))}{\log r} \quad \text{and} \quad \lim_{r \to 0} \frac{\sum_{k=0}^{n(\omega, r)-1} \log \tilde{\psi}(\sigma^k(\omega))}{\sum_{k=0}^{n(\omega, r)-1} \log |\tilde{a}(\sigma^k(\omega))|^{-1}}$$

exist simultaneously, and the proof of the lemma follows.   ■

Applying Lemma 3 we conclude that $\dim_H J_{\alpha(q)} = T(q) + q\alpha(q)$, and Statements 1–5 follow.

We now prove the final statement of the theorem. Given $r > 0$, consider a Moran cover $\mathfrak{U}_r = \{R_r^{(j)}\}$ of J. There are positive constants C_{12} and C_{13} independent of r such that for every j one can find a point $x_j \in R_r^{(j)}$ satisfying

$$B(x_j, C_{12} r) \subset R_r^{(j)} \subset B(x_j, C_{13} r). \tag{21.11}$$

Since the measure ν is diametrically regular (see Proposition 21.4) it follows from (21.11) that for every $q \in \mathbb{R}$,

$$\sum_j \nu(B(x_j, C_{13} r))^q \leq C_{14} \sum_j \nu(B(x_j, C_{12} r))^q \leq C_{14} \sum_j \nu(R_r^{(j)})^q, \tag{21.12}$$

where $C_{14} > 0$ is a constant independent of j and r.

Let $\mathcal{G}_r$ be a cover of the repeller J by balls $B(y_i, r)$. For each $j \geq 0$ there exists $B(y_{i_j}, r) \in \mathcal{G}_r$ such that $B(y_{i_j}, r) \cap R_r^{(j)} \neq \varnothing$. Consider the new cover of J by the balls $B_j = B(y_{i_j}, 2C_{13} r)$. By (21.11) each basic set $R_r^{(j)}$ is contained in at least one element of the new cover. Since the measure ν is diametrically regular we obtain by (21.11) that for any $q \in \mathbb{R}$,

$$\sum_{R_r^j \in \xi_k} \nu(R_r^{(j)})^q \leq C_{15} \nu(B_k)^q, \tag{21.13}$$

where $C_{15} > 0$ is a constant. Exploiting again the fact that the measure ν is diametrically regular we conclude using (21.13) that for any $q \in \mathbb{R}$,

$$\sum_{R_r^{(j)} \in \mathfrak{U}_r} \nu(R_r^j)^q \leq C_{15}^q \sum_k \nu(B_k)^q \leq C_{16} \sum_{B \in \mathcal{G}_r} \nu(B)^q,$$

where $C_{16} > 0$ is a constant. Statement 5 of the theorem follows immediately from (21.6), (21.12), and (21.13). ∎

Remarks.

(1) Assume that $\nu = m$ is the measure of full dimension. This implies that the functions $-s \log |a \circ \chi|$ and $\log \psi$ are cohomologous to each other (see Appendix V). Since $P_{\Sigma_A^+}(-T(q) \log |a \circ \chi| + q \log \psi) = 0$ it follows that $T(q) = (1 - q)s$ (thus, $T(q)$ is a *linear* function). Since the pointwise dimension of m is equal to s everywhere in J we have that $f_\nu(s) = s$ and $f_\nu(\alpha) = 0$ for all $\alpha \neq s$.

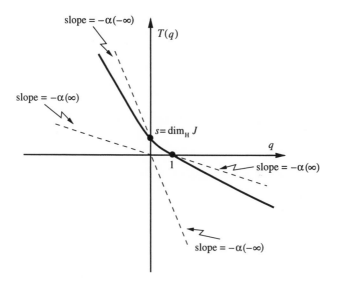

Figure 17a. "Typical" Graph of the Function $T(q)$.

(2) Assume that $\nu \neq m$. There is an interesting manifestation of the multifractal analysis which immediately follows from Statement 4 of Theorem 21.1: *for any $\varepsilon > 0$ there exists a distinct (singular) equilibrium measure μ on J such that $|\dim_H \mu - \dim_H \nu| \leq \varepsilon$.* In fact, using analyticity of the topological pressure and Statement 1 of Theorem 21.1 (see also (21.20) and (21.22) below) one can show that (we leave detail arguments to the reader): *the Hausdorff dimension*

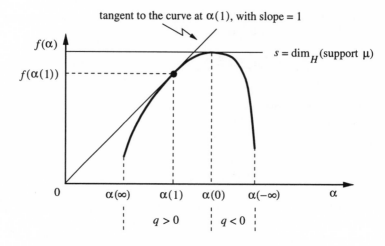

Figure 17b. "Typical" Graph of the Function $f_\nu(\alpha)$.

of an equilibrium measure on a conformal repeller (corresponding to a Hölder continuous potential) depends continuously on the potential.

(3)   Assume that $\nu \neq m$. The function $f_\nu(\alpha)$ is defined on the interval $[\alpha_1, \alpha_2]$ where

$$\alpha_1 = -\lim_{q \to +\infty} T'(q), \quad \alpha_2 = -\lim_{q \to -\infty} T'(q).$$

We have that $0 < \alpha_1 < \alpha_2 < \infty$. Moreover, $f_\nu(\alpha) < s = \dim_H J$ for any $\alpha \neq \alpha(0)$ and $f_\nu(\alpha(0)) = T(0) = s$ (since the function $f_\nu(\alpha)$ is strictly convex $\alpha(0)$ is the only value where this function attains its maximum s; see Figure 17b). Notice that $\alpha(0) < s$ and $\nu_0 = m$.

Differentiating the equality $f_\nu(\alpha(q)) = T(q) + q\alpha(q)$ with respect to q we obtain that $\frac{d}{d\alpha} f_\nu(\alpha(q)) = q$ for every $q \in \mathbb{R}$. This implies that

$$\lim_{\alpha \to \alpha_1} \frac{d}{d\alpha} f_\nu(\alpha) = +\infty, \quad \lim_{\alpha \to \alpha_2} \frac{d}{d\alpha} f_\nu(\alpha) = -\infty. \qquad (21.14)$$

Furthermore, $\frac{d}{d\alpha} f_\nu(\alpha(1)) = 1$. Since $T(1) = 0$ we have that $f_\nu(\alpha(1)) = \alpha(1)$. Taking into consideration that the function $f_\nu(\alpha)$ is strictly convex we conclude that the equation $f_\nu(\alpha) = \alpha$ has the unique root $\alpha(1)$. Moreover, the function $f_\nu(\alpha)$ is tangent to the line of slope 1 at $\alpha(1)$ (see Figure 17b). Notice that $\nu_1 = \nu$ and $\alpha = \alpha(1)$ is the only number for which $\nu(J_\alpha) = 1$.

(4)   Given a number $q_0 \neq 0$, consider the measure ν_{q_0}. We claim that its dimension spectrum can be computed by the following formula:

$$f_{\nu_{q_0}}(\alpha(q)) = f_\nu(\alpha(qq_0)). \qquad (21.15)$$

Indeed, since $P(\varphi_{q_0}) = 0$ we have that $\log \psi_{q_0} = \varphi_{q_0}$ and hence for every $q \in \mathbb{R}$,

$$\begin{aligned}
(\varphi_{q_0})_q(x) &= -T_{q_0}(q) \log |a(x)| + q \log \psi_{q_0}(x) \\
&= -(T_{q_0}(q) + T(q_0)q) \log |a(x)| + qq_0 \log \psi(x).
\end{aligned}$$

Since $P((\varphi_{q_0})_q) = 0$ we obtain that $T_{q_0}(q) + T(q_0)q = T(qq_0)$ for every $q \in \mathbb{R}$. Differentiating over q yields that

$$\alpha_{q_0}(q) = -T'_{q_0}(q) = \alpha(qq_0)q_0 + T(q_0)$$

and the desired formula follows. In particular, we conclude that for every $q \in \mathbb{R}$, $q \neq 0$,

$$\left\{ f_{\nu_q}(\alpha(+\infty)), f_{\nu_q}(\alpha(-\infty)) \right\} = \left\{ f_\nu(\alpha(+\infty)), f_\nu(\alpha(-\infty)) \right\}.$$

(5) Assume that ν_q is the measure of maximal entropy for some $q \neq 0$. This implies that the function φ_q is cohomologous to 0, and hence the functions $T(q) \log |a|$ and $q \log \psi$ are cohomologous. Therefore, the measure ν must be a unique equilibrium measure corresponding to the Hölder continuous function $t \log |a|$, where $t = \frac{T(q)}{q}$.

(6) In [Si], Simpelaere obtained a *variational* description of the function $T(q)$: namely,

$$T(q) = \inf \frac{h_\rho(f) - q \int_J \log \psi \, d\rho}{\int_J \log a \, d\rho}, \tag{21.16}$$

where the infimum is taken over all Borel ergodic measures ρ on J. It follows from Statements 1 and 3 of Theorem 21.1 and (21.3) that the infimum is attained at $\rho = \nu_q$.

Completeness of the $f_\nu(\alpha)$-Spectrum

Schmeling [Sch] showed that the $f_\nu(\alpha)$-spectrum is **complete**, i.e., for any number α outside the interval $[\alpha_1, \alpha_2]$ the set J_α is empty. More precisely, the following result holds.

Theorem 21.2. *We have that*

$$\alpha_1 = \inf_{x \in J} \underline{d}_\nu(x), \quad \alpha_2 = \sup_{x \in J} \overline{d}_\nu(x)$$

and hence $f_\nu(\alpha) = 0$ if and only if $\alpha \notin [\alpha_1, \alpha_2]$.

Proof. Using (21.16) we find that for every $q \geq 0$,

$$q\alpha_1 \leq -q \frac{\int_J \log \psi \, d\nu_q}{\int_J \log |a| \, d\nu_q} \leq \dim_H \nu - q \frac{\int_J \log \psi \, d\nu_q}{\int_J \log |a| \, d\nu_q} = T(q)$$

$$= \inf \left(\dim_H \rho - q \frac{\int_J \log \psi \, d\rho}{\int_J \log |a| \, d\rho} \right) \leq \dim_H J + q \inf - \frac{\int_J \log \psi \, d\rho}{\int_J \log |a| \, d\rho},$$

where the infimum is taken over all Borel ergodic measures ρ on J. Dividing by q and letting $q \to \infty$ yields that

$$\alpha_1 \leq + \inf - \frac{\int_J \log \psi \, d\rho}{\int_J \log |a| \, d\rho}. \tag{21.17}$$

Recall that $\alpha_1 = \lim_{q\to\infty} \alpha(q)$. Therefore, in view of Theorem 21.1, $\alpha_1 \geq \inf_{x\in J} \underline{d}_\nu(x)$. Assume that this inequality is strict. One can find $\varepsilon > 0$, $x \in X$, and a sequence of numbers n_k such that

$$-\sum_{j=0}^{n_k} \log \psi(x) \left(\sum_{j=0}^{n_k} \log |a(x)| \right)^{-1} \leq \alpha_1 - \varepsilon.$$

Let ρ be an accumulation measure of the sequence of measures $\rho_k = \frac{1}{n_k} \sum_{j=0}^{n_k-1} \delta_{f^j(x)}$ (here δ_y denotes the unit mass concentrated at the point y). We have that

$$-\int_J \log \psi \, d\rho_k \left(\int_J \log |a| \, d\rho_k \right)^{-1} \leq \alpha_1 - \varepsilon.$$

Passing to the limit as $k \to \infty$ yields that

$$-\int_J \log \psi \, d\rho \left(\int_J \log |a| \, d\rho \right)^{-1} \leq \alpha_1 - \varepsilon$$

which contradicts (21.17). This implies the first equality. The second one can be proved in a similar fashion.   ∎

Schmeling also proved that for any $\tau_1, \tau_2 \in [0, s]$ there exists an equilibrium measure ν on J corresponding to a Hölder continuous function for which $f_\nu(\alpha_1) = \tau_1$ and $f_\nu(\alpha_2) = \tau_2$. However, "generically" $f_\nu(\alpha_1) = f_\nu(\alpha_2) = 0$; more precisely, this holds for equilibrium measures whose potential functions belong to an open and dense set in the space of Hölder continuous functions on J — the phenomenon noticed previously in physical literature (most of the graphs of the function $f_\nu(\alpha)$ that one can find there end up on the x-axis). Note that in the latter case the measures $\nu_{\alpha(-\infty)}$ and $\nu_{\alpha(\infty)}$ have zero measure-theoretic entropy.

We have the following **multifractal decomposition** of a conformal repeller J associated with the pointwise dimension of an equilibrium measure supported on J and corresponding to a Hölder continuous function:

$$J = \hat{J} \cup \left(\bigcup_\alpha J_\alpha \right). \tag{21.18}$$

Here J_α is the set of points where the pointwise dimension takes on the value α. Each set J_α is everywhere dense in J and supports a measure ν_α (i.e., $\nu_\alpha(J_\alpha) = 1$), which is a unique equilibrium measure corresponding to the Hölder continuous function $\varphi_q(x) = -T(q) \log |a(x)| + q \log \psi(x)$, where q is chosen such that $\alpha = \alpha(q)$. The set $\hat{J}$ — the **irregular part** of the multifractal decomposition — consists of points with no pointwise dimension. We will see in Appendix IV that $\hat{J} \neq \varnothing$; moreover, it is everywhere dense in J and carries full Hausdorff dimension (i.e., $\dim_H \hat{J} = \dim_H J$) and full topological entropy (i.e., $h_{\hat{J}}(f) h_J(f)$).

Pointwise Dimension of Measures on Conformal Repellers

Given a point $x \in J$, we define the **Lyapunov exponent** at x by

$$\lambda(x) = \lim_{n \to \infty} \frac{\log \|df_x^n\|}{n} > 0 \tag{21.19}$$

(provided the limit exists; see Section 26 for more details). If ν is an f-invariant measure then by the Birkhoff ergodic theorem, the above limit exists ν-almost everywhere, and if ν is ergodic then it is constant almost everywhere. We denote the corresponding value by $\lambda_\nu > 0$.

Let ν be an equilibrium measure corresponding to a Hölder continuous function on J. It follows from Statement 1 of Theorem 21.1 that for ν-almost every $x \in J$,

$$d_\nu(x) = \frac{h_\nu(f)}{\lambda_\nu}, \tag{21.20}$$

where $h_\nu(f)$ is the measure-theoretic entropy of f and

$$\lambda_\nu = \lim_{n \to \infty} \frac{1}{n} \sum_{k=0}^{n} \log |a(f^k(x))| = \int_J \log |a(x)| \, d\nu(x). \tag{21.21}$$

Furthermore, let G_ν be the set of all forward generic points of ν (i.e., points for which the Birkhoff ergodic theorem holds for any continuous function on X; see Appendix II). It follows from Lemmas 2 and 3 in the proof of Theorem 21.1 (applied to the measure $\nu_1 = \nu$) that for *every* $x \in G_\nu$,

$$d_\nu(x) = \frac{h_\nu(f)}{\lambda_\nu}.$$

In view of Theorems 7.1 and 7.2, this implies that

$$\frac{h_\nu(f)}{\lambda_\nu} = \dim_H \nu = \dim_H G_\nu. \tag{21.22}$$

We extend (21.20) and (21.22) to any Borel ergodic measure on J of positive measure-theoretic entropy which is not necessarily an equilibrium measure.

Theorem 21.3. *For any Borel ergodic measure ν of positive measure-theoretic entropy supported on a conformal repeller J we have*

(1) *the equality (21.20) holds for almost every $x \in J$;*
(2) $\frac{h_\nu(f)}{\lambda_\nu} = \dim_H \nu = \dim_H G_\nu = \underline{\dim}_B G_\nu = \overline{\dim}_B G_\nu.$

Proof. Set $d_\nu = \frac{h_\nu(f)}{\lambda_\nu}$. We first show that $d_\nu(x) \geq d_\nu$. The proof is a slight modification of the proof of Statement 2 of Theorem 13.1. Consider a Markov partition $\mathcal{R} = \{R_1, \ldots, R_p\}$ for f and the corresponding symbolic model (Σ_A^+, σ) (see Section 20). Let μ be the pullback to Σ_A^+ of the measure ν by the coding map χ.

Fix $\varepsilon > 0$. It follows from the Shannon–McMillan–Breiman theorem that for μ-almost every $\omega \in \Sigma_A^+$ one can find $N_1(\omega) > 0$ such that for any $n \geq N_1(\omega)$,

$$\mu(C_{i_0\ldots i_n}(\omega)) \leq \exp(-(h - \varepsilon)n), \tag{21.23}$$

where $C_{i_0\ldots i_n}(\omega)$ is the cylinder set containing ω and $h = h_\mu(\sigma)$ is the measure-theoretic entropy of the shift σ.

It follows from the Birkhoff ergodic theorem, applied to the function $\log|a(x)|$, that for ν-almost every $x \in \mathcal{M}$ there exists $N_2(x)$ such that for any $n \geq N_2(x)$,

$$\int_{\mathcal{M}} \log|a(x)|d\nu \leq \frac{1}{n}\log\prod_{j=1}^n |a(f^j(x))| + \varepsilon. \tag{21.24}$$

In order to prove the desired lower bound for $d_\nu(x)$ it remains only to use (21.23) and (21.24) and to repeat readily the argument in the proof of Statement 2 of Theorem 13.1.

We now prove the opposite inequality. Fix $0 < r < 1$. By (20.1) it follows that $R_{i_0\ldots i_{n(x)+1}} \subset B(x, C_1 r)$, where $C_1 > 0$ is a constant. Therefore,

$$\nu(B(x_1 C r)) \geq \nu(R_{i_0\ldots i_{n(x)+1}}) \geq C_2 \exp(-(h + \varepsilon)n(x)),$$

where $C_2 > 0$ is a constant. By virtue of (20.1) we obtain for all $x \in J$,

$$\overline{d}_\nu(x) = \varlimsup_{r \to 0} \frac{\log \nu(B(x,r))}{\log r} \leq d_\nu + 2\varepsilon.$$

Since ε can be arbitrarily small this proves that $\underline{d}_\nu(x) \leq d_\nu$.

In order to prove the second statement we note that $d_\nu = \dim_H \nu \leq \dim_H G_\nu$. On the other hand, by (A2.16) (see Appendix II), we conclude that $0 = h_\nu(f) - \int_J d_\nu \lambda_\nu \, d\nu = P_{G_\nu}(-d_\nu \lambda_\nu)$, i.e., d_ν is the root of Bowen's equation. Repeating arguments in the proof of Theorem 20.1 (applied to the set G_ν instead of J) we obtain that $\overline{\dim}_B G_\nu \leq d_\nu$. This completes the proof of the theorem. ∎

Some results, similar to Theorem 21.3 for measures supported on conformal repellers of holomorphic maps, were obtained in [Ma2, PUZ].

Since the measure-theoretic entropy is a semi-continuous function we obtain as an immediate corollary of Theorem 21.3 and (21.21) that: *the Hausdorff dimension of a Borel ergodic measure ν on a conformal repeller is a semicontinuous function of ν.*

Information Dimension

We compute the information dimension of a Gibbs measure ν on a conformal repeller J. Applying Theorem 21.1 and taking into account that the function $T(q)$ is differentiable we obtain that the limit

$$\lim_{q \to 1} \frac{T(q)}{1 - q}$$

exists and is equal to $-T'(1) = \alpha(1)$. As we know the latter coincides with the Hausdorff dimension of ν (see (21.22) and Remark 3 above). This implies that

$$f_\nu(\alpha(1)) = \alpha(1) = -T'(1) = \underline{I}(\nu) = \overline{I}(\nu),$$

where $\underline{I}(\nu)$ and $\overline{I}(\nu)$ are the lower and upper information dimensions of ν (see Section 18).

We note that Statement 5 of Theorem 21.1 allows one to extend the Hentschel–Procaccia spectrum and Rényi spectrum for dimensions for any $q \neq 1$. Moreover, the above argument makes it possible to define these spectra even for $q = 1$ (as being equal to $\alpha(1)$).

Dimension Spectrum for Lyapunov Exponents

We consider the **multifractal decomposition** of the repeller J associated with the Lyapunov exponent $\lambda(x)$ (see (21.19)

$$J = \hat{L} \cup \left(\bigcup_{\beta \in \mathbb{R}} L_\beta \right), \tag{21.25}$$

where

$$\hat{L} = \{x \in J : \text{the limit in (21.19) does not exist}\}$$

is the **irregular part** and

$$L_\beta = \{x \in J : \text{ the limit in (21.19) exists and } \lambda(x) = \beta\}.$$

If ν is an ergodic measure for f we obtain that $\lambda(x) = \lambda_\nu$ for ν-almost every $x \in J$. Thus, the set $L_{\lambda_\nu} \neq \varnothing$. Moreover, if ν is an equilibrium measure corresponding to a Hölder continuous function the set L_{λ_ν} is everywhere dense (since in this case the support of ν is the set J). We note that if the set L_β is not empty then it supports an ergodic measure ν_β for which $\lambda_{\nu_\beta} = \beta$ (indeed, for every $x \in L_\beta$ the sequence of measures $\frac{1}{n} \sum_{k=0}^{n-1} \delta_{f^k(x)}$ has an accumulation measure whose ergodic components satisfy the above property).

There are several fundamental questions related to the above decomposition, for example,

(1) *Are there points x for which the limit in (21.19) does not exist, i.e., $\hat{L} \neq \varnothing$?*
(2) *How large is the range of values of $\lambda(x)$?*
(3) *Is there any number β for which any ergodic measure ν with $\nu(L_\beta) > 0$ is* **not** *an equilibrium measure?*

In order to characterize the above multifractal decomposition quantitatively, we introduce the **dimension spectrum for Lyapunov exponents** of f by

$$\ell(\beta) = \dim_H L_\beta.$$

This definition is inspired by the work of Eckmann and Procaccia [EP]. In [We], Weiss derived the complete study of the Lyapunov dimension spectrum for conformal expanding maps by establishing its relation to the $f_{\nu_{\max}}(\alpha)$-spectrum, where $\nu_{\max}$ is the measure of maximal entropy. Notice that the measure of maximal entropy is a unique equilibrium measure corresponding to the function $\varphi = 0$, and hence $\psi = \text{constant} = \exp(-h_J(f))$, where $h_J(f)$ is the topological entropy of f on J. Therefore, for *every* $x \in L_\beta$,

$$d_{\nu_{\max}}(x) = \frac{h_J(f)}{\beta}.$$

This implies the following result.

Theorem 21.4. [We] *Let J be a conformal repeller for a smooth expanding map f. Then*

(1) *If $\nu_{\max} \neq m$ (m is the measure of full dimension) then the dimension spectrum for Lyapunov exponents*

$$\ell(\beta) = f_{\nu_{\max}}\left(\frac{h_J(f)}{\beta}\right)$$

is a real analytic strictly convex function on an open interval $[\beta_1, \beta_2]$ containing the point $\beta = h_J(f)/s$.

(2) *If $\nu_{\max} = m$ then the dimension spectrum for Lyapunov exponents is a delta function, i.e.,*

$$\ell(\beta) = \begin{cases} s, & \text{for } \beta = h_J(f)/s \\ 0, & \text{for } \beta \neq h_J(f)/s. \end{cases}$$

As immediate consequences of Theorem 21.4 we obtain that *if $\nu_{\max} \neq m$ then the range of the function $\lambda(x)$ contains an open interval, and hence the Lyapunov exponent attains uncountably many distinct values.* On the contrary, *if the Lyapunov exponent $\lambda(x)$ attains only countably many values, then $\nu_{\max} = m$.*

There is an interesting application of this result to rational maps. In [Z], Zdunik proved that in the case of rational maps the coincidence $\nu_{\max} = m$ implies that the map must be of the form $z \to z^{\pm n}$. Therefore, we obtain the following rigidity theorem for rational maps.

Theorem 21.5. *If the Lyapunov exponent of a rational map with a hyperbolic Julia set attains only countably many values, then the map must be of the form $z \to z^{\pm n}$.*

We can now answer the above questions. Namely,

(1) *the set $\hat{L}$ is not empty and has full Hausdorff dimension (see Appendix IV; compare to (21.18));*

(2) *the range of values of $\lambda(x)$ is an interval $[\beta_1, \beta_2]$ and for any β outside this interval the set L_β is empty (i.e., the spectrum is **complete**);*

(3) *for any $\beta \in [\beta_1, \beta_2]$ there exists an equilibrium measure ν corresponding to a Hölder continuous function for which $\nu(L_\beta) = 1$.*

22. Hausdorff Dimension and Box Dimension
of Basic Sets for Axiom A Diffeomorphisms

Axiom A Diffeomorphisms

In this section we study the Hausdorff dimension and box dimension of sets invariant under smooth dynamical systems with strong hyperbolic behavior.

Let $\mathcal{M}$ be a smooth finite-dimensional Riemannian manifold and $f: M \to M$ a $C^{1+\alpha}$-diffeomorphism (i.e., f is a $C^{1+\alpha}$-invertible map whose inverse is of class $C^{1+\alpha}$). A compact f-invariant set $\Lambda \subset \mathcal{M}$ is said to be **hyperbolic** if there exist a continuous splitting of the tangent bundle $T_\Lambda \mathcal{M} = E^{(s)} \oplus E^{(u)}$ and constants $C > 0$ and $0 < \lambda < 1$ such that for every $x \in \Lambda$

(1) $df E^{(s)}(x) = E^{(s)}(f(x)),\ df E^{(u)}(x) = E^{(u)}(f(x));$

(2) for all $n \geq 0$

$$\|df^n v\| \leq C\lambda^n \|v\| \quad \text{if } v \in E^{(s)}(x),$$
$$\|df^{-n} v\| \leq C\lambda^n \|v\| \quad \text{if } v \in E^{(u)}(x).$$

The subspaces $E^{(s)}(x)$ and $E^{(u)}(x)$ are called **stable** and **unstable subspaces** at x respectively and they depend Hölder continuously on x.

It is well-known (see, for example, [KH]) that for every $x \in \Lambda$ one can construct **stable** and **unstable local manifolds**, $W_{\text{loc}}^{(s)}(x)$ and $W_{\text{loc}}^{(u)}(x)$. They have the following properties:

(3) $x \in W_{\text{loc}}^{(s)}(x),\ x \in W_{\text{loc}}^{(u)}(x);$

(4) $T_x W_{\text{loc}}^{(s)}(x) = E^{(s)}(x),\ T_x W_{\text{loc}}^{(u)}(x) = E^{(u)}(x);$

(5) $f(W_{\text{loc}}^{(s)}(x)) \subset W_{\text{loc}}^{(s)}(f(x)),\ f^{-1}(W_{\text{loc}}^{(u)}(x)) \subset W_{\text{loc}}^{(u)}(f^{-1}(x));$

(6) there exist $K > 0$ and $0 < \mu < 1$ such that for every $n \geq 0,$

$$\rho(f^n(y), f^n(x)) \leq K\mu^n \rho(y, x) \text{ for all } y \in W_{\text{loc}}^{(s)}(x) \tag{22.1}$$

and

$$\rho(f^{-n}(y), f^{-n}(x)) \leq K\mu^n \rho(y, x) \text{ for all } y \in W_{\text{loc}}^{(u)}(x), \tag{22.2}$$

where ρ is the distance in $\mathcal{M}$ induced by the Riemannian metric;

(7) there exist $\varepsilon > 0$ such that the intersection $W_{\text{loc}}^{(s)}(x) \cap B(x, \varepsilon)$ (respectively, $W_{\text{loc}}^{(u)}(x) \cap B(x, \varepsilon)$) consists of all points in $B(x, \varepsilon)$ that satisfy (22.1) (respectively, (22.2)).

A hyperbolic set Λ is called **locally maximal** if there exists a neighborhood U of Λ such that for any closed f-invariant subset $\Lambda' \subset U$ we have $\Lambda' \subset \Lambda$. In this case

$$\Lambda = \bigcap_{-\infty < n < \infty} f^n(U).$$

The set Λ is locally maximal if and only if the following property holds:

(8) there exists $\delta > 0$ such that for all $x, y \in \Lambda$ with $\rho(x, y) \le \delta$ the set $\mathcal{W}_{\mathrm{loc}}^{(s)}(x) \cap \mathcal{W}_{\mathrm{loc}}^{(u)}(y)$ consists of a single point $z \in \Lambda$, which we denote by $z = [x, y]$; moreover, the map

$$[\cdot, \cdot] \colon \{(x, y) \in \Lambda \times \Lambda : \rho(x, y) \le \delta\} \to \Lambda \tag{22.3}$$

is continuous.

We define **stable** and **unstable global manifolds** at $x \in \Lambda$ by

$$\mathcal{W}^{(s)}(x) = \bigcup_{n \ge 0} f^{-n}(\mathcal{W}_{\mathrm{loc}}^{(s)}(f^n(x))), \quad \mathcal{W}^{(u)}(x) = \bigcup_{n \ge 0} f^n(\mathcal{W}_{\mathrm{loc}}^{(u)}(f^{-n}(x))).$$

They can be characterized as follows:

$$\begin{aligned}
\mathcal{W}^{(s)}(x) &= \{y \in \Lambda : \rho(f^n(y), f^n(x)) \to 0 \quad \text{as } n \to \infty\}, \\
\mathcal{W}^{(u)}(x) &= \{y \in \Lambda : \rho(f^{-n}(y), f^{-n}(x)) \to 0 \quad \text{as } n \to \infty\}.
\end{aligned} \tag{22.4}$$

A diffeomorphism f is called an **Axiom A diffeomorphism** if its non-wandering set $\Omega(f)$ is a locally maximal hyperbolic set (see Section 20 for definition of a non-wandering set).

The **Spectral Decomposition Theorem** claims (see [KH]) that the set $\Omega(f)$ can be decomposed into finitely many disjoint closed f-invariant locally maximal hyperbolic sets, $\Omega(f) = \Lambda_1 \cup \cdots \cup \Lambda_m$, such that $f \mid \Lambda_i$ is topologically transitive. Moreover, for each i there exist a number n_i and a set $A_i \subset \Lambda_i$ such that the sets $f^k(A_i)$ are disjoint for $0 \le k < n_i$, their union is the set Λ_i, $f^{n_i}(A_i) = A_i$, and the map $f^{n_i} \mid A_i$ is topologically mixing.

Let Λ be a locally maximal hyperbolic set of f. From now on we assume that $f \mid \Lambda$ is topologically mixing. This assumption is technical and we require it to simplify arguments in the proofs. The general case can be reduced to this one by using the Spectral Decomposition Theorem.

A non-empty closed set $R \subset \Lambda$ is called a **rectangle** if $\operatorname{diam} R \le \delta$ (where δ is chosen as in (22.3)), $R = \overline{\operatorname{int} R}$, and $[x, y] \in R$ whenever $x, y \in R$. By Property (8) a rectangle R has "direct product structure", i.e., given $x \in R$, there exists a Hölder continuous homeomorphism

$$\theta \colon R \to R \cap \mathcal{W}_{\mathrm{loc}}^{(s)}(x) \times R \cap \mathcal{W}_{\mathrm{loc}}^{(u)}(x). \tag{22.5}$$

We remind the reader that a finite cover $\mathcal{R} = \{R_1, \ldots, R_p\}$ of Λ is called a **Markov partition** for f if

(1) $\operatorname{int} R_i \cap \operatorname{int} R_j = \varnothing$ unless $i = j$;
(2) for each $x \in \operatorname{int} R_i \cap f^{-1}(\operatorname{int} R_i)$ we have

$$\begin{aligned}
f(\mathcal{W}_{\mathrm{loc}}^{(s)}(x) \cap R_i) &\subset \mathcal{W}_{\mathrm{loc}}^{(s)}(f(x)) \cap R_j, \\
f(\mathcal{W}_{\mathrm{loc}}^{(u)}(x) \cap R_i) &\supset \mathcal{W}_{\mathrm{loc}}^{(u)}(f(x)) \cap R_j.
\end{aligned}$$

A Markov partition $\mathcal{R} = \{R_1, \ldots, R_p\}$ generates a symbolic model of Λ by a subshift of finite type (Σ_A, σ), where Σ_A is the set of two-sided infinite sequences of integers which are admissible with respect to the transfer matrix of the Markov partition $A = (a_{ij})$ (i.e., $a_{ij} = 1$ if $\mathrm{int}R_i \cap f^{-1}(\mathrm{int}R_j) \neq \varnothing$ and $a_{ij} = 0$ otherwise; see Appendix II). Namely, define

$$R^{(u)}_{i_0 \ldots i_n} = \bigcap_{j=0}^{n} f^{-j}(R_{i_j}), \quad R^{(s)}_{i_{-n} \ldots i_0} = \bigcap_{j=-n}^{0} f^{-j}(R_{i_j}),$$

$$R_{i_{-n} \ldots i_n} = R^{(s)}_{i_{-n} \ldots i_0} \cap R^{(u)}_{i_0 \ldots i_n}.$$

(22.6)

Now we define the **coding map** $\chi \colon \Sigma_A \to \Lambda$ by

$$\chi(\ldots i_{-n} \ldots i_0 \ldots i_n \ldots) = \bigcap_{n \geq 0} R_{i_{-n} \ldots i_n}.$$

Note that the diagram

$$
\begin{array}{ccc}
\Sigma_A & \xrightarrow{\ \sigma\ } & \Sigma_A \\
\chi \downarrow & & \downarrow \chi \\
\Lambda & \xrightarrow{\ f\ } & \Lambda
\end{array}
$$

is commutative. The map χ is Hölder continuous and injective on the set of points whose trajectories never hit the boundary of any element of the Markov partition.

For any point $\omega = (\ldots i_{-1} i_0 i_1 \ldots) \in \Sigma_A$ and any point $\omega' = (\ldots i'_{-1} i'_0 i'_1 \ldots) \in \Sigma_A$ with the same past as ω (i.e., $i'_j = i_j$ for any $j \leq 0$) we have that $\chi(\omega') \in \mathcal{W}^{(u)}_{\mathrm{loc}}(x) \cap R(x)$, where $x = \chi(\omega)$ and $R(x)$ is the element of an Markov partition containing x. Similarly, for any point $\omega'' = (\ldots i''_{-1} i''_0 i''_1 \ldots) \in \Sigma_A$ with the same future as ω (i.e., $i''_j = i_j$ for any $j \geq 0$) we have that $\chi(\omega'') \in \mathcal{W}^{(s)}_{\mathrm{loc}}(\chi(\omega)) \cap R(x)$. Thus, the set $\mathcal{W}^{(u)}_{\mathrm{loc}}(x) \cap R(x)$ can be identified via the coding map χ with the cylinder $C^+_{i_0}$ in Σ^+_A and the set $\mathcal{W}^{(s)}_{\mathrm{loc}}(x) \cap R(x)$ can be identified via the coding map χ with the cylinder $C^-_{i_0}$ in Σ^-_A.

It is well-known that a locally maximal hyperbolic set Λ admits Markov partitions of arbitrary small diameters (see [KH]).

Let ψ be a Hölder continuous function on Λ. The following result allows one to compare the values of this function along trajectories of two points lying on the same stable (or unstable) local manifold.

Proposition 22.1. *There exist positive constants $L_1 = L_1(\psi)$ and $L_2 = L_2(\psi)$ such that for any $x \in \Lambda$ and $n > 0$,*

$$L_1 \leq \prod_{i=0}^{n} \psi(f^i(y)) \left(\prod_{i=0}^{n} \psi(f^i(x)) \right)^{-1} \leq L_2 \ \text{ if } y \in \mathcal{W}^{(s)}_{\mathrm{loc}}(x),$$

$$L_1 \leq \prod_{i=0}^{n} \psi(f^{-i}(y)) \left(\prod_{i=0}^{n} \psi(f^{-i}(x)) \right)^{-1} \leq L_2 \ \text{ if } y \in \mathcal{W}^{(u)}_{\mathrm{loc}}(x).$$

Proof. See the proof of Proposition 20.1. ∎

Given $x \in \Lambda$, we set

$$J^{(s)}(x) = \operatorname{Jac}\left(df \mid E^{(s)}(x)\right), \quad J^{(u)}(x) = \operatorname{Jac}\left(df \mid E^{(u)}(x)\right).$$

$J^{(s)}(x)$ and $J^{(u)}(x)$ are Jacobians at x in the stable and unstable directions respectively. Proposition 22.1 applies to these functions.

Consider a Markov partition $\mathcal{R} = \{R_1, \ldots, R_p\}$ of Λ and a Hölder continuous function φ on Λ. The pull back by the coding map χ of φ is a Hölder continuous function $\tilde{\varphi}$ on Σ_A, i.e., $\tilde{\varphi}(\omega) = \varphi(\chi(\omega))$ for $\omega \in \Sigma_A$. If f is topologically mixing (and so is the shift σ) an equilibrium measure corresponding to this function, $\mu = \mu_{\tilde{\varphi}}$, is the unique Gibbs measure for σ (see Appendix II). Its push forward is a measure on Λ which is a unique equilibrium measure corresponding to φ. We denote it by $\nu = \nu_\varphi$. Following Appendix II we consider stable and unstable parts of the measure μ, i.e., the measures $\mu^{(s)}$ and $\mu^{(u)}$. Fix a point $x \in \Lambda$ and choose the element $R(x)$ of the Markov partition containing x (we assume that $x \in \operatorname{int}R(x)$). The sets $\mathcal{W}^{(u)}_{\mathrm{loc}}(y) \cap R(x)$, $y \in R(x)$ either are disjoint or coincide and hence comprise a measurable partition of $R(x)$ which we denote by $\xi^{(u)}$. Similarly the sets $\mathcal{W}^{(u)}_{\mathrm{loc}}(y) \cap R(x)$, $y \in R(x)$ comprise a measurable partition of $R(x)$ which we denote by $\xi^{(u)}$. Since μ is a Gibbs measure we have $\nu(R(x)) > 0$. For almost every $y \in R(x)$ the conditional measure $\nu^{(s)}(y)$ generated by ν on $\mathcal{W}^{(s)}_{\mathrm{loc}}(y) \cap R(x)$ (respectively, the conditional measure $\nu^{(u)}(y)$ generated by ν on $\mathcal{W}^{(u)}_{\mathrm{loc}}(y) \cap R(x)$) coincide with the push forward of the measure $\mu^{(s)}$ (respectively, $\mu^{(u)}$). By Proposition A2.2 (see Appendix II) we obtain the following result.

Proposition 22.2. *There are positive constants A_1 and A_2 such that for any Borel sets $E \in \mathcal{W}^{(s)}_{\mathrm{loc}}(x) \cap R(x)$ and $F \in \mathcal{W}^{(u)}_{\mathrm{loc}}(x) \cap R(x)$,*

$$A_1(\nu^{(s)}(E) \times \nu^{(u)}(F)) \leq \nu(E \times F) \leq A_2(\nu^{(s)}(E) \times \nu^{(u)}(F)).$$

In other words, the measure ν on $R(x)$ is equivalent to the direct product of measures $\nu^{(s)}(x)$ and $\nu^{(u)}(x)$.

Conformal Axiom A Diffeomorphisms

We say that a diffeomorphism f of a locally maximal hyperbolic set Λ is u-**conformal** (respectively, s-**conformal**) if there exists a continuous function $a^{(u)}(x)$ (respectively, $a^{(s)}(x)$) on Λ such that $df \mid E^{(u)}(x) = a^{(u)}(x) \operatorname{Isom}_x$ for every $x \in \Lambda$ (respectively, $df \mid E^{(s)}(x) = a^{(s)}(x) \operatorname{Isom}_x$; recall that Isom_x denotes an isometry of $E^{(u)}(x)$ or $E^{(s)}(x)$). Since the subspaces $E^{(u)}(x)$ and $E^{(s)}(x)$ depend Hölder continuously on x the functions $a^{(u)}(x)$ and $a^{(s)}(x)$ are also Hölder continuous. Note that $|a^{(u)}(x)| > 1$ and $|a^{(s)}(x)| < 1$ for every $x \in \Lambda$. A diffeomorphism f on Λ is called **conformal** if it is u-conformal and s-conformal as well.

Consider a Markov partition $\mathcal{R} = \{R_1, \ldots, R_p\}$ of Λ of a small diameter (which is less than the number ε in Property (7)). Given a point $x \in \Lambda$, define the sets

$$A^{(u)}(x) = \mathcal{W}^{(u)}_{\mathrm{loc}}(x) \cap R(x), \quad A^{(s)}(x) = \mathcal{W}^{(s)}_{\mathrm{loc}}(x) \cap R(x) \qquad (22.7)$$

(we assume that $x \in \operatorname{int} R(x)$). Note that by (22.5) the rectangle $R(x)$ has the direct product structure (defined up to a homeomorphism θ): $R(x) = A^{(u)}(x) \times A^{(s)}(x)$.

From now on we assume that f is u-conformal. Note that one can view $A^{(u)}(x)$ as the limit set for a geometric construction (CG1–CG2) with $R^{(u)}_{i_0 \ldots i_n}$ to be the basic sets. Namely,

$$A^{(u)}(x) = \bigcap_{n \geq 0} \bigcup_{(i_0 \ldots i_n)} R^{(u)}_{i_0 \ldots i_n} \cap \mathcal{W}^{(u)}_{\mathrm{loc}}(x),$$

where the union is taken over all admissible n-tuples $(i_0 \ldots i_n)$ with $R_{i_0} = R(x)$. Moreover, this geometric construction is a Moran-like geometric construction with non-stationary ratio coefficients since its basic sets satisfy Condition (B2) (see Section 15) as the following statement shows.

Proposition 22.3. *Let Λ be a locally maximal hyperbolic set for a u-conformal $C^{1+\alpha}$-diffeomorphism f. We have that*

(1) *each basic set $R^{(u)}_{i_0 \ldots i_n} \cap \mathcal{W}^{(u)}_{\mathrm{loc}}(x)$ contains a ball in $\mathcal{W}^{(u)}_{\mathrm{loc}}(x)$ of radius $\underline{r}_{i_0 \ldots i_n}$ and is contained in a ball in $\mathcal{W}^{(u)}_{\mathrm{loc}}(x)$ of radius $\overline{r}_{i_0 \ldots i_n}$;*

(2) *There exist positive constants K_1 and K_2 such that for every basic set $R^{(u)}_{i_0 \ldots i_n}$ and every $x \in R^{(u)}_{i_0 \ldots i_n}$,*

$$K_1 \prod_{j=0}^{n-1} |a^{(u)}(f^j(x))|^{-1} \leq \underline{r}_{i_1 \ldots i_n} \leq \overline{r}_{i_1 \ldots i_n} \leq K_2 \prod_{j=0}^{n-1} |a^{(u)}(f^j(x))|^{-1}.$$

Proof. See the proof of Proposition 20.2.   ∎

Using the analogy with geometric constructions described above we will construct a **Moran cover** of the set $A^{(u)}(x)$. It is comprised from basic sets and it allows us to build up an optimal cover for computing the Hausdorff dimension and lower and upper box dimensions of Λ.

Let $\hat{\omega} = (\ldots \hat{i}_{-1} \hat{i}_0 \hat{i}_1 \ldots) \in \Sigma_A$ is chosen such that $\chi(\hat{\omega}) = x$. We identify the set of points in Σ_A having the same past as $\hat{\omega}$ with the cylinder set $C^+_{\hat{i}_0} \subset \Sigma^+_A$. Given $r > 0$ and a point $\omega \in C^+_{\hat{i}_0}$, let $n(\omega)$ denote the unique positive integer such that

$$\prod_{k=0}^{n(\omega)} |a^{(u)}\left(\chi(\sigma^k(\omega))\right)|^{-1} > r, \qquad \prod_{k=0}^{n(\omega)+1} |a^{(u)}\left(\chi(\sigma^k(\omega))\right)|^{-1} \leq r. \qquad (22.8)$$

It is easy to see that $n(\omega) \to \infty$ as $r \to 0$ uniformly in ω. Fix $\omega = (i_0 i_1 \ldots) \in C^+_{\hat{i}_0}$ and consider the cylinder set $C_{i_0 \ldots i_{n(\omega)}} \subset C^+_{\hat{i}_0}$. We have $\omega \in C_{i_0 \ldots i_{n(\omega)}}$ and if $\omega' \in C_{i_0 \ldots i_{n(\omega)}}$ and $n(\omega') \geq n(\omega)$ then

$$C_{i_0 \ldots i_{n(\omega')}} \subset C_{i_0 \ldots i_{n(\omega)}}.$$

Let $C(\omega)$ be the largest cylinder set containing ω with the property that $C(\omega) = C_{i_0...i_{n(\omega'')}}$ for some $\omega'' \in C(\omega)$ and $C_{i_0...i_{n(\omega)}} \subset C(\omega)$ for any $\omega' \in C(\omega)$. The sets $C(\omega)$ corresponding to different $\omega \in C_{i_0}^+$ either coincide or are disjoint. We denote these sets by $C_j^{(u)}, j = 1, \ldots, N_r$. There exist points $\omega_j \in C_{i_0}^+$ such that $C_j^{(u)} = C_{i_0...i_{n(\omega_j)}}$. These sets form a disjoint cover of $C_{i_0}^+$ which we denote by $\mathfrak{U}^{(u)}$. The sets $R_j^{(u)} = \chi(C_j^{(u)}), j = 1, \ldots, N_r$ may overlap along their boundaries and comprise a cover of $A^{(u)}(x)$ (which we will denote by the same symbol $\mathfrak{U}_r^{(u)}$ if it does not cause any confusion). We have that $R_j^{(u)} = R_{i_0...i_{n(y_j)}}$ for some $y_j \in A^{(u)}(x)$.

Let $Q \subset C_{i_0}^+$ be a subset (not necessarily invariant). One can repeat the above arguments to construct a Moran cover of the set Q. It consists of cylinder sets $C_j^{(u)}, j = 1, \ldots, N_r$ for which there exist points $\omega_j \in Q$ such that $C_j^{(u)} = C_{i_0...i_{n(\omega_j)}}$ and the intersection $C_j^{(u)} \cap C_i^{(u)} \cap Q$ is empty as soon as $i \neq j$. We denote this cover by $\mathfrak{U}_{r,Q}^{(u)}$.

Moran covers have the following crucial property. Given a point $z \in A^{(u)}(x)$ and a sufficiently small $r > 0$, the number of basic sets $R_j^{(u)}$ in a Moran cover $\mathfrak{U}_r^{(u)}$ that have non-empty intersection with the ball $B^{(u)}(z, r)$ is bounded from above by a number M which is independent of z and r. We call this number a **Moran multiplicity factor**.

In order to establish this property let $\hat{r} = \max\{\text{diam}\, R_i : i = 1, \ldots, p\}$. Since the sets R_i are the closure of their interiors there exists a number $0 < r_1 \leq \hat{r}$ such that for every $x \in \Lambda$,

$$B^{(u)}(x, r_1) \subset A^{(u)}(x).$$

The desired property of Moran covers follows from Proposition 22.3.

Hausdorff Dimension and Box Dimension of Locally Maximal Hyperbolic Sets for u-Conformal and s-Conformal Diffeomorphisms

Let Λ be a locally maximal hyperbolic set for a $C^{1+\alpha}$-diffeomorphism f. Assume that f is topologically mixing.

We denote by $\kappa^{(u)}$ a unique equilibrium measure corresponding to the Hölder continuous function $-t^{(u)} \log |a^{(u)}(x)|$ on Λ, where $t^{(u)}$ is the unique root of Bowen's equation $P_\Lambda(-t \log |a^{(u)}|) = 0$. For every $y \in R(x)$ we also denote by $m^{(u)}(y)$ the conditional measure on $\mathcal{W}^{(u)}(y) \cap R(x)$ generated by $\kappa^{(u)}$ as explained above (see Proposition 22.2; recall that $m^{(u)}(y)$ is the push forward by the coding map of the unstable part of the measure $\kappa^{(u)}$).

We now state our main result assuming that f is u-conformal.

Theorem 22.1. *Let f be a u-conformal $C^{1+\alpha}$-diffeomorphism of a locally maximal hyperbolic set Λ. Then*

(1) *for any $x \in \Lambda$ and for any open set $U \subset \mathcal{W}^{(u)}(x)$ such that $U \cap \Lambda \neq \varnothing$,*

$$\dim_H(U \cap \Lambda) = \underline{\dim}_B(U \cap \Lambda) = \overline{\dim}_B(U \cap \Lambda) = t^{(u)};$$

(2)
$$t^{(u)} = \frac{h_{\kappa^{(u)}}(f)}{\int_\Lambda \log |a^{(u)}(x)| \, d\kappa^{(u)}(x)},$$

where $h_{\kappa^{(u)}}(f)$ is the measure-theoretic entropy of f with respect to the measure $\kappa^{(u)}$;

(3) the $t^{(u)}$-Hausdorff measure of $U \cap \Lambda$ is positive and finite; moreover, it is equivalent to the measure $m^{(u)}(x)|U$ for every $x \in \Lambda$;

(4) $t^{(u)} = \dim_H m^{(u)}(x)$ for every $x \in \Lambda$, i.e., the measure $m^{(u)}(x)$ is the measure of full dimension (see Section 5).

Proof. It is sufficient to prove the theorem assuming that $U = A^{(u)}(y)$ for some point $y \in \mathcal{W}^{(u)}(x)$. The arguments are very similar to the arguments in the proof of Theorem 20.1 and exploit the crucial property of Moran covers observed above.

Choose a point $\hat{\omega} = (\ldots \hat{i}_{-1} \hat{i}_0 \hat{i}_1 \ldots) \in \Sigma_A$ such that $\chi(\hat{\omega}) = y$ and set $d = \dim_H(A^{(u)}(y) \cap \Lambda)$. We show first that $t^{(u)} \le d$. Fix $\varepsilon > 0$. By the definition of the Hausdorff dimension there exists a number $r > 0$ and a cover of $A^{(u)}(y) \cap \Lambda$ by balls B_ℓ, $\ell = 1, 2, \ldots$ of radius $r_\ell \le r$ such that

$$\sum_\ell r_\ell^{d+\varepsilon} \le 1.$$

For every $\ell > 0$ consider a Moran cover $\mathfrak{U}_{r_\ell}^{(u)}$ of $A^{(u)}(y) \cap \Lambda$ and choose those basic sets from the cover that intersect B_ℓ. Denote them by $R_\ell^{(1)}, \ldots, R_\ell^{(m(\ell))}$. Note that $R_\ell^{(j)} = R_{i_0 \ldots i_{n(\ell,j)}}^{(u)}$ for some $(i_0 \ldots i_{n(\ell,j)})$. Using Proposition 22.1, the property of the Moran cover, and repeating the argument in the proof of Statement 1 of Theorem 20.1 one can show that

$$\sum_{R_\ell^{(j)} \in \mathcal{G}} \sup_{x \in R_\ell^{(j)}} \prod_{k=0}^{n(\ell,j)-1} |a^{(u)}(f^k(x))|^{-d-\varepsilon} \le C_1,$$

where $C_1 > 0$ is a constant. Given a number $N > 0$, choose r so small that $n(\ell, j) \ge N$ for all ℓ and j. We now have that for any $n > 0$ and $N > n$,

$$M(C_{i_0}^+, 0, \varphi, \mathcal{U}_n, N) \le \sum_{R_\ell^{(j)} \in \mathcal{G}} \exp \left(\sup_{\omega \in C_\ell^{(j)}} \sum_{k=0}^{n(\ell,j)} \varphi(\sigma^k(\omega)) \right)$$

$$= \sum_{R_\ell^{(j)} \in \mathcal{G}} \sup_{x \in R_\ell^{(j)}} \prod_{k=0}^{n(\ell,j)} |a^{(u)}(f^k(x))|^{-d-\varepsilon} \le C_1,$$

where $M(C_{i_0}^+, 0, \varphi, \mathcal{U}_n, N)$ is defined by (A2.19) (see Appendix II) with $\alpha = 0$ and

$$\varphi(\omega) = -(d + \varepsilon) \log |a^{(u)}(\chi(\omega))|.$$

This implies that

$$P_{C_{i_0}^+} \left(-(d+\varepsilon) \log |a^{(u)} \circ \chi| \right) \leq 0.$$

Notice that for any continuous function $\varphi \colon \Lambda \to \mathbb{R}$

$$P_\Lambda(\varphi) = P_{\Sigma_A}(\chi^* \varphi) = \sup_{n \geq 0} P_{\sigma^n(C_{i_0}^+)}(\chi^* \varphi) = P_{C_{i_0}^+}(\chi^* \varphi).$$

Hence,

$$P_\Lambda \left(-(d+\varepsilon) \log |a^{(u)}| \right) \leq 0.$$

Therefore, by Theorem A2.5 (see Appendix II), $t^{(u)} \leq d+\varepsilon$. Since this inequality holds for all ε we conclude that $t^{(u)} \leq d$.

Denote $\bar{d} = \overline{\dim}_B(A^{(u)}(y) \cap \Lambda)$. We now proceed with the upper estimate and show that $\bar{d} \leq t^{(u)}$. Fix $\varepsilon > 0$. Repeating arguments in the proof of Theorem 20.1 one can show that there exists a positive integer N such that for an arbitrary cover $\mathcal{G}$ of $C_{i_0}^+$ by cylinder sets $C_{i_0 \ldots i_N}$,

$$\sum_{C_{i_0 \ldots i_N} \in \mathcal{G}} \sup_{x \in R_\ell^{(j)}} \prod_{k=0}^N |a^{(u)}(f^k(x))|^{-\bar{d}+2\varepsilon} \geq C_2,$$

where $C_2 > 0$ is a constant. We now have that for any $n > 0$ and $N > n$,

$$R(C_{i_0}^+, 0, \varphi, \mathcal{U}_n, N) = \sum_{C_{i_0 \ldots i_N} \in \mathcal{G}} \exp \left(\sup_{\omega \in C_{i_0 \ldots i_N}} \sum_{k=0}^N \varphi(\sigma^k(\omega)) \right)$$

$$= \sum_{C_{i_0 \ldots i_N} \in \mathcal{G}} \sup_{x \in R_\ell^{(j)}} \prod_{k=0}^N |a^{(u)}(f^k(x))|^{-\bar{d}+2\varepsilon} \geq C_2,$$

where $R(C_{i_0}^+, 0, \varphi, \mathcal{U}_N, N)$ is defined by (A2.19') (see Appendix II) with $\alpha = 0$ and

$$\varphi(\omega) = -(\bar{d} - 2\varepsilon) \log |a^{(u)}(\chi(\omega))|.$$

This implies that

$$\overline{CP}_{C_{i_0}^+} \left(-(\bar{d} - 2\varepsilon) \log |a^{(u)} \circ \chi| \right) \geq 0.$$

Notice that for any continuous function $\varphi \colon \Lambda \to \mathbb{R}$

$$P_\Lambda(\varphi) = P_{\Sigma_A}(\chi^* \varphi) = \overline{CP}_{\Sigma_A}(\chi^* \varphi) \geq \sup_{n \geq 0} \overline{CP}_{\sigma^n(C_{i_0}^+)}(\chi^* \varphi) = \overline{CP}_{C_{i_0}^+}(\chi^* \varphi).$$

It follows that

$$P_\Lambda \left(-(\bar{d} - 2\varepsilon) \log |a^{(u)}| \right) \geq 0$$

and hence $\bar{d} - 2\varepsilon \leq t^{(u)}$. Since this inequality holds for all ε we conclude that $\bar{d} \leq t^{(u)}$. This completes the proof of the first statement.

Since the measure $\kappa^{(u)}$ is an equilibrium measure corresponding to the function $-t^{(u)} \log |a^{(u)}(x)|$ on Λ, whose topological pressure is zero, we obtain in view of the variational principle (see Appendix II) that

$$P_\Lambda(-t^{(u)} \log |a^{(u)}|) = h_{\kappa^{(u)}}(f) - t^{(u)} \int_\Lambda \log |a^{(u)}(x)| \, d\kappa^{(u)}(x) = 0.$$

This implies the second statement.

Since the measure $\mu^{(u)}$ is the stable part of the measure $\kappa^{(u)}$ (i.e., the projection of $\kappa^{(u)}$ to Σ_A^+) for every $x \in \Lambda$ and any basic set $R_{i_0 \ldots i_n}^{(u)}$ containing x we have that

$$m^{(u)}(x)(R_{i_0 \ldots i_n}^{(u)} \cap \mathcal{W}^{(u)}(x)) \leq \text{const} \prod_{k=0}^{n} |a^{(u)}(f^k(x))|^{-1}.$$

Fix $x \in \Lambda$, $r > 0$ and consider a Moran cover of the set $A^{(u)}(x)$ by basic sets $R^{(j)}$. It follows from (22.8) that

$$m^{(u)}(x)(R^{(j)}) \leq \text{const} \prod_{k=0}^{n} |a^{(u)}(f^k(x_j))|^{-t^{(u)}}.$$

Therefore, by the property of the Moran cover we find that

$$m^{(u)}(x)(B^{(u)}(x,r)) \leq \sum_j m^{(u)}(x)(R^{(j)}) \leq M r^{t^{(u)}},$$

where M is a Moran multiplicity factor (which does not depend on r). Thus, $m^{(u)}(x)$ satisfies the uniform mass distribution principle (see Section 7) and hence $\dim_H m^{(u)}(x) \geq t^{(u)}$. The opposite inequality follows from the first statement. This also shows that the $t^{(u)}$-Hausdorff measure of $U \cap \Lambda$ is positive and finite and is equivalent to the measure $m^{(u)}(x)|A^{(u)}(x)$. $\blacksquare$

Let Λ be a locally maximal hyperbolic set for a $C^{1+\alpha}$-diffeomorphism f. Assume that f is s-conformal. Denote by $\kappa^{(s)}$ a unique equilibrium measure corresponding to the Hölder continuous function $t^{(s)} \log |a^{(s)}(x)|$ on Λ, where $t^{(s)}$ is the unique root of Bowen's equation $P_\Lambda(t \log |a^{(s)}|) = 0$. Consider the system of conditional measures $m^{(s)}(y)$ on $\mathcal{W}^{(s)}(y) \cap R(x)$, $y \in R(x)$ generated by $\kappa^{(s)}$ (see Proposition 22.2; recall that $m^{(s)}(y)$ is the push forward by the coding map of the unstable part of the measure $\kappa^{(s)}$; see Appendix II).

Similarly to Theorem 22.1, one can prove that for any $x \in \Lambda$ and any open set $U \subset \mathcal{W}^{(s)}(x)$,

$$\dim_H(U \cap \Lambda) = \underline{\dim}_B(U \cap \Lambda) = \overline{\dim}_B(U \cap \Lambda) = t^{(s)}.$$

Moreover,

$$t^{(s)} = -\frac{h_{\kappa^{(s)}}(f)}{\int_\Lambda \log |a^{(s)}(x)| \, d\kappa^{(s)}(x)},$$

where $h_{\kappa^{(s)}}(f)$ is the measure-theoretic entropy of f with respect to the measure $\kappa^{(s)}$. In addition, the $t^{(s)}$-Hausdorff measure of $U \cap \Lambda$ is positive and finite.

We assume now that Λ is a locally maximal hyperbolic set for a $C^{1+\alpha}$-diffeomorphism f which is both s- and u-conformal. By results of Hasselblatt [Ha] in this case the stable and unstable distributions on Λ are smooth. Hence, given a point $x \in \Lambda$ and a rectangle R containing x, the homeomorphism θ in (22.5) is Lipschitz continuous. Therefore, using Theorems 6.5 and 22.1 we compute the Hausdorff dimension and box dimension of Λ.

Theorem 22.2. *We have*

$$\dim_H \Lambda = \underline{\dim}_B \Lambda = \overline{\dim}_B \Lambda = t^{(u)} + t^{(s)},$$

where $t^{(u)}$ and $t^{(s)}$ are unique roots of Bowen's equations

$$P_\Lambda(-t \log |a^{(u)}|) = 0, \quad P_\Lambda(t \log |a^{(s)}|) = 0$$

respectively and can be computed by the formulae

$$t^{(u)} = \frac{h_{\kappa^{(u)}}(f)}{\int_\Lambda \log |a^{(u)}(x)| \, d\kappa^{(u)}(x)}, \quad t^{(s)} = -\frac{h_{\kappa^{(s)}}(f)}{\int_\Lambda \log |a^{(s)}(x)| \, d\kappa^{(s)}(x)}. \quad (22.9)$$

This result applies and produces a formula for the Hausdorff dimension and box dimension of a basic set of an Axiom A $C^{1+\alpha}$-surface diffeomorphism, which is clearly seen to be both s- and u-conformal. For this case, McCluskey and Manning [MM] proved that $\dim_H \Lambda = t^{(u)} + t^{(s)}$. For diffeomorphisms of class C^2 Takens [T2] showed that the Hausdorff dimension of Λ coincides with its lower and upper box dimensions. Palis and Viana [PV] extended this result to diffeomorphisms of class C^1. Barreira [Bar2] obtained the same result using another approach based on the non-additive version of the thermodynamic formalism. This approach allowed him to extend Theorem 22.2 to "conformal" homeomorphisms possessing locally maximal "topologically hyperbolic" sets.

Consider the measures $m^{(u)}(x)$ and $m^{(s)}(x)$ for $x \in \Lambda$. By (22.9), Proposition 26.1, and Theorem 7.1,

$$\dim_H m^{(u)}(x) = t^{(u)} + d^{(s)}, \quad \dim_H m^{(s)}(x) = t^{(s)} + d^{(u)},$$

where $d^{(s)} \leq t^{(s)}$ and $d^{(u)} \leq t^{(u)}$. Moreover, the equalities hold if and only if

$$\kappa^{(u)} = \kappa^{(s)} \stackrel{\text{def}}{=} \kappa. \quad (22.10)$$

In this case, κ is the measure of full dimension. Condition (22.10) is a "rigidity" type condition. It holds if and only if the functions $-t^{(u)} \log |a^{(u)}(x)|$ and $t^{(s)} \log |a^{(s)}(x)|$ are cohomologous (see [KH] and Appendix V). One can show that this is the case if and only if for any periodic point $x \in \Lambda$ of period p,

$$\prod_{k=0}^{p-1} a^{(u)}(f^k(x))^{t^{(u)}} a^{(s)}(f^k(x))^{t^{(s)}} = 1.$$

Estimates of Hausdorff Dimension and Box Dimension of Locally Maximal Hyperbolic Sets: Non-conformal Case

Let Λ be a locally maximal hyperbolic set for a topologically transitive $C^{1+\alpha}$-diffeomorphism f of a compact manifold $\mathcal{M}$. In the multidimensional case the map f may *not* be either u- or s-conformal and the Hausdorff dimension of the intersection $U \cap \Lambda$ (where $U \subset \mathcal{W}^{(s)}(x)$ is an open set) may not coincide with its lower and upper box dimensions (see example of a multidimensional horseshoe in Section 23). Nevertheless, the above approach, which relied upon the thermodynamic formalism (or its non-additive version), can be still used to establish effective dimension estimates (i.e., estimates that can not be improved). We follow Barreira [Bar2].

Define functions $\underline{\varphi}^{(u)}$ and $\overline{\varphi}^{(u)}$ on Λ by

$$\underline{\varphi}^{(u)}(x) = -\log \|d_x f | E^{(u)}\|, \quad \overline{\varphi}^{(u)}(x) = \log \|(d_x f)^{-1} | E^{(u)}\|$$

and functions $\underline{\varphi}^{(s)}$ and $\overline{\varphi}^{(s)}$ on Λ by

$$\underline{\varphi}^{(s)} = -\log \|d_x f^{-1} | E^{(s)}\|, \quad \overline{\varphi}^{(s)} = \log \|(d_x f^{-1})^{-1} | E^{(u)}\|.$$

Let $\underline{t}^{(u)}(x)$ and $\overline{t}^{(u)}(x)$ be the unique roots of Bowen's equations

$$P_{A^{(u)}(x)}(t\underline{\varphi}^{(u)}) = 0, \quad P_{A^{(u)}(x)}(t\overline{\varphi}^{(u)}) = 0$$

and $\underline{t}^{(s)}(x)$ and $\overline{t}^{(s)}(x)$ be the unique roots of Bowen's equations

$$P_{A^{(s)}(x)}(t\underline{\varphi}^{(s)}) = 0, \quad P_{A^{(s)}(x)}(t\overline{\varphi}^{(s)}) = 0.$$

Proposition 22.4. *For any* $x \in \Lambda$,

$$\underline{t}^{(u)}(x) \leq \dim_H(\mathcal{W}_{\mathrm{loc}}^{(u)}(x) \cap \Lambda) \leq \underline{\dim}_B(\mathcal{W}_{\mathrm{loc}}^{(u)}(x) \cap \Lambda)$$
$$\leq \overline{\dim}_B(\mathcal{W}_{\mathrm{loc}}^{(u)}(x) \cap \Lambda) \leq \overline{t}^{(u)}(x),$$

$$\underline{t}^{(s)}(x) \leq \dim_H(\mathcal{W}_{\mathrm{loc}}^{(s)}(x) \cap \Lambda) \leq \underline{\dim}_B(\mathcal{W}_{\mathrm{loc}}^{(s)}(x) \cap \Lambda)$$
$$\leq \overline{\dim}_B(\mathcal{W}_{\mathrm{loc}}^{(s)}(x) \cap \Lambda) \leq \overline{t}^{(s)}(x).$$

One can obtain even sharper dimension estimates using another approach. Define two sequences of functions $\underline{\varphi}_n^{(u)}$ and $\overline{\varphi}_n^{(u)}$ on Λ by

$$\underline{\varphi}_n^{(u)}(x) = -\log \|d_x f^n | E^{(u)}\|, \quad \overline{\varphi}_n^{(u)}(x) = \log \|(d_x f^n)^{-1} | E^{(u)}\| \qquad (22.11)$$

and two other sequences of functions $\underline{\varphi}_n^{(s)}$ and $\overline{\varphi}_n^{(s)}$ on Λ by

$$\underline{\varphi}_n^{(s)}(x) = -\log \|d_x f^{-n} | E^{(s)}\|, \quad \overline{\varphi}_n^{(s)}(x) = \log \|(d_x f^{-n})^{-1} | E^{(s)}\| \qquad (22.12)$$

If the derivatives $df|E^{(u)}$ and $df^{-1}|E^{(s)}$ are δ-bunched (see Condition (20.9)), one can show that all four sequences of functions, (22.11) and (22.12), satisfy Condition (12.1). Thus, by Theorem A2.6 (see Appendix II), Bowen's equations

$$\overline{CP}_{A^{(u)}(x)}(r\,\underline{\varphi}^{(u)}) = 0, \quad P_{A^{(u)}(x)}(r\,\overline{\varphi}^{(u)}) = 0$$

have unique roots $\underline{r}^{(u)}(x)$ and $\overline{r}^{(u)}(x)$, and Bowen's equations

$$\overline{CP}_{A^{(s)}(x)}(r\,\underline{\varphi}^{(s)}) = 0, \quad P_{A^{(s)}(x)}(r\,\overline{\varphi}^{(s)}) = 0$$

have unique roots $\underline{r}^{(s)}(x)$ and $\overline{r}^{(s)}(x)$. We can now state dimension estimates.

Proposition 22.5. *If the derivatives $df|E^{(u)}$ and $df^{-1}|E^{(s)}$ are δ-bunched for some $\delta > 0$ then for any $x \in \Lambda$,*

$$\underline{r}^{(u)}(x) \le \dim_H(\mathcal{W}^{(u)}_{\mathrm{loc}}(x) \cap \Lambda) \le \underline{\dim}_B(\mathcal{W}^{(u)}_{\mathrm{loc}}(x) \cap \Lambda)$$
$$\le \overline{\dim}_B(\mathcal{W}^{(u)}_{\mathrm{loc}}(x) \cap \Lambda) \le \overline{r}^{(u)}(x),$$

$$\underline{r}^{(s)}(x) \le \dim_H(\mathcal{W}^{(s)}_{\mathrm{loc}}(x) \cap \Lambda) \le \underline{\dim}_B(\mathcal{W}^{(s)}_{\mathrm{loc}}(x) \cap \Lambda)$$
$$\le \overline{\dim}_B(\mathcal{W}^{(s)}_{\mathrm{loc}}(x) \cap \Lambda) \le \overline{r}^{(s)}(x).$$

23. Hausdorff Dimension of Horseshoes and Solenoids

Multidimensional Horseshoes

Let $B_1 \subset \mathbb{R}^k$ and $B_2 \subset \mathbb{R}^\ell$ be balls and $\Delta = B_1 \times B_2 \subset \mathbb{R}^k \oplus \mathbb{R}^\ell = \mathbb{R}^n$. We refer to $\mathbb{R}^k$ as "horizontal" and $\mathbb{R}^\ell$ as "vertical" directions in $\mathbb{R}^n$. Let also $U \subset \mathbb{R}^n$ be an open set and $f: U \to \mathbb{R}^n$ a C^r-diffeomorphism, $r \ge 1$. The set $\Delta = B_1 \times B_2 \subset U$ is called a **horseshoe** for f if the intersection $\Delta' = \Delta \cap f(\Delta)$ can be decomposed into simply connected closed disjoint sets $\Delta' = \Delta_1 \cup \cdots \cup \Delta_p$ such that

(1) $\pi_1(\Delta_i) = B_1$, $\pi_2(f^{-1}(\Delta_i)) = B_2$, where $\pi_1 : \mathbb{R}^n \to \mathbb{R}^k$ and $\pi_2 : \mathbb{R}^n \to \mathbb{R}^k$ are the canonical projections;

(2) $\pi_2|f^{-1}(\Delta_i \cap (B_1 \times \pi(z)))$ is a bijection onto B_2 for any $z \in \Delta_i$;

(3) the map df preserves and expands a vertical cone family $C_1(z, \alpha)$ and the map $df^{-1}|\Delta'$ preserves and expands a horizontal cone family $C_2(z, \alpha)$.

The latter means that $C_1(z, \alpha)$ and $C_2(z, \alpha)$ are cones in $\mathbb{R}^n$ at z of angle α around $\mathbb{R}^\ell$ and $\mathbb{R}^k$ respectively such that

$$df(C_1(z, \alpha)) \subset C_1(f(z), \alpha) \quad \text{if } z \in \Delta$$
$$df^{-1}(C_2(z, \alpha)) \subset C_2(f^{-1}(z), \alpha) \quad \text{if } z \in \Delta'$$

and

$$\|df_z v\| \ge \lambda \|v\| \qquad \text{for any } v \in C_1(z, \alpha)$$
$$\|df_z^{-1} v\| \ge \lambda \|v\| \qquad \text{for any } v \in C_2(z, \alpha),$$

where $\lambda > 1$ is a constant.

One can show that the set

$$\Lambda = \bigcap_{n \in \mathbb{Z}} f^n(\Delta')$$

is a locally maximal hyperbolic set with "almost" vertical unstable and "almost" horizontal stable subspaces (see [KH]). In the two-dimensional case the above construction was first introduced by Smale [Sm] and the set Λ is known as a "Smale horseshoe."

We describe the topological structure of Λ. Note that for every $n \geq 0$ the set $f^n(\Delta')$ consists of p^n simply connected close disjoint "almost" vertical components which we denote by $\Delta^{(u)}_{i_0 \ldots i_n}$. Similarly, for every $n \geq 0$ the set $f^{-n}(\Delta')$ consists of p^n simply connected close disjoint "almost" horizontal components which we denote by $\Delta^{(s)}_{i_0 \ldots i_n}$ (see Figure 18). Given a point $z \in \Lambda$, denote by $\Delta^{(u)}_{i_0 \ldots i_n}(z)$ and $\Delta^{(s)}_{i_0 \ldots i_n}(z)$ the vertical and, respectively, horizontal component that contains z (clearly, it is uniquely defined). One can show that

(1) for every $z \in \Lambda$ the set $\bigcap_{n \geq 0} \Delta^{(u)}_{i_0 \ldots i_n}(z)$ is a smooth ℓ-dimensional unstable submanifold that is isomorphic to B_2; similarly, the set $\bigcap_{n \geq 0} \Delta^{(s)}_{i_0 \ldots i_n}(z)$ is a smooth k-dimensional stable submanifold that is isomorphic to B_1;

(2) every point $z \in \Lambda$ can be coded by a two-sided infinite sequence of integers $(\ldots i_{-1} i_0 i_1 \ldots)$, $i_j = 1, \ldots, p$ such that

$$z = \left(\bigcap_{n \geq 0} \Delta^{(u)}_{i_0 \ldots i_n}(z) \right) \cap \left(\bigcap_{n \geq 0} \Delta^{(s)}_{i_0 \ldots i_{-n}}(z) \right);$$

moreover, the **coding map** $\chi : \Lambda \to \Sigma_p$ defined by $\chi(x) = (\ldots i_{-1} i_0 i_1 \ldots)$ is bijective and onto.

The map χ establishes the symbolic representation of the horseshoe by the full shift on p symbols. There is another description of the horseshoe which is more suitable for studying its dimension. Notice first that the sets

$$\Delta_{i_{-n} \ldots i_0 \ldots i_n} = \Delta^{(u)}_{i_0 \ldots i_n} \cap \Delta^{(s)}_{i_0 \ldots i_{-n}}$$

define a geometric construction in Δ (modeled by the full shift on $2p$ symbols) whose limit set is Λ. Furthermore, consider the sets

$$\Lambda^{(s)}_{i_0 \ldots i_n} = \Delta^{(s)}_{i_0 \ldots i_n} \cap B_1, \qquad \Lambda^{(u)}_{i_0 \ldots i_n} = \Delta^{(u)}_{i_0 \ldots i_n} \cap B_2 \,.$$

They define geometric constructions in B_1 and B_2 respectively (modeled by the full shift on p symbols). We denote these constructions by $CG^{(s)}$ and $CG^{(u)}$. If $F^{(s)}$ and $F^{(u)}$ are their limit sets then $\Lambda = F^{(s)} \times F^{(u)}$.

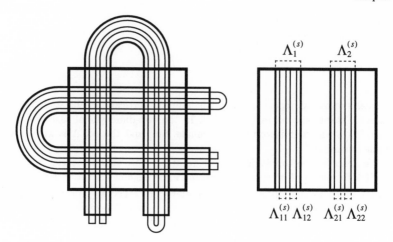

Figure 18. A SMALE HORSESHOE.

A horseshoe Λ is called **linear** if every component Δ_i is a vertical strip and every component $f^{-1}(\Delta_i)$ is a horizontal strip and the maps $h_i = f|f^{-1}(\Delta_i)$ are linear. In this case, the constructions $CG^{(s)}$ and $CG^{(u)}$ are generated by affine maps $g_i^{(s)}$ and $g_i^{(u)}$ respectively given as follows:

$$g_i^{(s)}(y) = \pi_1(f^{-1}(\pi_1^{-1}(y) \cap \Delta_i)), \quad g_i^{(u)}(x) = \pi_2(f^{-1}(\pi_2^{-1}(x) \cap \Delta_i)). \quad (23.1)$$

On the other hand, let $CG^{(s)}$ and $CG^{(u)}$ be two geometric constructions in B_1 and B_2 modeled by the full shift on p symbols and generated by affine maps $g_i^{(s)}$ and $g_i^{(u)}$ respectively. Let also $F^{(s)}$ and $F^{(u)}$ be their limit sets. One can build a linear horseshoe Λ for a map f such that $\Lambda = F^{(s)} \times F^{(u)}$ and relations (23.1) hold. If the maps $g_i^{(s)}$ and $g_i^{(u)}$ are conformal then by Theorem 22.2, we obtain that

$$\dim_H \Lambda = \underline{\dim}_B\Lambda = \overline{\dim}_B\Lambda = \dim_H F^{(s)} + \dim_H F^{(u)} .$$

On the other hand, using Example 16.1 we conclude that there exist a linear horseshoe in $\mathbb{R}^4$ for which

$$\dim_H \Lambda < \underline{\dim}_B\Lambda = \overline{\dim}_B\Lambda .$$

We also remark that if the ratio coefficients of the affine maps $g_i^{(s)}$ and $g_i^{(u)}$ are *equal* then the coding map is an isometry between the the horseshoe Λ and Σ_p. Thus, it preserves the Hausdorff dimension and box dimension (compare to Theorem A2.9 in Appendix II).

Three-Dimensional Solenoids

We follow Bothe [Bot]. Let S^1 be the unit circle and $\mathcal{D}^2$ the unit disk in $\mathbb{R}^2$. Then $V = S^1 \times \mathcal{D}^2$ is a solid torus. The projections $\pi : V \to S^1$ and $\rho_1, \rho_2 : V \to S^1 \times [-1, 1]$ are defined by

$$\pi(t, x, y) = t, \qquad \rho_1(t, x, y) = (t, x), \qquad \rho_2(t, x, y = (t, y) .$$

We also denote by $\mathcal{D}(t) = \{t\} \times \mathcal{D}^2 = \bar{u}^{-1}(t)$ (where $t \in S^1$).

Let $\mathfrak{F}$ be the space of all C^1-embeddings $f : V \to V$ of the form

$$f(t, x, y) = \big(\varphi(t),\, \lambda_1(t)x + z_1(t),\, \lambda_2(t)y + z_2(t)\big), \qquad (23.2)$$

where

$$\varphi : S^1 \to S^2, \qquad \lambda_1, \lambda_2 : S^1 \to (0, 1), \qquad z_1, z_2 : S^1 \to (-1, 1)$$

are C^1-maps, and φ is expanding, i.e.

$$\frac{d\varphi}{dt} > 1.$$

The last condition implies that the degree θ of φ is at least 2 and that f stretches the solid torus V in the direction of S^1. Since $0 < \lambda_1, \lambda_2 < 1$ the disks $\mathcal{D}(t)$ are contracted. This implies that the image $f(V)$ is thinner then V and it is wrapped around inside V exactly θ times. The set

$$\Lambda = \bigcap_{n \geq 0} f^n(V)$$

is called a **solenoid**. One can show that Λ is a hyperbolic attractor for f (see definition in Section 26). The local topological structure of Λ is described as follows. For each $t \in S^1$ the set $\Lambda(t) = \Lambda \cap \mathcal{D}(t)$ is a Cantor-like set; it is the limit set for a geometric construction in $\mathcal{D}(t)$ modeled by the full shift on θ symbols; its basic sets on the step n are mutually disjoint ellipses $\mathcal{D}_{i_0 \ldots i_n}$ (where $i_j = 1, \ldots, \theta$) which comprise the intersection $f^n(V) \cap \mathcal{D}(t)$ (see Figure 19). Furthermore, for any arc $B \subset S^1$ containing t there is a homeomorphism

$$h : B \times \Lambda(t) \to \Lambda \cap \pi^{-1}(B) \qquad (23.3)$$

which can be chosen such that for $q \in B$ and $x \in \Lambda(t)$,

$$\pi\, h(q, x) = q, \qquad h(t, x) = x.$$

For each $x \in \Lambda(t)$ the embedding $h_x = h(\cdot, x) : B \to V$ is of class C^1, and h_x depends continuously on x in the C^1-topology.

A map $f \in \mathfrak{F}$ is called *intrinsically transverse* (with respect to the projections ρ_1 and ρ_2) if for any arc $B \subset S^1$ and any two components $B_1, B_2 \subset \Lambda \cap \pi^{-1}(B)$ the arcs $\rho_i(B_1)$ and $\rho_i(B_2)$ are transverse in $S^1 \times [-1, 1]$ at each point of $\rho_i(B_1) \cap \rho_i(B_2)$. We denote by $\mathfrak{F}^*$ the set of all intrinsically transverse maps $f \in \mathfrak{F}$. For $i = 1, 2$ we also denote by

$$\mathfrak{F}_i = \left\{ f \in \mathfrak{F} : \sup \lambda_i < \inf \left(\frac{d\varphi}{dt} \right) \sup \left(\frac{d\varphi}{dt} \right)^{-4 \log \inf \lambda_i / \log \sup \lambda_i} \right\},$$

where the infimum and supremum are taken over $0 \leq t \leq 1$. Obviously, $\mathfrak{F}_i$ is open in $\mathfrak{F}$. In [Bot], Bothe proved the following statement.

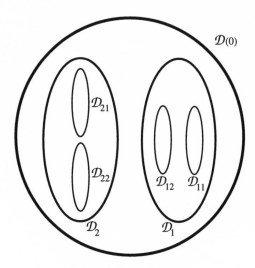

Figure 19. THE CROSS-SECTION OF A SMALE–WILLIAMS SOLENOID
FOR $t = 0$.

Proposition 23.1. *The set $\mathfrak{F}^* \cap \mathfrak{F}_i$ is open and dense in $\mathfrak{F}_i$.*

We now compute the Hausdorff dimension of the cross-sections $\Lambda(t)$. Since the map φ of the circle S^1 is smooth and expanding and the functions λ_i are of class C^1 for $i = 1, 2$ there exists a unique root of Bowen's equation

$$P_{S^1}(s_i \log \lambda_i) = 0$$

(where P_{S^1} is the topological pressure with respect to the map φ). One can show that s_i is the unique number of which the functional equation

$$\sum_{t' \in \varphi^{-1}(t)} \lambda_i(t')^{s_i} \xi(t') = \xi(t)$$

has a positive continuous solution $\xi : S^1 \to \mathbb{R}$ (see [Bot]).

Proposition 23.2.

(1) *For every $f \in \mathfrak{F}$ and every $t \in S^1$,*

$$\dim_H \Lambda(t) \leq \max\{s_1, s_2\} .$$

(2) *If $s_{i_0} = \max\{s_1, s_2\}$ then for every $f \in \mathfrak{F}_{i_0} \cap \mathfrak{F}^*$ and every $t \in S^1$,*

$$\dim_H \Lambda(t) = s_{i_0} .$$

Note that the stable foliation of the solid torus by disks $\mathcal{D}(t)$ is obviously smooth. However, the one-dimensional unstable distribution on Λ is, in general, only Hölder continuous. So is the homeomorphism h defined by (23.3). We set

$$\underline{\lambda} = \min_{0 \le t \le 1} \min\{\lambda_1(t), \lambda_2(t)\}, \quad \overline{\lambda} = \max_{0 \le t \le 1} \max\{\lambda_1(t), \lambda_2(t)\} \,.$$

It follows from results of Hasselblatt [Ha] that the homeomorphism h is Hölder continuous with the Hölder exponent α satisfying

$$\theta^\alpha \frac{\log \underline{\lambda}}{\log \overline{\lambda}} > 1 \,.$$

In particular, if we assume that

$$\theta \frac{\log \underline{\lambda}}{\log \overline{\lambda}} > 1 \tag{23.4}$$

then the map h is Lipschitz continuous. By Theorem 6.5 and Proposition 23.2 we conclude that under the assumption (23.4), if $s_{i_0} = \max\{s_1, s_2\}$, then for every $f \in \mathfrak{F}_{i_0} \cap \mathfrak{F}^*$,

$$\dim_H \Lambda = \max\{s_1, s_2\} + 1 \,.$$

We consider now the special case when

$$\varphi(t) = 2t \pmod 1, \quad \lambda_1(t) = \lambda_1, \quad \lambda_2(t) = \lambda_2,$$
$$z_1(t) = a \cos 2\pi t, \quad z_2(t) = a \sin 2\pi t,$$

where $1 > a > 0$ is a constant. If we assume that

$$0 < \lambda_1, \lambda_2 < \min\left\{\frac{1}{2}, a\right\}$$

then the map f (see (23.2)) is a C^1-embedding. The solenoid Λ for f is known as the Smale–Williams solenoid (see Figure 19). Assume that $\lambda_1 \le \lambda_2$. It is easy to check that $f \in \mathfrak{F}_2$ and

$$s_1 \le s_2 = -\frac{\log 2}{\log \lambda_2} \,.$$

In [Sim2], Simon showed that, in fact, $f \in \mathfrak{F}^*$. Thus, we obtain that

$$\dim_H \Lambda(t) = -\frac{\log 2}{\log \lambda_2} \,.$$

Moreover, it is easy to see that the map h is Lipschitz continuous. Therefore,

$$\dim_H \Lambda = -\frac{\log 2}{\log \lambda_2} + 1 \,. \tag{23.5}$$

Note that if $\lambda_1 = \lambda_2$ then for every $t \in S^1$ the set $\Lambda(t)$ is the limit set for a geometric construction (CPW1–CPW4) with basic sets on the nth step to be balls of radius λ_2^n. By Theorem 13.3 we have that the formula (23.5) holds for any $0 < \lambda_1 = \lambda_2 < \min\left\{\frac{1}{2}, a\right\}$. Using results in [Sim2] Simon extended this statement and proved that the formula (23.5) holds for any $0 < \lambda_1 \leq \lambda_2 < \min\left\{\frac{1}{2}, a\right\}$.

Baker's Transformations

Consider the map T of the square $S = [-1, 1] \times [-1, 1]$ given as follows

$$T(x, y) = \begin{cases} (\lambda_1 x + c_1, & 2y - 1) & \text{if } y \geq 0 \\ (\lambda_2 x + c_2, & 2y + 1) & \text{if } y < 0 \end{cases}$$

where we assume that

$$0 < |\lambda_1|, |\lambda_2| < 1 \tag{23.6}$$

and for $i = 1, 2$,

$$|\lambda_i + c_i| \leq 1, \qquad |-\lambda_i + c_i| \leq 1. \tag{23.7}$$

Conditions (23.6) and (23.7) assure that T maps S into itself. In fact, the set $\overline{T(S)}$ consists of two right rectangles both of vertical height 2, one of width $|\lambda_1|$ and the other of width $|\lambda_2|$. They may or may not overlap. The map T is called a **generalized baker's transformation**. In the case $\lambda_1 = \lambda_2 = \frac{1}{2}$, $c_1 = \frac{1}{2}$, $c_2 = -\frac{1}{2}$ the map T is the **classical baker's transformation**. Clearly, T is piecewise linear with the discontinuity set $[-1, 1] \times \{0\}$. For every $z \in S$ the map T is expanding by a factor of 2 in the vertical direction and is contracting by a factor λ_1 or λ_2 in the horizontal direction depending on whether z lies in the top part or bottom part of the square S.

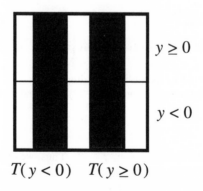

$y \geq 0$

$y < 0$

$T(y < 0) \quad T(y \geq 0)$

Figure 20. A Skinny Baker's Transformation.

The set

$$\Delta = \bigcap_{n \geq 0} \overline{T^n(S)}$$

is the attractor for T (i.e., it attracts all the trajectories in S). Clearly, the set Λ is the direct product of the interval $[-1, 1]$ in the y-direction and a Cantor-like set Δ in the x-direction. The latter is the limit set for a self-similar geometric construction on $[-1, 1]$ whose basic sets at a step n are not necessarily disjoint.

Alexander and York [AY] considered the case $\lambda_1 = \lambda_2 = \beta$, $c_1 = 1 - \beta$, and $c_2 = \beta - 1$. Following their work we call T a **skinny baker's transformation** if $|\lambda_1| + |\lambda_2| < 1$ (see Figure 20) and a **fat baker's transformation** otherwise.

We first study the case of a skinny baker's transformation T. Under an appropriate choice of numbers c_1 and c_2 the two rectangles that comprise $\overline{T(S)}$ are disjoint. The map T is clearly seen to be one-to-one. Moreover, basic sets at each step of the geometric construction, mentioned above, are disjoint. Therefore,

$$\dim_H \Delta = \underline{\dim}_B \Delta = \overline{\dim}_B \Delta = s,$$

where s is the unique root of the equation

$$|\lambda_1|^s + |\lambda_2|^s = 1. \tag{23.8}$$

Hence,

$$\dim_H \Lambda = \underline{\dim}_B \Lambda = \overline{\dim}_B \Lambda = 1 + s.$$

We now consider the case of a fat baker's transformation T assuming for simplicity that $\lambda_1 = \lambda_2 = \beta > \frac{1}{2}$, $c_1 = 1 - \beta$, and $c_2 = \beta - 1$. Note that T is non-invertible and area expanding and that the attractor for T is the whole square S.

Alexander and York [AY] proved that for every $1 \geq \beta > \frac{1}{2}$ the map $T = T_\beta$ possesses the Sinai–Bowen–Ruelle measure μ_β, that is the limit evolution of the Lebesgue measure mes on S, i.e.,

$$\mu_\beta = \lim_{n \to \infty} \frac{1}{n} \sum_{k=0}^{n-1} (T_\beta{}^k)_* \text{mes}.$$

The measure μ_β has a characteristic property that its conditional measures on vertical lines (where T_β expands) are equivalent to the linear Lebesgue measure. Let ν_β be the factor-measure induced by μ_β. It has the following "measure-arithmetic" interpretation and plays an important role in the old and not yet completely solved problem by Erdős.

Let ϵ_n, $n = 0, 1, 2, \ldots$ be a sequence of independent random variables, each with the values $+1$ and -1 with equal probabilities. The measure ν_β can be shown to be the measure with distribution function of the random variable

$$\sum_{n=0}^{\infty} \epsilon_n (1 - \beta) \beta^n.$$

In other words, for any interval (a, b) and any integer N let $\nu_{\beta,N}(a, b)$ be the proportion of points of the form $\sum_{n=0}^{N-1} \epsilon_n (1 - \beta) \beta^n$ that lie in (a, b), i.e.,

$$\nu_{\beta,N}(a, b) = 2^{-N} \text{card} \left\{ x : x = \sum_{n=0}^{N-1} \epsilon_n (1 - \beta) \beta^n, \quad a < x < b \right\},$$

where card(A) denotes the cardinality of the set A. Then

$$\nu_\beta(a,b) = \lim_{N\to\infty} \nu_{\beta,N}(a,b).$$

The measure ν_β is called an *infinitely convolved Bernoulli measure*.

It is easy to see that for $\beta < \frac{1}{2}$, ν_β has a Cantor distribution and for $\beta = \frac{1}{2}$ the uniform (Lebesgue) distribution. For $\beta > \frac{1}{2}$, ν_β is known to be continuous and always pure, i.e., either absolutely continuous or totally singular (see [JW]). For β the nth root of $\frac{1}{2}$, ν_β is absolutely continuous (and indeed, progressively smoother as n increases; see [W]). Erdös [Er2] proved that for almost all β sufficiently close to 1 the measure ν_β is absolutely continuous; on the contrary, he showed [Er1] that if β is the reciprocal of a PV (Pisot–Vijayarghavan) number (i.e., an algebraic integer whose conjugates lie inside the unit circle in the complex plane; an example of a PV number is the Golden mean $(-1+\sqrt{5})/2$; see Section 16), ν_β is singular.

If ν_β is absolutely continuous then its Hausdorff dimension is 1 and hence the Hausdorff dimension of the Sinai–Bowen–Ruelle measure μ_β is 2. Ledrappier and Porzio [LP] considered the case when β is the Golden mean (and hence ν_β is singular). They found an explicit formula for the Hausdorff dimension of ν_β which implies that it is strictly less than 1 (and hence the Hausdorff dimension of μ_β is strictly less than 2). Alexander and York [AY] explicitly computed the information dimension of μ_β provided β is the reciprocal of a PV number. They also showed that if β is the Golden mean then the information dimension is strictly less than 2.

In [F1], Falconer studied the so-called **slanting baker's transformation** of the square S given as follows

$$T(x,y) = \begin{cases} (\lambda_1 x + \mu_1 y + c_1,\, 2y-1) & \text{if } y \geq 0 \\ (\lambda_2 x + \mu_2 y + c_2,\, 2y-1) & \text{if } y < 0. \end{cases}$$

We assume that (23.6) and (23.7) hold and also for $i = 1, 2$,

$$|\lambda_i + \mu_i + c_i| < 1, \qquad |-\lambda_i + \mu_i + c_i| < 1. \tag{23.9}$$

This guarantees that T maps S into itself and hence possesses an attractor. Assuming that T is "skinny", i.e., $|\lambda_1|+|\lambda_2| < 1$, Falconer proved that for almost every (c_1, c_2) the Hausdorff dimension of the attractor for T is the solution s of the equation (23.8). In [Sim1], Simon found that if $\mu_1 \neq \mu_2$ and $0 < |\lambda_1|, |\lambda_2| < |\mu_1 - \mu_2|$ then for *every* pair (c_1, c_2) satisfying (23.7) and (23.9)) the Hausdorff dimension of the attractor is equal to s and that the result holds true for any sufficiently small perturbation of T in the first component (which may make T non-linear).

24. Multifractal Analysis of Equilibrium Measures on Basic Sets of Axiom A Diffeomorphisms

We undertake the complete multifractal analysis of Gibbs measures on a locally maximal hyperbolic set Λ of a $C^{1+\alpha}$-diffeomorphism assuming that f is both s- and u-conformal. We follow the approach suggested by Pesin and Weiss in [PW3].

We assume that f is topologically mixing. The general case can be reduced to this one using the Spectral Decomposition Theorem.

Thermodynamic Description of the Dimension Spectrum

Let φ be a Hölder continuous function on Λ and $\nu = \nu_\varphi$ an equilibrium measure for φ. Define the *normalized* function ψ on Λ by $\log \psi = \varphi - P_\Lambda(\varphi)$. Clearly, $P(\log \psi) = 0$ and ν is an equilibrium measure for ψ.

Consider the one-parameter family of functions $\varphi_q^{(s)}$, $q \in (-\infty, \infty)$ on Λ

$$\varphi_q^{(s)}(x) = T^{(s)}(q) \log |a^{(s)}(x)| + q \log \psi(x),$$

where $T^{(s)}(q)$ is chosen such that $P_\Lambda(\varphi_q^{(s)}) = 0$. Furthermore, consider the one-parameter family of functions $\varphi_q^{(u)}$, $q \in (-\infty, \infty)$ on Λ

$$\varphi_q^{(u)}(x) = -T^{(u)}(q) \log |a^{(u)}(x)| + q \log \psi(x),$$

where $T^{(u)}(q)$ is chosen such that $P_\Lambda(\varphi_q^{(u)}) = 0$. Finally, let us set

$$T(q) = T^{(s)}(q) + T^{(u)}(q).$$

The functions $T^{(s)}(q)$ and $T^{(u)}(q)$ admit another description in terms of the underlying symbolic dynamical system (Σ_A, σ). Namely, let $\tilde{\varphi}$, $\tilde{\psi}$, $\tilde{a}^{(s)}$, and $\tilde{a}^{(u)}$ be the pull back to Σ_A by the coding map of the functions φ, ψ, $a^{(s)}$, and $a^{(u)}$ respectively. Note that $\log \tilde{\psi} = \tilde{\varphi} - P_{\Sigma_A}(\tilde{\varphi})$. Therefore, the functions $T^{(s)}(q)$ and $T^{(u)}(q)$ satisfy

$$\tilde{\varphi}_q^{(s)}(\omega) = T^{(s)}(q) \log |\tilde{a}^{(s)}(\omega)| + q \log \tilde{\psi}(\omega),$$
$$\tilde{\varphi}_q^{(u)}(\omega) = -T^{(u)}(q) \log |\tilde{a}^{(u)}(\omega)| + q \log \tilde{\psi}(\omega),$$

where $\tilde{\varphi}_q^{(s)}$ and $\tilde{\varphi}_q^{(u)}$ are the pull back to Σ_A by the coding map of the functions $\varphi_q^{(s)}$ and $\varphi_q^{(u)}$ respectively. We have that

$$P_{\Sigma_A}(\tilde{\varphi}_q^{(s)}) = P_{\Sigma_A}(\tilde{\varphi}_q^{(u)}) = 0.$$

There is also a description of the functions $T^{(s)}(q)$ and $T^{(u)}(q)$ which uses the stable and unstable parts of the functions $\tilde{\psi}$, $\tilde{a}^{(s)}$, and $\tilde{a}^{(u)}$ (see definitions in Appendix II). Namely, we define the function $\tilde{\psi}^{(s)}$ on Σ_A^- by setting

$$\log \tilde{\psi}^{(s)}(\omega^-) = -\lim_{n \to \infty} \log \frac{\mu(C_{i_{-n} \ldots i_1})}{\mu(C_{i_{-n} \ldots i_0})},$$

where $\omega^- = (\ldots i_{-n} \ldots i_{-1} i_0) \in \Sigma_A^-$. We also define the function $\tilde{\psi}^{(u)}$ on Σ_A^+ by setting

$$\log \tilde{\psi}^{(u)}(\omega^+) = -\lim_{n \to \infty} \log \frac{\mu(C_{i_1 \ldots i_n})}{\mu(C_{i_0 \ldots i_n})},$$

where $\omega^+ = (i_0 i_1 \ldots i_n \ldots) \in \Sigma_A^+$. According to Appendix II the above limits exist and the functions $\log \tilde{\psi}^{(s)}$ and $\log \tilde{\psi}^{(u)}$ are Hölder continuous. Moreover, these functions are strictly cohomologous to the projections of the function $\log \tilde{\psi}$ to Σ_A^- and Σ_A^+ respectively, i.e., they are cohomologous and

$$P_{\Sigma_A^-}(\log \tilde{\psi}^{(s)}) = P_{\Sigma_A^+}(\log \tilde{\psi}^{(u)}) = 0$$

(see Appendix V). Let $\mu^{(s)}$ and $\mu^{(u)}$ be the Gibbs measures corresponding to the functions $\tilde{\psi}^{(s)}$ and $\tilde{\psi}^{(u)}$ respectively. As we saw in Appendix II, these measures are the stable and unstable parts of the Gibbs measure μ corresponding to the function $\tilde{\varphi}$ (recall that $\nu = \chi_* \mu$). In addition,

$$\int_{\Sigma_A^-} \log \tilde{\psi}^{(s)}(\omega^-) \, d\mu^{(s)}(\omega^-) = \int_{\Sigma_A^+} \log \tilde{\psi}^{(u)}(\omega^+) \, d\mu^{(u)}(\omega^+) = \int_{\Sigma_A} \log \tilde{\psi}(\omega) \, d\mu(\omega)$$

$$= h_{\mu^{(s)}}(\sigma | \Sigma_A^-) = h_{\mu^{(u)}}(\sigma | \Sigma_A^+) = h_\mu(\sigma) = h_\nu(f).$$

Arguing as above one can also find the function $\tilde{a}^{(ss)}$ defined on Σ_A^- and the function $\tilde{a}^{(uu)}$ defined on Σ_A^+ such that $\log \tilde{a}^{(ss)}$ is strictly cohomologous to the projection of the function $\tilde{a}^{(s)}$ to Σ_A^- and $\log \tilde{a}^{(uu)}$ is strictly cohomologous to the projection of the function $\tilde{a}^{(u)}$ to Σ_A^+. Therefore, the function

$$\tilde{\varphi}_q^{(ss)}(\omega^-) = T^{(s)}(q) \log |\tilde{a}^{(ss)}(\omega^-)| + q \log \tilde{\psi}^{(s)}(\omega^-)$$

is strictly cohomologous to the projection of the function $\tilde{\varphi}_q^{(s)}(\omega)$ to Σ_A^- and the function

$$\tilde{\varphi}_q^{(uu)}(\omega^+) = -T^{(u)}(q) \log |\tilde{a}^{(uu)}(\omega^+)| + q \log \tilde{\psi}^{(u)}(\omega^+)$$

is strictly cohomologous to the projection of the function $\tilde{\varphi}_q^{(u)}(\omega)$ to Σ_A^+. This observation allows us to use results in Section 21 to study the function $T(q)$.

It follows from Proposition 21.1 that the function $T(q)$ is real analytic. In order to compute its first derivative we define for each $q \in \mathbb{R}$,

$$\alpha^{(s)}(q) = \frac{\int_{\Sigma_A^-} \log(\tilde{\psi}^{(s)}(\omega^-)) \, d\mu_q^{(s)}}{\int_{\Sigma_A^-} \log |\tilde{a}^{(ss)}(\omega^-)| \, d\mu_q^{(s)}}, \quad \alpha^{(u)}(q) = -\frac{\int_{\Sigma_A^+} \log(\tilde{\psi}^{(u)}(\omega^+)) \, d\mu_q^{(u)}}{\int_{\Sigma_A^+} \log |\tilde{a}^{(uu)}(\omega^+)| \, d\mu_q^{(u)}},$$

$$\alpha(q) = \alpha^{(s)}(q) + \alpha^{(s)}(q),$$

where $\mu_q^{(s)}$ and $\mu_q^{(u)}$ are Gibbs measures corresponding to Hölder continuous functions $\tilde{\varphi}_q^{(ss)}$ on Σ_A^- and $\tilde{\varphi}_q^{(uu)}$ on Σ_A^+.

It follows from Propositions 21.2 and 21.3 that $\alpha(q) = -T'(q)$ and the function $T(q)$ is convex, i.e., $T''(q) \geq 0$. Moreover, one can show that the function

$T(q)$ is strictly convex if and only if the measure $\nu|R(x)$ is not equivalent to the measure $m^{(s)}(x) \times m^{(u)}(x)$ for every $x \in \mathcal{M}$. We recall that, according to Section 22, $m^{(u)}(x)$ and $m^{(s)}(x))$ denote the conditional measures on $A^{(u)}(x)$ and $A^{(s)}(x)$ generated by the measures $\kappa^{(u)}$ and $\kappa^{(s)}$ on Λ respectively. The latter are equilibrium measures corresponding to the Hölder continuous functions $t^{(u)} \log |a^{(u)}(x)|$ and $-t^{(s)} \log |a^{(s)}(x))|)$, where $t^{(u)}$ and $t^{(s)}$ are the unique roots of Bowen's equations $P_\Lambda(t \log |a^{(u)}(x)|) = 0$ and $P_\Lambda(-t \log |a^{(s)}(x)|) = 0$ respectively.

We leave it as an easy exercise to the reader to show that if the measure $\nu|R(x)$ is equivalent to the measure $m^{(s)}(x) \times m^{(u)}(x)$ for *some* $x \in \mathcal{M}$ then it holds for *every* $x \in \mathcal{M}$ (note that f is topologically mixing). The equivalence also means that ν is the measure of full dimension (see Condition (22.10)).

Diametrical Regularity of Equilibrium Measures

Our approach to the multifractal analysis of equilibrium measures is based upon the diametrically regular property of these measures (see Condition (8.15)).

Proposition 24.1. *Let φ be a Hölder continuous function on Λ. Then the equilibrium measure $\nu = \nu_\varphi$ for φ is diametrically regular and so are the measure $\nu^{(s)}$ on $A^{(s)}$ and the measure $\nu^{(u)}$ on $A^{(u)}$.*

Proof. The fact that the measures $\nu^{(s)}$ and $\nu^{(u)}$ are diametrically regular uses Proposition 22.1 and is a slight modification of arguments in the proof of Proposition 21.4. The diametrical regularity of the measure ν follows from Proposition 22.2. ∎

Multifractal Analysis of Equilibrium Measures

We now establish a complete multifractal analysis of equilibrium measures (corresponding to Hölder continuous functions) supported on locally maximal conformal hyperbolic sets.

For $\alpha \geq 0$ consider the sets Λ_α defined by

$$\Lambda_\alpha = \{x \in \Lambda : d_\nu(x) = \alpha\}$$

(compare with (18.5)) and the $f_\nu(\alpha)$-spectrum for dimensions $f_\nu(\alpha) = \dim_H \Lambda_\alpha$.

Theorem 24.1. [PW3]

(1) *The pointwise dimension $d_\nu(x)$ exists for ν-almost every $x \in \Lambda$ and*

$$d_\nu(x) = \frac{\int_{\Sigma_A} \log \tilde{\psi}(\omega)\, d\mu(\omega)}{\int_{\Sigma_A} \log |\tilde{a}^{(u)}(\omega)|\, d\mu(\omega)} - \frac{\int_{\Sigma_A} \log \tilde{\psi}(\omega)\, d\mu(\omega)}{\int_{\Sigma_A} \log |\tilde{a}^{(s)}(\omega)|\, d\mu(\omega)}$$

$$= h_\nu(f)\left(\frac{1}{\lambda_\nu^+} - \frac{1}{\lambda_\nu^-}\right), \tag{24.1}$$

where $h_\nu(f)$ is the measure-theoretic entropy of f and λ_ν^+, λ_ν^- are positive and negative values of the Lyapunov exponent of ν (see definition of the Lyapunov exponent in Section 26).

(2) The function $f_\nu(\alpha)$ is defined on the interval $[\alpha_1, \alpha_2]$, which is the range of the function $\alpha(q)$ (i.e., $0 \le \alpha_1 \le \alpha_2 < \infty$, $\alpha_1 = \alpha(\infty)$ and $\alpha_2 = \alpha(-\infty)$); this function is real analytic and $f_\nu(\alpha(q)) = T(q) + q\alpha(q)$ (see Figure 17b).

(3) For any $q \in \mathbb{R}$ we have that $\nu_q(\Lambda_{\alpha(q)} \cap R(x)) = 1$ for every $x \in \Lambda$ and $\underline{d}_{\nu_q}(x) = \overline{d}_{\nu_q}(x) = T(q) + q\alpha(q)$ for ν_q-almost every $x \in \Lambda_{\alpha(q)} \cap R(x)$.

(4) If ν is not the measure of full dimension then the functions $f_\nu(\alpha)$ and $T(q)$ are strictly convex and form a Legendre transform pair (see Appendix V).

(5) The ν-measure of any open ball centered at points in Λ is positive and for any $q \in \mathbb{R}$ we have

$$T(q) = -\lim_{r \to 0} \frac{\log \inf_{\mathcal{G}_r} \sum_{B \in \mathcal{G}_r} \nu(B)^q}{\log r},$$

where the infimum is taken over all finite covers $\mathcal{G}_r$ of Λ by open balls of radius r. In particular, for every $q > 1$,

$$\frac{T(q)}{1-q} = \underline{HP}_q(\nu) = \overline{HP}_q(\nu) = \underline{R}_q(\nu) = \overline{R}_q(\nu).$$

Proof. First we note that

$$T(0) = T^{(s)}(0) + T^{(u)}(0) = t^{(s)} + t^{(u)} = \dim_H \Lambda.$$

Repeating arguments in the proof of Theorem 21.1 we obtain that (see (21.6))

$$T^{(s)}(q) = -\lim_{r \to 0} \frac{\log \sum_{C^{(j)} \in \mathfrak{U}_r^{(s)}} \mu^{(s)}(C^{(j)})^q}{\log r},$$
$$T^{(u)}(q) = -\lim_{r \to 0} \frac{\log \sum_{C^{(j)} \in \mathfrak{U}_r^{(u)}} \mu^{(u)}(C^{(j)})^q}{\log r}, \tag{24.2}$$

where $\mathfrak{U}_r^{(s)}$ and $\mathfrak{U}_r^{(u)}$ are Moran covers of Σ_A^- and Σ_A^+ respectively. In particular, $T(1) = T^{(s)}(1) + T^{(u)}(1) = 0$.

Fix an element of the Markov partition $R(x)$, where $x \in \text{int}R(x)$. Note that $R(x) = \chi(C_{i_0})$.

Given $0 < r < 1$ and $\omega = (\ldots i_{-1}i_0i_1 \ldots)$, choose $n^- = n^-(\omega, r)$ and $n^+ = n^+(\omega, r)$ such that

$$\prod_{k=1-n^-}^{0} |\tilde{a}^{(ss)}\left(\sigma^k(\omega^-)\right)| > r, \qquad \prod_{k=1-n^-}^{0} |\tilde{a}^{(ss)}\left(\sigma^k(\omega^-)\right)| \le r,$$
$$\prod_{k=0}^{n^+-1} |\tilde{a}^{(uu)}\left(\sigma^k(\omega^+)\right)|^{-1} > r, \qquad \prod_{k=0}^{n^+-1} |\tilde{a}^{(uu)}\left(\sigma^k(\omega^+)\right)|^{-1} \le r. \tag{24.3}$$

Fix a number $\alpha \ge 0$ and let $\tilde{\Lambda}_\alpha$ be the set of points $\omega = (\ldots i_{-1}i_0i_1 \ldots)$ for which the limit

$$\lim_{r \to 0} \left(\frac{\sum\limits_{k=1-n^-}^{0} \log \tilde{\psi}^{(s)}(\sigma^k(\omega^-))}{\sum\limits_{k=1-n^-}^{0} \log |\tilde{a}^{(ss)}(\sigma^k(\omega^-))|} - \frac{\sum\limits_{k=0}^{n^+-1} \log \tilde{\psi}^{(u)}(\sigma^k(\omega^+))}{\sum\limits_{k=0}^{n^+-1} \log |\tilde{a}^{(uu)}(\sigma^k(\omega^+))|} \right) \tag{24.4}$$

exists and is equal to α (the "symbolic" level set). Define the "symbolic" spectrum

$$\tilde{f}_\nu(\alpha) = \dim_H \tilde{\Lambda}_\alpha. \tag{24.5}$$

Given $x \in \Lambda$, we consider the measures $\nu_q^{(s)} = \chi_* \mu_q^{(s)} | C_{i_0}^-$ on $A^s(x)$ and $\nu_q^{(u)} = \chi_* \mu_q^{(u)} | C_{i_0}^+$ on $A^{(u)}(x)$ and let $\nu_q = \nu_q^{(s)} \times \nu_q^{(u)}$ be the measure on $R(x)$.

Lemma 1. *For every $q \in \mathbb{R}$ we have*

(1) $\nu_q(\chi(\tilde{\Lambda}_{\alpha(q)}) \cap R(x)) = 1$;

(2) $d_{\nu_q}(y) = T(q) + q\alpha(q)$ *for ν_q-almost all $y \in \chi(\tilde{\Lambda}_{\alpha(q)}) \cap R(x)$;*

(3) $\bar{d}_{\nu_q}(y) \leq T(q) + q\alpha(q)$ *for all $y \in \chi(\tilde{\Lambda}_{\alpha(q)}) \cap R(x)$;*

(4) $\dim_H \chi(\tilde{\Lambda}_{\alpha(q)}) \cap R(x) = T(q) + q\alpha(q)$.

Proof of the lemma. Given $\omega = (\ldots i_{-1} i_0 i_1 \ldots) \in \Sigma_A$, consider the functions $\omega^- \mapsto \log |\tilde{a}^{(ss)}(\omega^-)|$ and $\omega^- \mapsto \log \tilde{\psi}^{(s)}(\omega^-)$ on Σ_A^- and the functions $\omega^+ \mapsto \log |\tilde{a}^{(uu)}(\omega^+)|$ and $\omega^+ \mapsto \log \tilde{\psi}^{(u)}(\omega^+)$ on Σ_A^+. Since the measures $\mu_q^{(s)}$ and $\mu_q^{(u)}$ are ergodic the Birkhoff ergodic theorem yields that for $\mu_q^{(s)}$-almost every $\omega \in \Sigma_A^-$ and $\mu_q^{(u)}$-almost every $\omega \in \Sigma_A^+$,

$$\lim_{r \to 0} \frac{\sum_{k=0}^{1-n^-} \log \tilde{\psi}^{(s)}(\sigma^k(\omega^-))}{\sum_{k=0}^{1-n^-} \log |\tilde{a}^{(ss)}(\sigma^k(\omega^-))|} = \alpha^{(s)}(q),$$

$$\lim_{r \to 0} \frac{\sum_{k=0}^{n^+-1} \log \tilde{\psi}^{(u)}(\sigma^k(\omega^+))}{\sum_{k=0}^{n^+-1} \log |\tilde{a}^{(uu)}(\sigma^k(\omega^+))|} = -\alpha^{(u)}(q). \tag{24.6}$$

This implies the first statement.

By (24.6) for any $\varepsilon > 0$ and every $\omega \in \tilde{\Lambda}_{\alpha(q)}$ there exists $r(\omega)$ such that for any $r \leq r(\omega)$,

$$\alpha^{(s)}(q) - \varepsilon \leq \frac{\sum_{k=0}^{1-n^-} \log \tilde{\psi}^{(s)}(\sigma^k(\omega^-))}{\sum_{k=0}^{1-n^-} \log |\tilde{a}^{(ss)}(\sigma^k(\omega^-))|} \leq \alpha^{(s)}(q) + \varepsilon,$$

$$\alpha^{(u)}(q) - \varepsilon \leq \frac{\sum_{k=0}^{n^+-1} \log \tilde{\psi}^{(u)}(\sigma^k(\omega^+))}{\sum_{k=0}^{n^+-1} \log |\tilde{a}^{(uu)}(\sigma^k(\omega^+))|} \leq \alpha^{(u)}(q) + \varepsilon. \tag{24.7}$$

Given $\ell > 0$, denote by $Q_\ell = \{\omega \in \tilde{\Lambda}_{\alpha(q)} : r(\omega) \leq \frac{1}{\ell}\}$. It is easy to see that $Q_\ell \subset Q_{\ell+1}$ and $\tilde{\Lambda}_{\alpha(q)} = \bigcup_{\ell=1}^\infty Q_\ell$. Thus, there exists $\ell_0 > 0$ such that $\eta_q(Q_\ell) > 0$ if $\ell \geq \ell_0$. Let us choose $\ell \geq \ell_0$.

Since $\mu_q^{(s)}$ and $\mu_q^{(u)}$ are Gibbs measures we obtain by (A2.20) (see Appendix II) that for every $\omega^- = (\ldots i_{-1}i_0) \in \Sigma_A^-$ and $\omega^+ = (i_0i_1\ldots) \in \Sigma_A^+$,

$$C_1 \le \frac{\mu_q^{(s)}(C_{i_{-n+1}\ldots i_0})}{\prod\limits_{k=0}^{1-n} |\tilde{a}^{(ss)}(\sigma^k(\omega^-)))|^{-T^{(s)}(q)} \tilde{\psi}^{(s)}(\sigma^k(\omega^-))^q} \le C_2,$$

$$C_1 \le \frac{\mu_q^{(u)}(C_{i_0\ldots i_n})}{\prod\limits_{k=0}^{n^+-1} |\tilde{a}^{(uu)}(\sigma^k(\omega^+)))|^{-T^{(u)}(q)} \tilde{\psi}^{(u)}(\sigma^k(\omega^+))^q} \le C_2,$$

$$(24.8)$$

where C_1 and C_2 are positive constants. Repeating arguments in the proof of Lemma 2 of Theorem 21.1 one can show that for any $\ell > \ell_0$,

$$\underline{d}_{\nu_q^{(s)}}(z) \ge T^{(s)}(q) + q(\alpha^{(s)}(q) - \varepsilon) \text{ if } z \in A^{(s)}(x) \cap \chi(Q_\ell),$$

$$\underline{d}_{\nu_q^{(u)}}(z) \ge T^{(u)}(q) + q(\alpha^{(u)}(q) - \varepsilon) \text{ if } z \in A^{(u)}(x) \cap \chi(Q_\ell).$$

This implies that for almost every $y \in \chi(Q_\ell)$,

$$\underline{d}_{\nu_q}(y) \ge T(q) + q(\alpha(q) - \varepsilon).$$

Since sets Q_ℓ are nested and exhaust the set Q we obtain that $\underline{d}_{\nu_q}(y) \ge T(q) + q(\alpha(q) - \varepsilon)$ for ν_q-almost every $y \in \tilde{\Lambda}_{\alpha(q)}$. Since ε is arbitrary this implies that $\underline{d}_{\nu_q}(y) \ge T(q) + q\alpha(q)$ for ν_q-almost every $y \in \chi(\tilde{\Lambda}_{\alpha(q)})$. The second statement of the lemma follows. By Theorem 7.1 we also obtain that $\dim_H \chi(\tilde{\Lambda}_{\alpha(q)}) \ge T(q) + q\alpha(q)$.

Fix $0 < r < 1$. It follows from (24.3) that $R_{i_{-n^-}\ldots i_{n^+}} \subset B(y, Kr)$, where $y = \chi(\omega)$ and $K > 0$ is a constant independent of y and r. By virtue of (24.7) and (24.8) for all $\omega \in Q_\ell$,

$$\nu_q(B(y, Kr)) \ge \nu_q(R_{i_{-n^-}\ldots i_{n^+}})$$

$$\ge (C_2)^2 \prod_{k=-n^-+1}^{0} |\tilde{a}^{(ss)}(\sigma^k(\omega^-))|^{-T^{(s)}(q)} \tilde{\psi}^{(s)}(\sigma^k(\omega^-))^q$$

$$\times \prod_{k=0}^{n^+-1} |\tilde{a}^{(uu)}(\sigma^k(\omega^+))|^{-T^{(u)}(q)} \tilde{\psi}^{(u)}(\sigma^k(\omega^+))^q$$

$$\ge (C_2)^2 \prod_{k=-n^-+1}^{0} |\tilde{a}^{(ss)}(\sigma^k(\omega^-))|^{-T^{(s)}(q)-q(\alpha^{(s)}(q)+\varepsilon)}$$

$$\times \prod_{k=0}^{n^+-1} |\tilde{a}^{(uu)}(\sigma^k(\omega^+))|^{-T^{(u)}(q)-q(\alpha^{(u)}(q)+\varepsilon)} \ge (C_2)^2 r^{T(q)+q(\alpha(q)+\varepsilon)}.$$

It follows that for all $y \in \chi(Q_\ell)$,

$$\overline{d}_{\nu_q}(y) = \varlimsup_{r \to 0} \frac{\log \nu_q(B(y, r))}{\log r} \le T(q) + q(\alpha(q) + \varepsilon).$$

Since ε is arbitrary this implies that $\overline{d}_{\nu_q}(y) \leq T(q) + q\alpha(q)$ for every $y \in \chi(\tilde{\Lambda}_{\alpha(q)}) \cap R(x)$. The third statement of the lemma follows. Moreover, by Theorem 7.2 we obtain that $\dim_H(\chi(\tilde{\Lambda}_{\alpha(q)}) \cap R(x)) \leq T(q) + q\alpha(q)$. This completes the proof of the lemma.   ∎

We also need the following statement.

Lemma 2. $\Lambda_\alpha = \chi(\tilde{\Lambda}_\alpha)$.

Proof of the lemma. Notice that the measures $\nu^{(s)}$ and $\nu^{(u)}$ are diametrically regular (see Proposition 24.1) and that the measure ν is locally equivalent to their direct product (see Proposition 22.2). One can now apply arguments in the proof of lemma 3 of Theorem 21.1 and conclude that the limit

$$\lim_{r \to 0} \frac{\log \nu(B(x,r))}{\log r}$$

exists simultaneously with the limit (24.4). This completes the proof of the lemma.   ∎

Since $T(1) = 0$ we have that $\mu|R(x) = \eta_1$. The first statement of the theorem now follows from Lemmas 1 and 2 and the following obvious observation:

$$\alpha^{(s)}(0) = \frac{h_\nu(f)}{\lambda_\nu^-}, \quad \alpha^{(u)}(0) = \frac{h_\nu(f)}{\lambda_\nu^+}.$$

Note that by Lemma 2, $\Lambda_{\alpha(q)} = \chi(\tilde{\Lambda}_{\alpha(q)})$. Hence, $\dim_H \Lambda_{\alpha(q)} = T(q) + q\alpha(q)$. This implies Statements 2, 3, and 4. Since the measure ν is diametrically regular and has local product structure one can use (24.2) to prove Statement 5 by repeating arguments in the proof of the last statement of Theorem 21.1.   ∎

Remarks.

(1)   Assume that ν is the measure of full dimension. One can show (see Remark 1 in Section 21) that $T(q) = (1 - q)\dim_H \Lambda$ (thus, $T(q)$ is a linear function) and that $f_\nu(\dim_H \Lambda) = \dim_H \Lambda$ and $f_\nu(\alpha) = 0$ for all $\alpha \neq \dim_H \Lambda$.

(2)   (see Remark 2 in Section 21). Assume that $\nu \neq m$. Then *for any* $\varepsilon > 0$ *there exists a distinct (singular) equilibrium measure* μ *on* Λ *such that* $|\dim_H \mu - \dim_H \nu| \leq \varepsilon$. One can also show that: *the Hausdorff dimension of an equilibrium measure on a* Λ *(corresponding to a Hölder continuous potential) depends continuously on the potential.*

(3)   The function $f_\nu(\alpha)$ has properties described in Remarks 3 and 4 in Section 21.

(4)   We have the following **multifractal decomposition** of the hyperbolic set Λ associated with the pointwise dimension of an equilibrium measure on Λ corresponding to a Hölder continuous function (compare to the case of conformal repellers of smooth expanding maps; see (21.18)):

$$\Lambda = \hat{\Lambda} \cup \left(\bigcup_\alpha \Lambda_\alpha \right), \tag{24.9}$$

where Λ_α is the set of points for which the pointwise dimension takes on the value α and the **irregular part** $\hat\Lambda$ is the set of points with no pointwise dimension. It follows from results in [BS] that $\hat\Lambda \neq \varnothing$; moreover, it is everywhere dense in Λ and $\dim_H \hat\Lambda = \dim_H \Lambda$. We also have that each set Λ_α is everywhere dense in Λ.

Pointwise Dimension of Measures on Hyperbolic Sets

Let ν be an equilibrium measure corresponding to a Hölder continuous function on Λ. According to Statement 1 of Theorem 24.1 for ν-almost every $x \in \Lambda$,

$$d_\nu(x) = h_\nu(f)\left(\frac{1}{\lambda_\nu^+} - \frac{1}{\lambda_\nu^-}\right). \tag{24.10}$$

It follows from a result by Young [Y2] that this formula holds for an arbitrary ergodic measure ν on Λ which is not necessarily an equilibrium measure (see Proposition 26.3; it is a particular case of Theorem 26.1). We present a straightforward proof of this result based upon the approach which was used in the proof of Theorem 21.3.

Theorem 24.2. *Let ν be a Borel ergodic measure on Λ. Then for ν-almost every $x \in \Lambda$ the equality (24.10) holds.*

Proof. Set $d_\nu = h_\nu(f)\left(\frac{1}{\lambda_\nu^+} - \frac{1}{\lambda_\nu^-}\right)$. Let $\mathcal{R} = \{R_1,\ldots,R_p\}$ be a Markov partition for f, (Σ_A,σ) the corresponding symbolic model, and χ the corresponding coding map from Σ_A to Λ (see Section 20). Also, let μ be the pullback of ν by χ.

Fix $\varepsilon > 0$. It follows from the Shannon–McMillan–Breiman theorem that for μ-almost every $\omega \in \Sigma_A$ one can find $N_1(\omega) > 0$ such that for any $n, m \geq N_1(\omega)$,

$$\mu(C_{i_{-m}\ldots i_n}(\omega)) \leq \exp(-(h-\varepsilon)(n+m)),$$

where $C_{i_{-m}\ldots i_n}(\omega)$ is the cylinder set containing ω and $h = h_\mu(\sigma)$ is the measure-theoretic entropy.

It follows from the Birkhoff ergodic theorem, applied to the functions $\log|a^{(u)}(x)|$ and $\log|a^{(s)}(x)|$, that for ν-almost every $x \in \mathcal{M}$ there exists $N_2(x)$ such that for any $n, m \geq N_2(x)$,

$$\int_\mathcal{M} \log|a^{(u)}(x)|d\nu \leq \frac{1}{n}\log\prod_{j=0}^n |a^{(u)}(f^j(x))| + \varepsilon.$$

$$\int_\mathcal{M} \log|a^{(s)}(x)|d\nu \leq \frac{1}{n}\log\prod_{j=-m}^0 |a^{(s)}(f^j(x))| + \varepsilon.$$

For $\omega \in \Sigma_A$ we set $N_2(\omega) = N_2(\chi(\omega))$.

Given $\ell > 0$, denote by $Q_\ell = \{\omega \in \Sigma_A : N_1(\omega) \leq \ell$ and $N_2(\omega) \leq \ell\}$. It is easy to see that the sets Q_ℓ are nested and exhaust Σ_A. Thus, there exists $\ell_0 > 0$ such that $\mu(Q_\ell) > 0$ if $\ell \geq \ell_0$. Let us choose $\ell \geq \ell_0$.

Given $0 < r < 1$, consider a Moran cover $\mathfrak{U}_{r,Q_\ell}$ of the set $\chi(Q_\ell)$. It consists of sets $R_\ell^{(j)}$, $j = 1, \ldots, N_{r,\ell}$ for which there exist points $x_j \in \Lambda$ such that $R_\ell^{(j)} = R_{i_{-m(x_j)} \cdots i_{n(x_j)}}$.

Consider the open Euclidean ball $B(x, r)$ of radius r centered at a point x. Let $N(x, r, \ell)$ denote the number of sets $R_\ell^{(j)}$ that have non-empty intersection with $B(x, r)$. We have that $N(x, r, \ell) \leq M$, where M is a Moran multiplicity factor (which is independent of r and l). It now follows that

$$\nu(B(x,r) \cap \chi(Q_\ell)) \leq \sum_{j=1}^{N(x,r,\ell)} \nu(R_\ell^{(j)}) \leq M \exp(-(h - \varepsilon)(n(x_j) + m(x_j))).$$

Since $\nu(\chi(Q_\ell)) > 0$ by the Borel Density Lemma (see Appendix V) for ν-almost every $x \in \chi(Q_\ell)$ there exists a number $r_0 = r_0(x)$ such that for every $0 < r \leq r_0$ we have

$$\nu(B(x,r)) \leq 2\nu(B(x,r) \cap \chi(Q_\ell)).$$

By virtue of (22.8) this implies that for any $\ell > \ell_0$ and ν-almost every $x \in \chi(Q_\ell)$,

$$\underline{d}_\nu(x) = \varliminf_{r \to 0} \frac{\log \nu(B(x,r))}{\log r} \geq \varliminf_{r \to 0} \frac{\log \nu(B(x,r) \cap \chi(Q_\ell))}{\log r} \geq d_\nu - 2\varepsilon.$$

Since sets Q_ℓ are nested and exhaust the set Q we obtain that $\underline{d}_\nu(x) \geq d_\nu - 2\varepsilon$ for ν-almost every $x \in \Lambda$. Since ε can be arbitrarily small this proves that $\underline{d}_\nu(x) \geq d_\nu$.

We now prove the opposite inequality. Fix $0 < r < 1$. By (22.8) it follows that $R_{i_{-(m(x)+1)} \cdots i_{n(x)+1}} \subset B(x, C_1 r)$, where $C_1 > 0$ is a constant. This implies that

$$\nu(B(x_1, C_1 r)) \geq \nu(R_{i_{-(m(x)+1)} \cdots i_{n(x)+1}}) \geq C_2 \exp(-(h + \varepsilon)(n(x) + m(x))),$$

where $C_2 > 0$ is a constant. By virtue of (22.8) we obtain for all $x \in \Lambda$,

$$\overline{d}_\nu(x) = \varlimsup_{r \to 0} \frac{\log \nu(B(x,r))}{\log r} \leq d_\nu + 2\varepsilon.$$

Since ε can be arbitrarily small this proves that $\underline{d}_\nu(x) \leq d_\nu$. ∎

Since the measure-theoretic entropy is a semi-continuous function we obtain as an immediate corollary of Theorem 22.2 and (24.10) that: *the Hausdorff dimension of a Borel ergodic measure ν on Λ is a semi-continuous function of ν.*

Let ν be again an equilibrium measure corresponding to a Hölder continuous function on Λ and G_μ the set of all forward generic points of ν (i.e., points for which the Birkhoff ergodic theorem holds for any continuous function on X). It follows from Lemma 1 in the proof of Theorem 24.1 (applied to the measure $\nu_1 = \nu$) that for *every* $x \in G_\mu$,

$$d_\nu(x) = \frac{h_\nu(f)}{\lambda_\nu}.$$

In view of Theorems 7.1 and 7.2 this implies that

$$\frac{h_\nu(f)}{\lambda_\nu^-} = \dim_H(G_\nu \cap \mathcal{W}_{\text{loc}}^{(s)}(x)), \quad \frac{h_\nu(f)}{\lambda_\nu^+} = \dim_H(G_\nu \cap \mathcal{W}_{\text{loc}}^{(u)}(x)).$$

We extended this result to any Borel ergodic measure on Λ which is not necessarily an equilibrium measure.

Theorem 24.3. *Let ν be a Borel ergodic measure on Λ. Then for ν-almost every $x \in \Lambda$,*

$$\frac{h_\nu(f)}{\lambda_\nu^-} = \dim_H(G_\nu \cap \mathcal{W}_{\text{loc}}^{(s)}(x)) = \underline{\dim}_B(G_\nu \cap \mathcal{W}_{\text{loc}}^{(s)}(x)) = \overline{\dim}_B(G_\nu \cap \mathcal{W}_{\text{loc}}^{(s)}(x)),$$

$$\frac{h_\nu(f)}{\lambda_\nu^+} = \dim_H(G_\nu \cap \mathcal{W}_{\text{loc}}^{(u)}(x)) = \underline{\dim}_B(G_\nu \cap \mathcal{W}_{\text{loc}}^{(u)}(x)) = \overline{\dim}_B(G_\nu \cap \mathcal{W}_{\text{loc}}^{(u)}(x)).$$

Proof readily repeats arguments in the proof of Theorem 21.3. ∎

Manning [Ma1] proved the part of Theorem 24.2 involving the Hausdorff dimension of the sets $G_\nu \cap \mathcal{W}_{\text{loc}}^{(s)}(x)$ and $G_\nu \cap \mathcal{W}_{\text{loc}}^{(u)}(x)$.

Information Dimension

We study the information dimension of the measure ν. As in the case of conformal repellers one can show that

$$f_\nu(\alpha(1)) = \alpha(1) = -T'(1) = \underline{I}(\nu) = \overline{I}(\nu) = \dim_H \nu,$$

where $\underline{I}(\nu)$ and $\overline{I}(\nu)$ are the lower and upper information dimensions of ν (see Section 18).

We note that Statement 6 of Theorem 24.1 allows us to extend the notion of the Hentschel–Procaccia spectrum and Rényi spectrum for dimensions for any $q \neq 1$ and the above argument defines these spectra for $q = 1$.

Dimension Spectrum for Lyapunov Exponents

Consider the following **multifractal decomposition** of the set Λ associated with positive values of the Lyapunov exponent $\lambda^+(x)$ at points $x \in \Lambda$ (see Section 26):

$$\Lambda = \hat{L}^+ \cup \left(\bigcup_{\beta \in \mathbb{R}} L_\beta^+ \right), \tag{24.11}$$

where

$$\hat{L}^+ = \{x \in \Lambda : \text{the limit in (26.1) does not exist for any } v \in E^{(u)}(x)\}$$

is the **irregular part** and

$$L_\beta^+ = \{x \in \Lambda : \lambda^+(x) = \beta\}.$$

If ν is an ergodic measure for f we obtain that $\lambda^+(x) = \lambda_\nu^{(u)}$ for ν-almost every $x \in \Lambda$. Thus, the set $L_{\lambda_\nu^{(u)}}^+ \neq \varnothing$. Moreover, if ν is an equilibrium measure corresponding to a Hölder continuous function this set is everywhere dense (since in this case the support of ν is the set Λ). We note that if a set L_β is not empty then it supports an ergodic measure ν_β for which $\lambda_{\nu_\beta} = \beta$ (indeed, for every $x \in L_\beta$ the sequence of measures $\frac{1}{n} \sum_{k=0}^{n-1} \delta_{f^k(x)}$ has an accumulation measure whose ergodic components satisfy the above property).

As in the case of conformal repellers there are several fundamental questions related to the above multifractal decomposition, for example:

(1) *Are there points x for which the limit in (26.1) does not exist for any $v \in E^{(u)}(x)$, i.e., $\hat{L}^+ \neq \varnothing$?*
(2) *How large is the range of values of $\lambda^+(x)$?*
(3) *Is there any number β for which any ergodic measure ν with $\nu(L_\beta) > 0$ is* **not** *an equilibrium measure?*

We introduce the **dimension spectrum for (positive) Lyapunov exponents** of f by

$$\ell^+(\beta) = \dim_H L_\beta^+.$$

In [PW3], Pesin and Weiss established the relation between the Lyapunov dimension spectrum and the $f_{\nu_{\max}}(\alpha)$-spectrum, where $\nu_{\max}$ is the measure of maximal entropy. Notice that the measure of maximal entropy is a unique equilibrium measure corresponding to the function $\varphi = 0$ and hence $\psi = \text{constant} = \exp(-h_\Lambda(f))$, where $h_\Lambda(f)$ is the topological entropy of f on Λ. Therefore, for every $x \in L_\beta^+$

$$d_{\nu_{\max}^{(u)}}(x) = \frac{h_\Lambda(f)}{\beta}$$

(recall that $\nu_{\max}^{(u)}$ denotes the conditional measure induced by $\nu_{\max}$ on local unstable manifolds). This implies the following result.

Theorem 24.4. [PW3]

(1) *If $\nu_{\max}^{(u)}|R(x)$ is not equivalent to the measure $m^{(u)}(x)$ for some $x \in \Lambda$ then the Lyapunov spectrum*

$$\ell^+(\beta) = f_{\nu_{\max}^{(u)}}\left(\frac{h_\Lambda(f)}{\beta}\right)$$

is a real analytic strictly convex function on an interval $[\beta_1, \beta_2]$ containing the point $\beta = h_\Lambda(f)/\dim_H(\Lambda \cap \mathcal{W}_{\text{loc}}^{(u)}(x))$.
(2) *If $\nu_{\max}^{(u)}|R(x)$ is equivalent to $m^{(u)}|R(x)$ for some $x \in \Lambda$ then the Lyapunov spectrum is a delta function, i.e.,*

$$\ell^+(\beta) = \begin{cases} \dim_H \Lambda, & \text{for } \beta = h_\Lambda(f)/\dim_H(\Lambda \cap \mathcal{W}_{\text{loc}}^{(u)}(x)) \\ 0, & \text{for } \beta \neq h_\Lambda(f)/\dim_H(\Lambda \cap \mathcal{W}_{\text{loc}}^{(u)}(x)). \end{cases}$$

As immediate consequences of this result we obtain that *if the measure* $\nu_{\max}^{(u)}|R(x)$ *is not equivalent to the measure* $m^{(u)}(x)$ *for some* $x \in \Lambda$ *then the range of the function* $\lambda^+(x)$ *contains an open interval, and hence, the Lyapunov exponent attains uncountably many distinct values.* On the contrary, *if the Lyapunov exponent* $\lambda^+(x)$ *attains only countably many values then* $\nu_{\max}^{(u)}|R(x)$ *is equivalent to* $m^{(u)}|R(x)$ *for some* $x \in \Lambda$.

We can now answer the above questions. Namely,

(1) *The set* $\hat{L}^+$ *is not empty and has full Hausdorff dimension (see Remark (3) in this Section).*

(2) *The range of values of* $\lambda^+(x)$ *is an interval* $[\beta_1, \beta_2]$ *and for any* β *outside this interval the set* L_β^+ *is empty (i.e., the spectrum is **complete**).*

(3) *For any* $\beta \in [\beta_1, \beta_2]$ *there exists an equilibrium measure* ν *corresponding to a Hölder continuous function for which* $\nu(L_\beta) = 1$.

Similar statements hold true for **dimension spectrum for (negative) Lyapunov exponents** of f corresponding to negative values of the Lyapunov exponent $\lambda^-(x)$ at points $x \in \Lambda$.

Appendix IV

A General Concept of Multifractal Spectra; Multifractal Rigidity

Multifractal Spectra

In Sections 19, 21, and 24 we observed two dimension spectra — the dimension spectrum for pointwise dimensions and the dimension spectrum for Lyapunov exponents. The first one captures information about various dimensions associated with the dynamics (including the Hausdorff dimension, correlation dimension, and information dimension of invariant measures) while the second one yields integrated information on the instability of trajectories. These spectra are examples of so-called multifractal spectra which were introduced by Barreira, Pesin, and Schmeling in [BPS2] in an attempt to obtain a refined quantitative description of various multifractal structures generated by dynamical systems. The formal description follows.

Let X be a set, $Y \subset X$ a subset, and $g: Y \to [-\infty, +\infty]$ a function. The *level sets*

$$K_\alpha^g = \{x \in X : g(x) = \alpha\}, \quad -\infty \le \alpha \le +\infty$$

are disjoint and produce a **multifractal decomposition** of X,

$$X = \bigcup_{-\infty \le \alpha \le +\infty} K_\alpha^g \cup \hat{X},$$

where the set $\hat{X} = X \setminus Y$ is called the **irregular part**.

If g is a smooth function on a Euclidean space then the non-empty sets K_α^g are smooth hypersurfaces except for some critical values α. We will be interested in the case when g is not even a continuous (but Borel) function on a metric space so that K_α^g may have a very complicated topological structure.

Now, let G be a set function, i.e., a real function that is defined on subsets of X. Assume that $G(Z_1) \le G(Z_2)$ if $Z_1 \subset Z_2$. We introduce the **multifractal spectrum** specified by the pair of functions (g, G) (or simply the (g, G)-**multifractal spectrum**) as the function $\mathcal{F}: [-\infty, +\infty] \to \mathbb{R}$ defined by

$$\mathcal{F}(\alpha) = G(K_\alpha^g).$$

The function g generates a special structure on X, called the **multifractal structure**, and the function $\mathcal{F}$ captures important information about this structure.

259

Given α, let ν_α be a probability measure on K_α^g. If

$$\mathcal{F}(\alpha) = \inf\{G(Z) : Z \subset K_\alpha^g, \; \nu_\alpha(Z) = 1\}$$

then we call ν_α a (g, G)-**full measure**. Constructing a one-parameter family of (g, G)-full probability measures ν_α seems an effective way of studying multifractal decompositions (see examples of ν_α-measures below).

We consider the case when X is a complete separable metric space. Let $f\colon X \to X$ be a continuous map. There are two *natural* set functions on X. The first one is generated by the metric structure on X:

$$G_D(Z) = \dim_H Z,$$

and the second one is generated by the dynamics on X:

$$G_E(Z) = h_Z(f).$$

Multifractal spectrum generated by the function G_D is called the **dimension spectrum** while multifractal spectrum generated by the function G_E is called the **entropy spectrum**.

There are also three *natural* ways to choose the function g.

(1) Let μ be a Borel finite measure on X. Consider the subset $Y \subset X$ consisting of all points $x \in X$ for which the limit

$$d_\mu(x) = \lim_{r \to 0} \frac{\log \mu(B(x, r))}{\log r}$$

exists (i.e., the lower and upper pointwise dimensions coincide). We set

$$g(x) = d_\mu(x) \stackrel{\text{def}}{=} g_D(x), \quad x \in Y.$$

This leads to two multifractal spectra $\mathcal{D}_D = \mathcal{D}_D^{(\mu)}$ and $\mathcal{D}_E = \mathcal{D}_E^{(\mu)}$ specified respectively by the pairs of functions (g_D, G_D) and (g_D, G_E). We call them the **multifractal spectra for (pointwise) dimensions** (note that $\mathcal{D}_D$ is just the $f_\mu(\alpha)$-spectrum studied before). We stress that these spectra do not depend on the map f.

(2) Assume that the measure μ is invariant with respect to f. Consider a finite measurable partition ξ of X. For every $n > 0$, we write $\xi_n = \xi \vee f^{-1}\xi \vee \cdots \vee f^{-n}\xi$, and denote by $C_{\xi_n}(x)$ the element of the partition ξ_n that contains the point x. Consider the set $Y = Y_\xi \subset X$ consisting of all points $x \in X$ for which the limit

$$h_\mu(f, \xi, x) = \lim_{n \to \infty} -\frac{1}{n} \log \mu(C_{\xi_n}(x))$$

exists. We call $h_\mu(f, \xi, x)$ the **local entropy** of f at the point x (with respect to ξ). Clearly, Y is f-invariant and $h_\mu(f, \xi, f(x)) = h_\mu(f, \xi, x)$ for every $x \in Y$.

By the Shannon–McMillan–Breiman theorem, $\mu(Y) = 1$. In addition, if ξ is a generating partition and μ is ergodic then

$$h_\mu(f) = h_\mu(f, \xi, x)$$

for μ-almost all $x \in X$. Set

$$g(x) = h_\mu(f, \xi, x) \stackrel{\text{def}}{=} g_E(x), \quad x \in Y.$$

We emphasize that g_E may depend on ξ. We obtain two multifractal spectra $\mathcal{E}_D = \mathcal{E}_D^{(\mu)}$ and $\mathcal{E}_E = \mathcal{E}_E^{(\mu)}$ specified respectively by the pairs of functions (g_E, G_D) and (g_E, G_E). These spectra are called **multifractal spectra for (local) entropies** (see similar approach and discussion in [V2]). These spectra provide integrated information on the deviation of local entropy in the Shannon–McMillan–Breiman theorem from its mean value that is the entropy of the map.

(3) Let X be a differentiable manifold and $f: X \to X$ a C^1-map. Consider the subset $Y \subset X$ consisting of all points $x \in X$ for which the limit

$$\lambda(x) = \lim_{n \to +\infty} \frac{1}{n} \log \|d_x f^n\|$$

exists. The function $\lambda(x)$ is measurable and invariant under f. By Kingman's sub-additive ergodic theorem, if μ is an f-invariant Borel probability measure, then $\mu(Y) = 1$. We set

$$g(x) = \lambda(x) \stackrel{\text{def}}{=} g_L(x), \quad x \in Y.$$

This produces two multifractal spectra $\mathcal{L}_D$ and $\mathcal{L}_E$ specified respectively by the pairs of functions (g_L, G_D) and (g_L, G_E). These spectra are called **multifractal spectra for Lyapunov exponents**. It is worth emphasizing that these spectra do not depend on the measure μ.

If μ is ergodic the function $\lambda(x)$ is constant almost everywhere. Let λ_μ denote its value. Multifractal spectra for Lyapunov exponents provide integrated information on the deviation of Lyapunov exponent in Kingman's sub-additive ergodic theorem from its mean value λ_μ.

We consider some examples. First, we analyze four multifractal spectra $\mathcal{D}_D$, $\mathcal{D}_E$, $\mathcal{E}_D$, and $\mathcal{E}_E$ for a subshift of finite type (Σ_A^+, σ).

Denote by $\mathcal{P}$ the class of finite partitions of Σ_A^+ into disjoint cylinder sets (not necessarily all at the same level). Clearly, each $\xi \in \mathcal{P}$ is a generating partition. We use it to define the spectra for entropies $\mathcal{E}_D$ and $\mathcal{E}_E$.

The following theorem establishes the relations between the multifractal spectra for dimensions and entropies.

Theorem A4.1. [BPS2] *For every $\alpha \in \mathbb{R}$, we have*

$$\mathcal{E}_E(\alpha) = \mathcal{E}_D(\alpha) \log \beta = \mathcal{D}_E(\alpha/\log \beta) = \mathcal{D}_D(\alpha/\log \beta) \log \beta \qquad \text{(A4.1)}$$

(recall that $\beta > 1$ is the coefficient in the d_β metric on Σ_A^+; see (A2.18) in Appendix II). In particular, the common value is independent of the partition $\xi \in \mathcal{P}$. Moreover, the multifractal decompositions of the spectra $\mathcal{E}_E$, $\mathcal{E}_D$, $\mathcal{D}_E$, and $\mathcal{D}_D$ coincide, i.e., the families of level sets for these four spectra are equal up to the parameterizations given by (A4.1).

Proof. Notice that there exist constants $C_1 > 0$ and $C_2 > 0$ such that

$$C_1 a^{-n} \leq \operatorname{diam} \xi_n(x) \leq C_2 a^{-n}$$

for every $\xi \in \mathcal{P}$, $x \in \Sigma_A^+$, and $n \geq 1$. If the pointwise dimension $d_\mu(x)$ or the local entropy $h_\mu(f, \xi, x)$ exists for some $x \in \Sigma_A^+$ then

$$h_\mu(f, \xi, x) = \lim_{n\to\infty} -\frac{1}{n} \log \mu(\xi_n(x)) = \log a \lim_{n\to\infty} \frac{\log \mu(\xi_n(x))}{\operatorname{diam} \xi_n(x)} = \log a \cdot d_\mu(x).$$

This shows that $d_\mu(x)$ exists at a point x if and only if $h_\mu(f, \xi, x)$ exists. One can also see that $h_Z(f) = \dim_H Z \cdot \log a$ for any subset $Z \subset \Sigma_A^+$. The desired result follows. ∎

One can now obtain the complete description of all four spectra using Theorem 19.1 (note that the map σ is obviously a continuous weakly-conformal expanding map). We observe that these spectra are complete (see Section 21).

We describe four multifractal spectra $\mathcal{D}_D$, $\mathcal{E}_E$, $\mathcal{L}_D$, and $\mathcal{L}_E$ for equilibrium measures supported on conformal repellers of smooth expanding maps.

Let J be a repeller of a conformal $C^{1+\alpha}$-expanding map f. Consider a Markov partition ξ of J. It is clear that ξ is a generating partition. The same holds true for any partition of J by rectangles obtained from ξ (not necessarily all at the same level) and corresponding to disjoint cylinder sets in Σ_A^+. We denote the class of such partitions by $\mathcal{P}_f$. It is easy to check that $\chi_*^{-1}\xi \in \mathcal{P}$ for every partition $\xi \in \mathcal{P}_f$.

Let φ be a Hölder continuous function on J and μ the corresponding Gibbs measure with respect to f. Write $\log \psi = \varphi - P(\varphi)$.

For each $q, p \in \mathbb{R}$ consider the functions

$$\varphi_{D,q} = -T_D(q) \log |a| + q \log \psi, \quad \varphi_{E,p} = -T_E(p) + p \log \psi, \qquad (A4.2)$$

where the numbers $T_D(q)$ and $T_E(p)$ are chosen such that

$$P(\varphi_{D,q}) = P(\varphi_{E,p}) = 0.$$

Clearly, $T_E(p) = P(p \log \psi)$. As in Section 21 one can show that the functions $T_D(q)$ and $T_E(p)$ are real analytic decreasing and convex. Set

$$\alpha_D(q) = -T_D'(q), \quad \alpha_E(p) = -T_E'(p).$$

The complete description of the spectra $\mathcal{D}_D$ and $\mathcal{L}_D$ is given by Theorems 21.1 and 21.4 and the corresponding multifractal decompositions are (21.18) and (21.25) respectively. Since the coding map preserves the entropy the complete description of the spectrum $\mathcal{E}_E$ follows immediately from Theorem A4.1. More precisely, the following statement holds.

Theorem A4.2.

(1) *There exists a set $S \subset X$ with $\mu(S) = 1$ such that for every partition $\xi \in \mathcal{P}_f$ and every $x \in S$, the local entropy of μ at x exists, does not depend on x and ξ, and*

$$g_E(x) = h_\mu(f, \xi, x) = -\int_J \log \psi \, d\mu.$$

(2) *The function T_E is real analytic, and satisfies $T_E'(p) \leq 0$ and $T_E''(p) \geq 0$ for every $p \in \mathbb{R}$. We have $T_E(0) = h_J(f)$ and $T_E(1) = 0$.*

(3) *The domain of the function $\alpha \mapsto \mathcal{E}_E(\alpha)$ is a closed interval in $[0, +\infty)$ and coincides with the range of the function $\alpha_E(p)$. For every $p \in \mathbb{R}$, we have*

$$\mathcal{E}_E(\alpha_E(p)) = T_E(p) + p\alpha_E(p).$$

(4) *If μ is not the measure of maximal entropy then $\mathcal{E}_E$ and T_E are analytic strictly convex functions, and hence $(\mathcal{E}_E, T_E)$ form a Legendre transfer pair with respect to the variables α, q.*

(5) *If μ is the measure of maximal entropy then $\mathcal{E}_E$ is the delta function*

$$\mathcal{E}_E(\alpha) = \begin{cases} h & \text{if } \alpha = h \\ 0 & \text{if } \alpha \neq h \end{cases}.$$

We now give a complete description of the spectrum $\mathcal{L}_E$.

Theorem A4.3. [BPS2] *For every $\alpha \in \mathbb{R}$ we have*

$$\mathcal{L}_E(\alpha) = \mathcal{E}_E^{(m)}(\alpha \dim_H J),$$

where m is the measure of full dimension (see Theorem 20.1).

Proof. Since m is the Gibbs measure corresponding to $-s \log |a|$ (where $s = -\dim_H J$), by Proposition 20.2, for every $x = \chi(i_0 i_1 \dots) \in J$ we have

$$d_m(x) = \lim_{n \to \infty} \frac{\log m(R_{i_0 \dots i_n})}{\log \prod_{k=0}^n |a(f^k(x))|^{-1}} = \dim_H J.$$

Therefore, for every $x \in J \cap K_\alpha^{g_L}$ and $\xi \in \mathcal{P}_f$,

$$h_m(f, \xi, x) = \lim_{n \to \infty} -\frac{1}{n} \log m(R_{i_0 \dots i_n}) = s \lim_{n \to \infty} -\frac{1}{n} \log \operatorname{diam} R_{i_0 \dots i_n} = s\alpha$$

and hence $x \in K_{d\alpha}^{g_E}$. This implies that $\mathcal{L}_E(\alpha) = \mathcal{E}_E^{(m)}(\alpha \dim_H J)$. ∎

Let us notice that the Gibbs measures corresponding to the functions $\varphi_{D,q}$ and $\varphi_{E,p}$ (see (A4.2)) are (g, G)-full measures of the corresponding spectra.

It is an open problem to obtain a complete description of the spectra $\mathcal{D}_E$ and $\mathcal{E}_D$ for equilibrium measures on conformal repellers of smooth expanding maps.

Irregular Parts of Multifractal Decompositions

Consider the set

$$\mathcal{I} = \left\{ x \in J : \lim_{n \to \infty} \frac{1}{n} S_n g(x) \text{ does not exist for some } g \in C(J) \right\} = J \setminus \bigcup_{\mu \in \mathfrak{M}(J)} G_\mu$$

where $S_n g = \sum_{k=0}^{n-1} g \circ f^k$ (recall that G_μ is the set of all forward generic points of the measure μ; see (A2.16); $\mathfrak{M}(J)$ is the set of all Borel ergodic measures on J). This set is called **metrically irregular**. Note that $\mathcal{I}$ has zero measure with respect to any f-invariant measure. In [BS], Barreira and Schmeling demonstrated the surprising phenomenon that the metrically irregular set is *observable*, i.e., it has *full* Hausdorff dimension and carries *full* topological entropy. More precisely, they proved the following result.

Theorem A4.4. *Let J be a repeller for a $C^{1+\alpha}$-conformal expanding map f. Then $\dim_H \mathcal{I} = \dim_H J$ and $h_\mathcal{I}(f) = h_J(f)$.*

In the numerical study of dynamical systems one usually collects the data by observing *typical* trajectories, i.e., trajectories that correspond to points in G_μ for some invariant measure μ. This is based on the common belief that a computer always picks points "randomly" and thus, reproduces only "typical" orbits. Theorem A4.4 demonstrates that some fundamental characteristics of dynamical systems can be in fact computed while dealing with *non-typical* trajectories. Although such trajectories cannot be observed by random scanning of the phase space they can be obtained by "mixing up" two typical trajectories corresponding to two *distinct* invariant measures (in the case of the full shift on two symbols the procedure of "mixing up" two typical trajectories is explained in the proof of Proposition A2.1; for general subshifts a more sophisticated procedure is described in [BS]).

Metrically irregular sets are closely related to irregular parts of multifractal decompositions generated by the dimension spectrum for pointwise dimensions (i.e., the set $\hat{J}$ in (21.18)) and the dimension spectrum for Lyapunov exponents (i.e., the set $\hat{L}$ in (21.25)). Barreira and Schmeling showed that these irregular parts are *observable*, i.e., they have *full* Hausdorff dimension and carry *full* topological entropy.

Theorem A4.5. *Let J be a repeller for a $C^{1+\alpha}$-conformal expanding map f.*

(1) *Let μ be an equilibrium measure corresponding to a Hölder continuous function. Assume that μ is not the measure of full dimension. Then $\dim_H \hat{J} = \dim_H J$ and $h_{\hat{J}}(f) = h_J(f)$.*

(2) *If the measure of full dimension and the measure of maximal entropy are distinct then $\dim_H \hat{L} = \dim_H J$ and $h_{\hat{L}}(f) = h_J(f)$.*

The fact that the set $\hat{J}$ has positive Hausdorff dimension was first observed by Shereshevsky [Sh]. The proofs of Theorems A4.4 and A4.5 exploit the notion of φ-dimension of invariant measures (see Appendix III). We sketch these proofs.

Given a collection of sequences of continuous functions

$$G^{(i)} = \{g_n^{(i)} \colon J \to \mathbb{R}^+\}_{n \in \mathbb{N}}$$

for $i = 1, \ldots, m$, we define the corresponding **metrically irregular** set

$$\mathcal{I}(G^{(1)}, \ldots, G^{(m)}) = \left\{ x \in J : \varliminf_{n \to \infty} g_n^{(i)}(x) < \varlimsup_{n \to \infty} g_n^{(i)}(x) \text{ for } i = 1, \ldots, m \right\}.$$

We say that a collection of measures $\mu^{(i)}$, $i = 1, \ldots, k$, $k \leq 2m$ *distinguishes* the collection $\{G^{(1)}, \ldots, G^{(m)}\}$ if for every $1 \leq i \leq m$, there exist distinct integers $j_1 = j_1(i)$, $j_2 = j_2(i) \in [1, k]$ and numbers $a_{j_1}^{(i)} \neq a_{j_2}^{(i)}$ such that

$$\lim_{n \to \infty} g_n^{(i)}(x) = a_{j_1}^{(i)} \text{ for } \mu_{j_1}\text{-almost every } x,$$

$$\lim_{n \to \infty} g_n^{(i)}(x) = a_{j_2}^{(i)} \text{ for } \mu_{j_2}\text{-almost every } x.$$

The main result in [BS] claims the following.

Lemma.

(1) *Let ψ_i, $i = 1, \ldots, m$ be Hölder continuous functions on J. Assume that each ψ_i is cohomologous neither to the function $t = -s \log |a|$ (where $s = \dim_H J$) nor to the function $t = \text{const}$. Then for any $\varepsilon > 0$ there exists a collection of measures $\mu^{(i)}$, $i = 1, \ldots, 2m$ which satisfies the following conditions:*

   a) it distinguishes the collection of sequences of functions

$$G^{(i)} = \left\{ g_n^{(i)} = \frac{S_n \psi_i}{S_n t} \right\}_{n \in \mathbb{N}}, \ i = 1, \ldots, m;$$

   b)

$$\min_{1 \leq i \leq k} BS_\varphi(\mu_i) \geq BS_\varphi(J) - \varepsilon,$$

   where φ is a strictly positive Hölder continuous function and $BS_\varphi(\nu)$ (respectively, $BS_\varphi(J)$) is the Barreira–Schmeling dimension of the measure ν (respectively, of the repeller J); see Appendix III.

(2) *Let $\mu^{(i)}$, $i = 1, \ldots, k$ be a collection of measures which distinguishes a collection of sequences of functions $\{G^{(1)}, \ldots, G^{(m)}\}$, $k \leq 2m$; then for every Hölder continuous function φ on J we have*

$$BS_\varphi(\mathcal{I}(G^{(1)}, \ldots, G^{(m)})) \geq \min_{1 \leq i \leq k} BS_\varphi(\mu_i).$$

We proceed with the proof of Theorem A4.4. Fix $\varepsilon > 0$. Since Hölder continuous functions are dense in the space of continuous functions we obtain that the set $\mathcal{I}$ consists of all points $x \in J$ for which there exists a Hölder continuous function ψ satisfying

$$\varliminf_{n \to \infty} \frac{1}{n} S_n \psi(x) < \varlimsup_{n \to \infty} \frac{1}{n} S_n \psi(x).$$

Therefore, $\mathcal{I}(\Psi) \subset \mathcal{I}$, where $\Psi = \{S_n\psi\}_{n \in \mathbb{N}}$. One can choose a Hölder continuous function ψ which is not cohomologous either to the function $t = -s \log |a|$ or to the function $t = \text{const}$. Applying Lemma with $t = 1$, $\varphi = 1$, $m = 1$, and $\psi_1 = \psi$ we obtain in view of Remark 1 in Appendix III that

$$h_{\mathcal{I}}(f) \geq h_{\mathcal{I}(\Psi)}(f) = BS_\varphi(\mathcal{I}(\Psi)) \geq BS_\varphi(J) - \varepsilon = h_J(f) - \varepsilon.$$

The result about the topological entropy follows since ε is arbitrary.

Set $\varphi = \log |a|$. In view of Remark 2 in Appendix III we obtain that

$$\dim_H J = BS_\varphi(J).$$

Using Remark in Section 20 we also conclude that

$$\dim_H \mathcal{I}(\Psi) = BS_\varphi(\mathcal{I}(\Psi)).$$

Applying Lemma again with $t = 1$, $\varphi = \log |a|$, $m = 1$, and $\psi_1 = \psi$ we conclude that

$$\dim_H \mathcal{I} \geq \dim_H \mathcal{I}(\Psi) = BS_\varphi(\mathcal{I}(\Psi)) \geq BS_\varphi(J) - \varepsilon = \dim_H J - \varepsilon.$$

Since ε is arbitrary this yields the statement about the Hausdorff dimension and concludes the proof of Theorem A4.4.

We leave the proof of Theorem A4.5 to the reader. Hint: to prove the first statement choose the function ψ such that $\log \psi = \xi - P_J(\xi)$, where ξ is the potential of the measure μ; then apply Lemma with $t = -s \log |a|$, $m = 1$, $\psi_1 = \psi$ and set $\varphi = 1$ to establish equality of topological entropies and $\varphi = \log |a|$ to establish equality of Hausdorff dimensions; to prove the second statement choose $\psi = \log |a|$ and apply Lemma with $t = 1$, $m = 1$, $\psi_1 = \psi$ and set $\varphi = \log |a|$ or $\varphi = 1$.

Multifractal Rigidity

The dimension and entropy multifractal spectra capture important information about dynamics including information on the geometry of invariant sets and on invariant measures. Therefore, they can be used to identify main *macro-characteristics* of the system (such as distributions of Lyapunov exponents and topological entropy, dimension of invariant sets, etc.) — the phenomenon called **multifractal rigidity**. In other words, one can view multifractal spectra as "generalized degrees of freedom" to be used to "restore" the dynamics.

This leads to a new type of classification of dynamical systems which takes care of various aspects of the dynamics (chaotic behavior, instability, geometry, etc.) and fits better with the physical intuition of the equivalence of dynamical systems.

It is well-known that dynamical systems can be classified topologically (up to homeomorphisms) or measure-theoretically (up to measure-preserving automorphisms). From a physical point of view, these classifications trace separate "independent" characteristics of the dynamics. One can use multifractal spectra for **multifractal classification** of dynamical systems which combines features of each of the above classifications. The new classification has a strong physical content and identifies two systems up to a change of variables. It is much more rigid than the topological and measure-theoretic classifications and establishes the coincidence of dimension characteristics as well as the correspondence between invariant measures.

We illustrate the multifractal rigidity phenomenon by considering two one-dimensional linear Markov maps of the unit interval, modeled by the full shift on two symbols. Recall that this means that there are linear maps f_1 and f_2 defined respectively on two disjoint closed intervals I_1, $I_2 \subset [0,1]$ such that $f_1(I_1) = f_2(I_2) = [0,1]$, and the map $f: I_1 \cup I_2 \to \mathbb{R}$ is given by $f(x) = f_i(x)$ whenever $x \in I_i$, for $i = 1, 2$ (see Figure 15).

Let J be the repeller for f. The partition $\{J \cap I_1, J \cap I_2\}$ is a Markov partition of J and $f|J$ is topologically conjugate to the full shift $\sigma|\Sigma_2^+$.

We consider the Bernoulli measure on Σ_2^+ with probabilities β_1 and $\beta_2 = 1 - \beta_1$ (which is a Gibbs measure), and let $c_i = |f'|I_i| = |f_i'|I_i|$ for $i = 1, 2$.

We define the functions a and ϕ on J by

$$a(x) = c_i \text{ and } \phi(x) = \log \beta_i \quad \text{if } x \in I_i,$$

for $i = 1, 2$. For every $p, q \in \mathbb{R}$, the functions $T_E(p)$ and $T_D(q)$ satisfy the identities

$$e^{-T_E(p)}(\beta_1{}^p + \beta_2{}^p) = 1$$

and

$$c_1^{-T_D(q)}\beta_1{}^q + c_2^{-T_D(q)}\beta_2{}^q = 1. \tag{A4.3}$$

One can explicitly compute the measures $\nu_p^{\mathcal{E}_E}$ and $\nu_q^{\mathcal{D}_D}$: they are the Bernoulli measures with probabilities $e^{-T_E(p)}\beta_1{}^p$, $e^{-T_E(p)}\beta_2{}^p$ and $c_1^{-T_D(q)}\beta_1{}^q$, $c_2^{-T_D(q)}\beta_2{}^q$ respectively.

Let f and $\widehat{f}$ be two one-dimensional linear Markov maps of the unit interval as above with conformal repellers J and $\widehat{J}$ respectively. Let also $\chi: \Sigma_2^+ \to J$ and $\widehat{\chi}: \Sigma_2^+ \to \widehat{J}$ be the corresponding coding maps. We consider two Bernoulli measures μ and $\widehat{\mu}$ on Σ_2^+ with probabilities β_1, β_2 and $\widehat{\beta}_1$, $\widehat{\beta}_2$ respectively, where $\beta_1 + \beta_2 = \widehat{\beta}_1 + \widehat{\beta}_2 = 1$. We also consider the numbers c_1, c_2 and $\widehat{c}_1$, $\widehat{c}_2$ which are the absolute values of the derivatives of the linear pieces of f and $\widehat{f}$ respectively.

Define the functions a and ϕ on J as well as the functions $\widehat{a}$ and $\widehat{\phi}$ on $\widehat{J}$ as above.

Recall that an automorphism ρ of Σ_2^+ is a homeomorphism $\rho: \Sigma_2^+ \to \Sigma_2^+$ which commutes with the shift map σ. The involution automorphism is defined by $\rho(i_1 i_2 \ldots) = (i_1' i_2' \ldots)$, where $i_n' = 2$ if $i_n = 1$ and $i_n' = 1$ if $i_n = 2$ for each integer $n \geq 1$.

Since χ and $\widehat{\chi}$ are invertible one can define a homeomorphism $\theta: J \to \widehat{J}$ by $\theta = \widehat{\chi} \circ \chi^{-1}$. We note that $\theta \circ f = \widehat{f} \circ \theta$ on J and hence θ is a topological conjugacy between $f|J$ and $\widehat{f}|\widehat{J}$. If ρ is an automorphism of Σ_2^+, then the homeomorphism $\theta' = \widehat{\chi} \circ \rho \circ \chi^{-1}$ is also a topological conjugacy between $f|J$ and $\widehat{f}|\widehat{J}$, and all topological conjugacies are of this form.

We consider the spectrum $\mathcal{D}_D = \mathcal{D}_D^{(\mu)}$ specified by the measure μ as well as the spectrum $\widehat{\mathcal{D}}_D = \mathcal{D}_D^{(\widehat{\mu})}$ specified by the measure $\widehat{\mu}$.

Theorem A4.6. [BPS2] *If $\mathcal{D}_D(\alpha) = \widehat{\mathcal{D}}_D(\alpha)$ for every α and these spectra are not delta functions, then there is a homeomorphism $\zeta: J \to \widehat{J}$ such that:*

(1) *$\zeta \circ f = \widehat{f} \circ \zeta$ on J, that is, ζ is a topological conjugacy between $f|J$ and $\widehat{f}|\widehat{J}$;*

(2) *the automorphism ρ of Σ_2^+ satisfying $\chi \circ \zeta = \rho \circ \widehat{\chi}$ is either the identity or the involution automorphism;*

(3) *$a = \widehat{a} \circ \zeta$, i.e., either $c_1 = \widehat{c}_1$ and $c_2 = \widehat{c}_2$ or $c_1 = \widehat{c}_2$ and $c_2 = \widehat{c}_1$;*

(4) *$\phi = \widehat{\phi} \circ \zeta$, and $\mu = \widehat{\mu} \circ \zeta$.*

Proof. It is sufficient to prove that the spectrum $\mathcal{D}_D$ uniquely determines the numbers β_1, β_2, c_1, and c_2 up to a permutation of the indices 1 and 2.

By the uniqueness of the Legendre transform the spectrum $\mathcal{D}_D$ uniquely determines $T_D(q)$ for every $q \in \mathbb{R}$. Therefore, it remains to show that one can determine the numbers β_1, β_2, c_1, and c_2 uniquely up to a permutation of the indices 1 and 2 from the equation (A4.3).

One can verify that the numbers $\alpha_\pm = \alpha_D(\pm\infty)$ can be computed by

$$\alpha_\pm = - \lim_{q \to \pm\infty} (T_D(q)/q).$$

We observe that since the spectrum $\mathcal{D}_D$ is not a delta function, $T_D(q)$ is not linear and is strictly convex. Hence, $\alpha_+ < \dim_H J < \alpha_-$. Therefore, raising both sides of the equation (A4.3) to the power $1/q$ and letting $q \to \pm\infty$ we obtain

$$\max\{\beta_1 c_1^{\alpha_+}, \beta_2 c_2^{\alpha_+}\} = \min\{\beta_1 c_1^{\alpha_-}, \beta_2 c_2^{\alpha_-}\} = 1.$$

We assume that $\beta_1 c_1^{\alpha_+} = 1$ (the case $\beta_2 c_2^{\alpha_+} = 1$ can be treated in a similar way; in this case, ρ is the involution automorphism). Since $\alpha_+ < \alpha_-$ we must have $\beta_2 c_2^{\alpha_-} = 1$.

Setting $q = 0$ and $q = 1$ in the equation (A4.3) we obtain respectively,

$$c_1^{-\dim_H J} + c_2^{-\dim_H J} = 1 \quad \text{and} \quad \beta_1 + \beta_2 = 1.$$

Set $x = c_1^{-\dim_H J}$, $\gamma = \alpha_+/\dim_H J < 1$, and $b = \alpha_-/\dim_H J > 1$. Then one can easily derive the equation

$$x^\gamma + (1 - x)^b = 1.$$

We leave it as an easy exercise to the reader to show that this equation has a unique solution $x \in (0, 1)$ which uniquely determines the numbers c_1 and c_2 and hence also the numbers β_1 and β_2.   ∎

Remarks.

(1) We have shown that for a one-dimensional linear Markov map of the unit interval one can determine the four numbers β_1, β_2, c_1, and c_2 using the spectrum $\mathcal{D}_D$. If instead the spectrum $\mathcal{E}_E$ is used then only the numbers β_1 and β_2 can be recovered: one can show that if $\beta_1 \geq \beta_2$, then

$$\beta_1 = \exp \lim_{p \to +\infty} (T_E(p)/p) \quad \text{and} \quad \beta_2 = \exp \lim_{p \to -\infty} (T_E(p)/p).$$

In a similar way, using one of the spectra $\mathcal{L}_D$ or $\mathcal{L}_E$ one can determine only the numbers c_1 and c_2.

(2) The multifractal rigidity for two-dimensional horseshoes is demonstrated in [BPS3].

Chapter 8

Relations between Dimension, Entropy, and Lyapunov Exponents

In the previous chapters of the book we have seen that the pointwise dimension is a useful tool in computing the Hausdorff dimension and box dimension of measures and sets. The key idea is to establish whether a measure is exact dimensional (i.e., its lower and upper pointwise dimensions coincide and are constant almost everywhere). If this is the case then by Theorem 7.1, the Hausdorff dimension and lower and upper box dimensions of the measure coincide. This also gives an effective lower bound of the Hausdorff dimension of the set which supports the measure. One can obtain an effective upper bound of the Hausdorff dimension of the set by estimating the lower pointwise dimension at every point of the set and applying Theorem 7.2.

In the previous chapters of the book we described several classes of measures, invariant under dynamical systems, which are exact dimensional. We also obtained formulae for their pointwise dimension. These classes include Gibbs measures concentrated on limit sets of geometric constructions (CPW1–CPW4) (see Theorem 15.4), invariant ergodic measures supported on conformal repellers (see Theorem 21.3) or two-dimensional locally maximal hyperbolic sets (see Theorem 24.2).

In [ER], Eckmann and Ruelle discussed dimension of hyperbolic measures (i.e., measures invariant under diffeomorphisms with non-zero Lyapunov exponents almost everywhere). This led to the problem of whether a hyperbolic ergodic measure is exact dimensional. This problem has later become known as the Eckmann–Ruelle conjecture and has been acknowledged as one of the main problems in the interface of dimension theory and dynamical systems. Its role in the dimension theory of dynamical systems is similar to the role of the Shannon–McMillan–Breiman theorem in ergodic theory.

In [Y2], Young showed that hyperbolic measures invariant under surface diffeomorphisms are exact dimensional. Later Ledrappier [L] proved theis result for general Sinai–Ruelle–Bowen measures. In [PY], Pesin and Yue extended his approach to hyperbolic measures satisfying the so-called semi-local product structure (see below; this class includes, for example, Gibbs measures on locally maximal hyperbolic sets). Barreira, Pesin, and Schmeling [BPS1] obtained the complete affirmative solution of the Eckmann–Ruelle conjecture which we present in this chapter.

We also demonstrate that neither of the assumptions in the Eckmann–Ruelle conjecture can be omitted. Ledrappier and Misiurewicz [LM] constructed an example of a smooth map of a circle preserving an ergodic measure with zero Lyapunov exponent which is not exact dimensional (see Example 25.3 below). In [PW1], Pesin and Weiss presented an example of a Hölder homeomorphism whose measure of maximal entropy is not exact dimensional (see Example 25.1 below). Barreira and Schmeling [BS] showed that for "almost" any Gibbs measure on a two-dimensional hyperbolic set the set of points, where the lower and upper pointwise dimensions do not coincide, has full Hausdorff dimension (see Appendix IV).

25. Existence and Non-existence of Pointwise Dimension for Invariant Measures

We begin with examples that illustrate some problems of the existence of pointwise dimension.

Our first example shows that the pointwise dimension may not exist even for measures of maximal entropy invariant under a *continuous* map.

Example 25.1. [PW1] *There exists a geometric construction with rectangles modeled by the full shift on two symbols (see Section 16) such that the corresponding induced map G on the limit set F of the construction is a Hölder continuous endomorphism for which the unique measure of maximal entropy m satisfies $\underline{d}_m(x) < \overline{d}_m(x)$ for almost every $x \in F$.*

Proof. We begin with the geometric construction presented in Example 16.2. Note that for this construction

$$\underline{s} \stackrel{\text{def}}{=} s_{\underline{\lambda}} = \frac{\log 2}{-\log \underline{\lambda}}, \quad \overline{s} \stackrel{\text{def}}{=} s_{\overline{\lambda}} = \frac{\log 2}{-\log \overline{\lambda}}.$$

Hence, the functions $\varphi(\omega) = \underline{s}\log\underline{\lambda}$ and $\overline{\varphi}(\omega) = \overline{s}\log\overline{\lambda}$ defined on Σ_2^+ coincide and are constant ($= \log 2$). This implies that the Gibbs measures with respect to the shift, corresponding to these functions, also coincide: $\underline{\mu}_{\underline{\lambda}} = \overline{\mu}_{\lambda} \stackrel{\text{def}}{=} \mu$; moreover, μ is the measure of maximal entropy for the full shift σ (see Appendix II). Consider the induced map G. It is a Hölder continuous endomorphism. Obviously, the measure m, which is the push forward of μ to F, is the invariant measure for G of maximal entropy. We describe its lower and upper pointwise dimensions.

Lemma. *For m-almost every $x \in F$ we have*

$$\underline{d}_m(x) = \underline{s} = \frac{\log 2}{-\log \underline{\lambda}}, \quad \overline{d}_m(x) = \overline{s} = \frac{\log 2}{-\log \overline{\lambda}}.$$

Proof of the lemma. The fact that $\dim_H F = \underline{s}$ immediately implies that $\underline{d}_m(x) \leq \underline{s}$ for m-almost every $x \in F$. Otherwise there would exist a set A of positive m-measure with $\underline{d}_m(x) \geq \underline{s} + \varepsilon$ for any $x \in A$. The non-uniform mass distribution principle would then imply that $\dim_H F \geq \dim_H A \geq \underline{s} + \varepsilon$. The first equality now follows from Statement 2 of Theorem 16.1.

In order to prove the second statement consider $r_k = \overline{\lambda}^{n_{3k+1}}$ and denote by $\Delta_k(x)$ the unique cylinder set $\Delta_{i_1 \ldots i_{n_{3k+1}}}$ that contains $x \in F$. It is easy to see that $B(x, r_k) \cap F \subset \Delta_k(x) \cap F$. Since the measure μ is a Gibbs measure the inequalities (13.6) imply that for all $x \in F$,

$$m(B(x, r_k) \cap F) \leq m(\Delta_k(x) \cap F) \leq D_1 \overline{\lambda}^{\overline{s} n_{3k+1}} = D_1 r_k^{\overline{s}}$$

and hence

$$\overline{d}_m(x) \geq \varlimsup_{k \to \infty} \frac{\log m(B(x, r_k) \cap F)}{\log r_k} \geq \overline{s}.$$

The second equality now follows from Statement 2 of Theorem 16.1. ∎

As we have mentioned the map G constructed above is a Hölder continuous endomorphism but not a one-to-one map. Using the map G we now construct a Hölder continuous homeomorphism $\tilde{G}$ for which the measure of maximal entropy is not exact dimensional.

Example 25.1′. [PW1] *There exists a Hölder continuous homeomorphism of a compact subset in $\mathbb{R}^4$ with positive topological entropy for which the unique measure of maximal entropy has different lower and upper pointwise dimensions at almost every point; moreover, the homeomorphism is Hölder continuously conjugate to the full shift on two symbols.*

Proof. Consider the set $\tilde{F} = F \times F$ endowed with the metric

$$\tilde{\rho}((x_1, y_1), (x_2, y_2)) = \rho(x_1, x_2) + \rho(y_1, y_2), \qquad x_1, x_2, y_1, y_2 \in F$$

and the coding map $\tilde{\chi} \colon \Sigma_p \to \tilde{F}$ defined by $(x, y) = \tilde{\chi}(\ldots i_{-1} i_0 i_1 \ldots)$ where Σ_p denotes the space of two-sided sequences $(\ldots i_{-1} i_0 i_1 \ldots)$, $i_j = 1, \ldots, p$ and $x = \chi(\ldots i_{-1})$, $y = \chi(i_0 i_1 \ldots)$. Set $\tilde{G} = \tilde{\chi} \circ \sigma \circ \tilde{\chi}^{-1}$. It is easy to see that $\tilde{G}$ is a Hölder continuous homeomorphism and that for any $(x, y) \in \tilde{F}$,

$$\pi_1 \tilde{G}(x, y) = G(x), \quad \pi_2 \tilde{G}^{-1}(x, y) = G(y)$$

where π_1, π_2 are the projections defined by $\pi_1(x, y) = x$ and $\pi_2(x, y) = y$. Moreover, since G is expanding the map $\tilde{G}$ is *topologically hyperbolic*. Consider the measure $\tilde{m} = \tilde{\chi}^* \tilde{\mu}$ where $\tilde{\mu}$ is the measure on Σ_p^+ defined by

$$\tilde{\mu}(\Delta_{i_k \ldots i_n}) = \underline{\lambda}^{(n-k)\underline{s}} = \overline{\lambda}^{(n-k)\overline{s}} = 2^{-(n-k)}.$$

By virtue of Theorem 13.3 $\tilde{\mu}$ is invariant under σ and hence $\tilde{m}$ is invariant under $\tilde{G}$. It is easy to see that $\tilde{m} = m \times m$. It follows that for $\tilde{m}$-almost every (x, y)

$$\underline{d}_{\tilde{m}}(x, y) = \underline{d}_m(x) + \underline{d}_m(y) = \underline{s}, \quad \overline{d}_{\tilde{m}}(x, y) = \overline{d}_m(x) + \overline{d}_m(y) = \overline{s}.$$

It is not difficult to check that the measure $\tilde{m}$ (after a proper normalization) is the measure of maximal entropy for $\tilde{G}$. ∎

We remark that the Hölder exponent of the map G is $\log\overline{\lambda}/\log\underline{\lambda}$ (see Example 16.2) and hence can be chosen arbitrarily close to 1.

The following example illustrates that the pointwise dimension may exist almost everywhere but the measure, nevertheless, may not be exact-dimensional (see Section 7). We remark that this phenomenon may happen for measures invariant under continuous maps but never occurs for measures invariant under smooth maps (see Theorem 7.3).

Example 25.2.

(1) *There exists a geometric construction (CB1–CB3) on $[0,1]$ modeled by a subshift of finite type and determined by a sequence of affine maps such that*
   (a) the basic sets at each step of the construction are disjoint;
   (b) the induced map G on the limit set F is Hölder continuous;
   (c) there exists a G-invariant ergodic measure on F of positive entropy whose pointwise dimension exists almost everywhere but is not constant.
(2) *There exists a Hölder homeomorphism of a compact subset in $\mathbb{R}^4$ that possesses an invariant ergodic measure with positive entropy whose pointwise dimension exists almost everywhere but is not constant.*

Proof. Our approach follows Pesin and Weiss [PW1] and is a refined version of the approach by Cutler [Cu]. Consider a geometric construction (CB1–CB3) on $[0,1]$ with non-stationary ratio coefficients modeled by a subshift of finite type. We assume that $p = 3$ and the ratio coefficients $\lambda_{i,n}$, $i = 1, 2, 3$ and $n = 1, 2, 3, \ldots$ depend only on the steps of the constructions and are given by

$$\lambda_{1,n} = \lambda_{3,n} = \begin{cases} \alpha & \text{if } n \text{ is even} \\ \beta & \text{if } n \text{ is odd,} \end{cases}$$

$$\lambda_{2,n} = \begin{cases} \gamma, & \text{if } n \text{ is even} \\ \delta, & \text{if } n \text{ is odd,} \end{cases}$$

where $0 < \alpha \leq \beta < \frac{1}{3}$, $0 < \gamma \leq \delta < \frac{1}{3}$ and $\alpha\delta \neq \gamma\beta$. We also assume that the transfer matrix A is given by

$$A = \begin{pmatrix} 0 & 1 & 0 \\ 1 & 0 & 1 \\ 0 & 1 & 0 \end{pmatrix}.$$

We need the following lemma.

Lemma. *Let $\{\lambda_{i,n}\}$, $i = 1, \ldots, p$, $n = 1, 2, \ldots$ be sequences of numbers satisfying: $0 < a \leq \lambda_{i,n} \leq b < 1$ and for any n*

$$\sum_{i=1}^{p} \lambda_{i,n} = \lambda < 1.$$

Then there exist a sequence of affine maps $\{h_{i,n}\}$ and a geometric construction (CB1–CB3) on $[0, 1]$ modeled by a given symbolic dynamical system (Q, σ) such that

(1) *each basic set $\Delta_{i_0\ldots i_n} = h_{i_0,n} \circ \cdots \circ h_{i_n,n}([0,1])$;*
(2) *$\Delta_{i_0\ldots i_n} \cap \Delta_{j_0\ldots j_n} = \varnothing$ if $(i_0\ldots i_n) \neq (j_0\ldots j_n)$;*
(3) *the induced map G on the limit set F is Hölder continuous.*

Proof of the lemma. For each $n = 1, 2, \ldots$ define affine maps $h_{i,n}(x) = \lambda_{i,n}x + a_{i,n}$ and choose $a_{i,n}$ such that the sets $h_{i,n}([0,1])$ are at least $\frac{\lambda}{2p}$ apart. The result follows. ∎

It is easy to check that Conditions (15.3) and (15.5) hold. One can also see that the following limit exists:

$$\log \lambda_i \overset{\text{def}}{=} \lim_{n\to\infty} \frac{1}{n} \sum_{k=1}^{n} \log \lambda_{i,k} = \begin{cases} \frac{1}{2}\log(\alpha\delta), & \text{if } i = 1, 3 \\ \frac{1}{2}\log(\beta\gamma), & \text{if } i = 2. \end{cases} \quad (25.1)$$

Let μ be a Gibbs measure on Σ_A^+ corresponding to a Hölder continuous function and $\nu = \chi^*\mu$. Consider the sets

$$\mathcal{A} = \{x \in F : x = \chi(i_0 i_1 \ldots) \text{ with } i_1 = 1 \text{ or } 3\}$$

and

$$\mathcal{B} = \{x \in F : x = \chi(i_0 i_1 \ldots) \text{ with } i_1 = 2\}.$$

These sets are disjoint and comprise the space Σ_A^+. They are *not* invariant under G. Since μ is a Gibbs measure and the sets $\mathcal{A}$ and $\mathcal{B}$ are open we have that $\nu(\mathcal{A}) > 0$ and $\nu(\mathcal{B}) > 0$. One can check that for every $x \in \mathcal{A}$,

$$\lim_{n\to\infty} \frac{\log|\Delta_n(x)|}{n} = \frac{1}{2}\log(\gamma\beta)$$

and for every $x \in \mathcal{B}$,

$$\lim_{n\to\infty} \frac{\log|\Delta_n(x)|}{n} = \frac{1}{2}\log(\alpha\delta).$$

Since $\chi^{-1}(\Delta_n(x))$ is a cylinder set the Shannon–McMillan–Breiman theorem implies that for ν-almost every $x \in F$,

$$-\lim_{n\to\infty} \frac{\log\nu(\Delta_n(x))}{n} = h_\mu(\sigma) > 0.$$

Thus, by Theorem 15.3

$$\underline{d}(x) = \overline{d}(x) = -\frac{h_\mu(\sigma)}{\lim_{n\to\infty}\left(\frac{\log|\Delta_n(x)|}{n}\right)} = -\frac{2h_\mu(\sigma)}{\log(\beta\gamma)} \quad \text{for almost every } x \in \mathcal{A}$$

and

$$\underline{d}(x) = \overline{d}(x) = -\frac{h_\mu(\sigma)}{\lim_{n\to\infty}\left(\frac{\log|\Delta_n(x)|}{n}\right)} = -\frac{2h_\mu(\sigma)}{\log(\alpha\delta)} \quad \text{for almost every } x \in \mathcal{B}.$$

The first statement follows now from Theorem 15.3. Repeating the arguments in Section 25.1 one can define the Hölder homeomorphism $\tilde{G}$ and show that it possesses an invariant Borel ergodic measure $\tilde{\nu}$ with respect to which $d_{\tilde{\nu}}(x, y) = \underline{d}_{\tilde{\nu}}(x, y) = \overline{d}_{\tilde{\nu}}(x, y)$ for $\tilde{\nu}$-almost every (x, y); however, the function $d_{\tilde{\nu}}(x, y)$ is not essentially constant. This implies the second statement and completes the construction of the example. ∎

Remark.

The geometric construction described in Example 25.2 is called an **asymptotic Moran-like geometric construction** since its ratio coefficients admit asymptotic behavior (25.1). It is proved in [PW1] that $s_\lambda \leq \dim_H F$, where s_λ is a unique root of Bowen's equation $P_{\Sigma_A^+}(s \log \lambda_{i_1}) = 0$. Example 25.2 illustrates that the strict inequality can occur. Indeed, one can easily check, using Statement 1 of Theorem 13.3, that $s_\lambda = \log 2 / \log(\alpha\beta\gamma\delta)$. On the other hand, one can compute, using Theorem 15.2, that

$$\dim_H F = \max \left\{ -\frac{\log 2}{\log(\beta\gamma)}, \ -\frac{\log 2}{\log(\alpha\delta)} \right\}.$$

The following example demonstrates non-existence of pointwise dimension for smooth maps. It was constructed by Ledrappier and Misiurewicz in [LM].

Example 25.3. *For every integer $r \geq 1$ there exist a C^r-map f of the interval $[0,1]$ and f-invariant ergodic measure μ for which $\underline{d}_\mu(x) < \overline{d}_\mu(x)$ for almost all points x.*

Proof. We outline the proof in the case $r = 1$. Choose numbers A, B, and C such that $A > 1$ and $0 < B < C < \frac{A}{A+1}$. Let θ_n be a sequence of numbers satisfying $\theta_1 = 1$ and $B \leq \theta_{n+1}/\theta_n \leq C$. Since $C < 1$ the sequence θ_n decreases towards 0. We define two sequences of points $\{a_n\}$, $\{b_n\}$ of the interval $[0,1]$

$$0 = a_2 < b_2 < a_4 < b_4 < a_6 < b_6 < \ldots$$
$$< b_7 < a_7 < b_5 < a_5 < b_3 < a_3 < b_1 < a_1 = 1$$

such that

$$|L_n| = |a_n - b_n| = \frac{\theta_{n+1}}{A}$$

and

$$|M_n| = |a_{n+2} - b_n| = \theta_n - \theta_{n+1} - \frac{\theta_{n+1}}{A},$$

where L_n and M_n are the intervals with endpoints a_n, b_n and b_n, a_{n+2} respectively. We have

$$\frac{|M_n|}{\theta_n} \geq 1 - C\frac{A+1}{A} > 0$$

and

$$\sum_{n=1}^{\infty}(|L_n| + |M_n|) = \sum_{n=1}^{\infty}(\theta_n - \theta_{n+1}) = \theta_1 = 1.$$

Therefore, all points with even indices lie to the left of all points with odd indices and there exists a common limit $c = \lim_{n\to\infty} a_n = \lim_{n\to\infty} b_n$.

We now define the map f. First we set:

$f(c) = 1$,
$f(a_n) = 1 - \gamma_n$, where $\gamma_n = \theta_n/A^{n-1}$,
$f(b_n) = 1 - \delta_n$, where $\delta_n = (\theta_n - \theta_{n+1})/A^{n-1}$.

Since

$$\frac{\theta_n - \theta_{n+1}}{A^{n-1}} - \frac{\theta_{n+1}}{A^n} = \frac{|M_n|}{A^{n-1}},$$

we have $\delta_n > \delta_{n+1}$ and thus, $1 = \gamma_1 > \delta_1 > \delta_2 > \cdots \to 0$. Therefore, we can define f as a linear map on each interval L_n with the slope λ_n given as

$$\lambda_n = \left| \frac{f(b_n) - f(a_n)}{b_n - a_n} \right| = \frac{\gamma_n - \delta_n}{|L_n|} = A^{2-n}.$$

Note that

$$[0,1] = \{c\} \cup \left(\bigcup_{n \geq 1} L_n \right) \cup \left(\bigcup_{n \geq 1} M_n \right)$$

and it remains to define f on the intervals M_n. Denote

$$w_n = \frac{f(a_{n+2}) - f(b_n)}{a_{n+2} - b_n}, \quad \alpha_n = 2w_n - \frac{1}{2}(\lambda_n + \lambda_{n+2}).$$

For n even we set

$$f\left(b_n + \frac{1}{2}|M_n| \right) = f(b_n) + \frac{1}{4}|M_n|(\lambda_n + \alpha_n),$$

$$f'\left(b_n + \frac{1}{2}|M_n| \right) = \alpha_n.$$

One can now define f separately on $[b_n, b_n + \frac{1}{2}|M_n|)$ and $[b_n + \frac{1}{2}|M_n|, a_{n+2})$ to obtain a C^∞-function on $[b_n, a_{n+2}]$ (see details in [Mi]). For n odd one can define f analogously. This gives a map f on $[0,1]$. The condition on f to be of class C^1 is

$$\lim_{n \to \infty} \max(|\lambda_n|, |\alpha_n|) = 0.$$

We have $|\lambda_n| = A^{2-n} \to 0$ as $n \to \infty$. Furthermore,

$$|w_n| = \frac{f(a_{n+2}) - f(b_n)}{|M_n|} < \frac{\gamma_n}{|M_n|} \leq \frac{A^{2-n}}{(A+1)\left(\frac{A}{A+1} - C \right)}.$$

Thus, $|w_n| \to 0$ as $n \to \infty$ and hence

$$|\alpha_n| \leq 2|w_n| + \frac{1}{2}|\lambda_n| + \frac{1}{2}|\lambda_{n+2}| \to 0$$

as $n \to \infty$.

Let K_n be the interval $[a_n, a_{n+1}]$ or $[a_{n+1}, a_n]$. We need the following properties of sets K_n (see [Mi]).

Lemma 1.

(1) *The sequence of sets $(K_n)_{n=1}^{\infty}$ is decreasing;*
(2) $f^{2^{n-1}}(K_n) = K_n$, $f^{2^{n-1}}(K_{n+1}) = L_n$;
(3) *The sets $f^i(K_n)$, $i = 1, \ldots, 2^{n-1}$ are disjoint;*
(4) $f^{2^{n-1}+i}(K_{n+1}) \cap f^i(K_{n+1}) = \varnothing$ *for* $i = 1, \ldots, 2^{n-1}$;
(5) *The map $f^{i-1}|f(K_n)$ is linear for $i = 1, \ldots, 2^{n-1}$.*

For each $i = 1, \ldots, 2^{n-1}$ we write $i = 1 + \sum_{k=j}^{n-1} \varepsilon_k(i)2^{k-1}$ with $\varepsilon_k(i) = 0$ or 1.

Let

$$\zeta_k(0) = \frac{1}{A}\frac{\theta_{k+1}}{\theta_k} \quad , \quad \zeta_k(1) = \frac{\theta_{k+1}}{\theta_k} .$$

Lemma 2. *For all $n \leq 1$ and $i = 1, \ldots, 2^{n-1}$ we have*

$$|f^i(K_n)| = \prod_{k=1}^{n-1} \zeta_k(\varepsilon_k(i)).$$

Proof of the lemma. We use induction on n. For $n = 1$ the result is obvious. Assume that the statement holds for $n = m$. We shall prove it for $n = m + 1$.

Observe that for $i = 1, \ldots, 2^{m-1}$ the set $f^i(K_m)$ is a disjoint union (modulo endpoints) of sets $f^i(K_{m+1})$, $f^i(L_m) = f^{2^{m-1}+i}(K_{m+1})$, and a remaining gap $G_{i,m}$.

Observe also that the map f^{i-1} is linear on $f(K_m)$ so that the lengths of these intervals are in the same proportions as

$$\frac{|f(K_{m+1})|}{|f(K_m)|} = \frac{\gamma_{m+1}}{\gamma_m} = \frac{1}{A}\frac{\theta_{m+1}}{\theta_m}, \quad \frac{|f(L_m)|}{|f(K_m)|} = \frac{\gamma_m - \delta_m}{\gamma_m} = \frac{\theta_{m+1}}{\theta_m} .$$

The induction follows clearly from the above two observations.   ∎

This lemma gives us information on the lengths of intervals building the attractor for the map f. The following lemma provides estimates of the lengths of the gaps $G_{i,m}$. Denote

$$\beta = \left(\frac{1}{C} - \frac{A+1}{A}\right)^{-1} .$$

Clearly, $\beta > 0$.

Lemma 3. *For all $m \geq 2$ and $i = 1, \ldots, 2^{m-1}$ we have*

$$|f^i(K_{m+1})| < |f^i(L_m)| \leq \beta|G_{i,m}|.$$

Proof of the lemma. Note that by linearity of $f^{i-1}|f(K_m)$ the above inequalities are equivalent to

$$\frac{|f(K_{m+1})|}{|f(K_m)|} < \frac{|f(L_m)|}{|f(K_m)|} \leq \beta\left[1 - \frac{|f(K_m)|}{|f(K_m)|} - \frac{|f(L_m)|}{|f(K_m)|}\right]$$

or

$$\frac{1}{A}\frac{\theta_{m+1}}{\theta_m} < \frac{\theta_{m+1}}{\theta_m} \le \beta \left[1 - \left(1 + \frac{1}{A}\right)\frac{\theta_{m+1}}{\theta_m}\right].$$

The first inequality holds since $A > 1$. The second inequality follows from the inequality

$$\beta^{-1} \le \frac{\theta_m}{\theta_{m+1}} - \frac{A+1}{A}$$

which is clearly true since $1/C \le \theta_m/\theta_{m+1}$.   ∎

The attractor Λ for the map f is defined as

$$\Lambda = \bigcap_{n=1}^{\infty} \bigcup_{i=1}^{2^{n-1}} f^i(K_n).$$

By Lemmas 1 and 2 Λ is a Cantor-like set of zero Lebesgue measure. We define a probability measure μ on Λ by assigning to every set $f^i(K_m)$, $i = 1, \ldots, 2^{m-1}$ the measure $\mu(f^i(K_m)) = 2^{1-m}$. The measure μ is clearly non-atomic and invariant.

Each point $x \in \Lambda$ admits a symbolic coding $x \mapsto \omega = (\varepsilon_n(x))$, where $\varepsilon_n(x) = 0$ or 1. Namely, for $x \in f^i(K_{n+1})$, $1 \le i \le 2^n$ we set $\varepsilon_n(x) = 0$ if $i \le 2^{n-1}$ and $\varepsilon_n(x) = 1$ if $i > 2^{n-1}$. Using this coding map one can show that the measure μ is ergodic and has zero measure-theoretic entropy (see [CE]).

Lemma 4. *For all* $x \in \Lambda$

$$\underline{d}_\mu(x) = \varliminf_{n\to\infty} \frac{-n\log 2}{\sum_{j=1}^n \log \zeta_j(\varepsilon_j(x))}, \quad \overline{d}_\mu(x) = \varlimsup_{n\to\infty} \frac{-n\log 2}{\sum_{j=1}^n \log \zeta_j(\varepsilon_j(x))}.$$

Proof of the lemma. For $x \in f^i(K_{n+1})$, $i = 1, \ldots, 2^n$ we set

$$\eta_n(x) = |f^i(K_{n+1})| = \prod_{k+1}^n \zeta_k(\varepsilon_k(x))$$

(see Lemma 2). Fix n_0 such that $\beta < C^{-n_0}$. Clearly, the interval $[x - \eta_n(x), x + \eta_n(x)]$ contains $f^i(K_{n+1})$. For each $k < n$ there exists $i(k)$ such that $1 \le i(k) \le 2^k$ and $x \in f^{i(k)}(K_{n+1})$. By Lemma 2 we have

$$|f^{i(k)}(K_{k+1})| \ge C^{-(n-k)}|f^i(K_{n+1})|$$

and hence if $k \le n - n_0$ then

$$|f^{i(k)}(K_{k+1})| \ge \beta\eta_n(x).$$

By Lemma 3 we obtain that if $k \le n - n_0$ then

$$|G_{i(k),k}| \ge h_n(x).$$

Therefore, since the sets $G_{i(k),k}$ are gaps and hence are disjoint from Λ the set $[x - \eta_n(x), x + \eta_n(x)] \cap \Lambda$ is contained in $f^{i(n-n_0)}(K_{n-n_0+1})$. It follows that

$$2^{-n} \le \mu([x - \eta_n(x), x + \eta_n(x)]) \le 2^{-(n-n_0)}.$$

The desired result follows immediately from this relation and the inequalities

$$0 < BA^{-1} \le \frac{\eta_{n+1}(x)}{\eta_n(x)} \le C < 1.$$

The lemma is proved.   ∎

Note that

$$\log \zeta_j(0) = \log \frac{\theta_{j+1}}{\theta_j} - \log A , \quad \log \zeta_j(1) = \log \frac{\theta_{j+1}}{\theta_j} .$$

Since $\varepsilon_j(x)$ are independent stationary sequences for μ-almost every $x \in A$ we have

$$\lim_{n\to\infty} \frac{1}{n} \sum_{j=1}^{n} \log \zeta_j(\varepsilon_j(x)) = \lim_{n\to\infty} \frac{1}{n} \log \theta_n - \frac{1}{2} \log A,$$

$$\overline{\lim}_{n\to\infty} \frac{1}{n} \sum_{j=1}^{n} \log \zeta_j(\varepsilon_j(x)) = \overline{\lim}_{n\to\infty} \frac{1}{n} \log \theta_n - \frac{1}{2} \log A.$$

Let us choose the sequence θ_n such that $B \leq \frac{\theta_{n+1}}{\theta_n} \leq C$ and

$$\lim_{n\to\infty} \frac{1}{n} \log \theta_n = B , \quad \overline{\lim}_{n\to\infty} \frac{1}{n} \log \theta_n = C.$$

From Lemma 4 it follows that for μ-almost every $x \in \Lambda$,

$$\underline{d}_\mu(x) = \frac{\log 2}{\log(\sqrt{A}/B)} < \frac{\log 2}{\log(\sqrt{A}/C)} = \overline{d}_\mu(x).$$

This completes the construction of the example. ∎

26. Dimension of Measures with Non-zero Lyapunov Exponents; The Eckmann–Ruelle Conjecture

We assume that $\mathcal{M}$ is a compact smooth Riemannian p-dimensional manifold and $f: \mathcal{M} \to \mathcal{M}$ is a $C^{1+\alpha}$-diffeomorphism. We recall some basic notions of the theory of dynamical systems with non-zero Lyapunov exponents (see [KH], [M] for more details).

Given $x \in \mathcal{M}$ and $v \in T_x\mathcal{M}$, define the **Lyapunov exponent** of v at x by the formula

$$\lambda(x, v) = \overline{\lim}_{n\to\infty} \frac{\log \|df_x^n v\|}{n}. \qquad (26.1)$$

If x is fixed then the function $\lambda(x, \cdot)$ can take on only finitely many distinct values $\lambda^{(1)}(x) > \cdots > \lambda^{(q)}(x)$, where $q = q(x)$ and $1 \leq q \leq p$. Let $k_i(x)$ be the multiplicity of the value $\lambda^{(i)}(x)$, $i = 1, \ldots, q$. The functions $q(x)$, $\lambda^{(i)}(x)$, and $k_i(x)$, $i = 1, \ldots, q$ are measurable and invariant under f.

Let μ be a Borel f-invariant measure on $\mathcal{M}$. We will always assume that μ is ergodic. Then the function $q(x) = q$ is constant μ-almost everywhere and so are the functions $\lambda^{(i)}(x)$ and $k_i(x)$, $i = 1, \ldots, q$. We denote the corresponding values by $\lambda_\mu^{(i)}$ and $k_\mu^{(i)}$. A measure μ is said to be **hyperbolic** if

$$\lambda_\mu^{(1)} > \cdots > \lambda_\mu^{(k)} > 0 > \lambda_\mu^{(k+1)} > \cdots > \lambda_\mu^{(q)}$$

for some k, $1 \leq k < q$. If μ is such a measure then for μ-almost every point $x \in \mathcal{M}$ there exist stable and unstable subspaces $E^{(s)}(x)$, $E^{(u)}(x) \subset T_x\mathcal{M}$ such that

(1) $E^{(s)}(x) \oplus E^{(u)}(x) = T_x\mathcal{M}$, $df_x E^{(s)}(x) = E^{(s)}(f(x))$, $df_x E^{(u)}(x) = E^{(u)}(f(x))$;

(2) for any $n \geq 0$,

$$\|df_x^n v\| \leq C_1(x)\gamma^n\|v\| \quad \text{if } v \in E^{(s)}(x),$$
$$\|df_x^{-n} v\| \leq C_1(x)\gamma^n\|v\| \quad \text{if } v \in E^{(u)}(x),$$

where $0 < \gamma < 1$ is a constant and $C_1(x) > 0$ is a measurable function;

(3) $\angle(E^{(s)}(x), E^{(u)}(x)) \geq C_2(x) > 0$, where $C_2(x)$ is a measurable function and $\angle$ denotes the angle between subspaces $E^{(s)}(x)$ and $E^{(u)}(x)$;

(4) $C_1(f^n(x)) \leq C_1(x)e^{n\delta}$, $C_2(f^n(x)) \geq C_2(x)e^{-n\delta}$ for any $n \geq 0$, where $\delta > 0$ is a constant which is sufficiently small compare to $1 - \gamma$.

Denote by Λ_ℓ, $\ell \geq 1$ the set of points x for which $C_1(x) \leq \ell$, $C_2(x) \geq \frac{1}{\ell}$. The sets Λ_ℓ are closed, $\Lambda_\ell \subset \Lambda_{\ell+1}$, and the set $\Lambda = \bigcup_{\ell \geq 1} \Lambda_\ell$ is f-invariant and coincides with $\mathcal{M}$ up to a set of μ-measure zero.

For any $x \in \Lambda$ one can construct stable and unstable smooth local manifolds which we denote by $\mathcal{W}_{\text{loc}}^{(s)}(x)$ and $\mathcal{W}_{\text{loc}}^{(u)}(x)$ respectively. They have the following properties:

(5) $x \in \mathcal{W}_{\text{loc}}^{(s)}(x)$ and $x \in \mathcal{W}_{\text{loc}}^{(u)}(x)$; $T_x\mathcal{W}_{\text{loc}}^{(s)}(x) = E^{(s)}(x)$ and $T_x\mathcal{W}_{\text{loc}}^{(u)}(x) = E^{(u)}(x)$;

(6) for any $n \geq 0$,

$$\rho(f^n(x), f^n(y)) \leq C_3(x)\gamma^n\rho(x,y) \quad \text{if } y \in \mathcal{W}_{\text{loc}}^{(s)}(x),$$
$$\rho(f^{-n}(x), f^{-n}(y)) \leq C_3(x)\gamma^n\rho(x,y) \quad \text{if } y \in \mathcal{W}_{\text{loc}}^{(u)}(x),$$

where ρ is the distance in $\mathcal{M}$ generated by the Riemannian metric and $C_3(x) > 0$ is a measurable function satisfying $C_3(f^n(x)) \leq C_3(x)e^{10n\delta}$ for every $n \geq 0$;

(7) $\mathcal{W}_{\text{loc}}^{(s)}(x)$ and $\mathcal{W}_{\text{loc}}^{(u)}(x)$ depend continuously on $x \in \Lambda_\ell$ in the C^1-topology and have *size* at least $r_\ell > 0$; this means that they contain respectively balls $B^{(s)}(x, r_\ell)$ and $B^{(u)}(x, r_\ell)$ centered at x of radius r_ℓ in the intrinsic topology generated by the Riemannian metric;

(8) there exists a function $\ell_1 = \psi(\ell) < \infty$, such that for any $x, y \in \Lambda_\ell$, if the intersection $B^{(s)}(x, r_\ell) \cap B^{(u)}(y, r_\ell) \neq \varnothing$, then it consists of a single point $z \in \Lambda_{\ell_1}$;

(9) there exists $K > 0$ such that for any $x \in \Lambda_\ell$,

$$\rho^{(u)}(y, z) \geq K\rho(y, z) \quad \text{if } y, z \in \mathcal{W}_{\text{loc}}^{(u)}(x),$$
$$\rho^{(s)}(y, z) \geq K\rho(y, z) \quad \text{if } y, z \in \mathcal{W}_{\text{loc}}^{(s)}(x),$$

where and $\rho^{(u)}$ and $\rho^{(s)}$ are the intrinsic distances in the local unstable and stable manifolds induced by the Riemannian metric.

A closed set $\Pi \subset \Lambda$ is called a **rectangle** if :

a) there exists $\ell > 1$ such that $\Pi \cap \Lambda_\ell$ is an open subset of Λ_ℓ ℓ (with respect to the induced topology of Λ_ℓ); moreover, for any $x \in \Pi$ there exists $y \in \Pi \cap \Lambda_\ell$ such that $x \in B^{(s)}(y, r_\ell)$ or $x \in B^{(u)}(y, r_\ell)$;

b) for any $x, y \in \Pi$ the intersection $B^{(s)}(x, r_\ell) \cap B^{(u)}(y, r_\ell)$ is not empty and consists of a single point $z \in \Pi$ (one can see that $\Pi \subset \Lambda_{\ell_1}$).

Given a rectangle Π and points $x, y \in \Pi$, define the u-**holonomy map** $H_{x,y}^{(u)} : B^{(s)}(x, r_\ell) \cap \Pi \to B^{(s)}(y, r_\ell) \cap \Pi$ by $H_{x,y}^{(u)}(z) = B^{(u)}(z, r_\ell) \cap B^{(s)}(y, r_\ell)$. One can similarly define the s-**holonomy map** $H_{x,y}^{(s)}$. For each rectangle Π such that $\mu(\Pi) > 0$ let $\xi^{(u)}$ and $\xi^{(s)}$ be the measurable partitions of Π : $\xi^{(u)}(x) = \Pi \cap B^{(u)}(x, r_\ell)$ and $\xi^{(s)}(x) = \Pi \cap B^{(s)}(x, r_\ell)$ (see [KH]). Let $\{\mu_x^{(u)}\}_{x \in \Pi}$ and $\{\mu_x^{(s)}\}_{x \in \Pi}$ be the canonical families of conditional measures associated with the partitions $\xi^{(u)}$ and $\xi^{(s)}$ respectively.

We say that a measure μ has the **local semi-product structure** if for any rectangle Π with $\mu(\Pi) > 0$ and μ-almost every $x, y \in \Pi$ the associated holonomy map $H_{x,y}^{(s)}$ is absolutely continuous with respect to the conditional measures $\mu_x^{(u)}$ and $\mu_y^{(u)}$, i.e., the measure $(H_{x,y}^{(s)})_* \mu_x^{(u)}$ is absolutely continuous with respect to the measure $\mu_y^{(u)}$ (or if the same is true for the holonomy maps $H_{x,y}^{(u)}$ and the conditional measures $\mu_x^{(s)}, \mu_y^{(u)}$ for μ-almost every $x, y \in \Pi$). We say that a measure μ has the **local product structure** if for any rectangle Π with $\mu(\Pi) > 0$ and μ-almost every $x, y \in \Pi$ the associated holonomy maps $H_{x,y}^{(u)}, H_{x,y}^{(s)}$ both are absolutely continuous with respect to the conditional measures $\mu_x^{(s)}, \mu_y^{(s)}$ and $\mu_x^{(u)}, \mu_y^{(u)}$, respectively.

There are two special cases of measures which have local product structure.

A hyperbolic measure μ is called a **Sinai–Ruelle–Bowen measure** (or SRB-measure) if for every rectangle Π with $\mu(\Pi) > 0$ and μ-almost every $x \in \Pi$ the conditional measure $\mu_x^{(u)}$ is absolutely continuous with respect to the Riemannian volume on $W_{\text{loc}}^{(u)}(x)$. SRB-measures have semi-local product structure.

Let Λ be a locally maximal hyperbolic set for a $C^{1+\alpha}$-diffeomorphism on a compact smooth manifold $\mathcal{M}$ (see definition in Section 22) and $\mu = \mu_\varphi$ a unique equilibrium measure corresponding to a Hölder continuous function φ on Λ. By Proposition 22.2 μ has local product structure.

Given $x \in \Lambda$ and a sufficiently small $r > 0$, define

$$d_\mu^{(s)}(x) = \lim_{r \to 0} \frac{\log \mu_x^{(s)}(B^{(s)}(x, r))}{\log r}, \quad d_\mu^{(u)}(x) = \lim_{r \to 0} \frac{\log \mu_x^{(u)}(B^{(u)}(x, r))}{\log r} \quad (26.2)$$

(assuming that the limits exist). One can show that the functions $d_\mu^{(s)}(x)$ and $d_\mu^{(u)}(x)$ are measurable and invariant under f. Since the measure μ is ergodic these functions are almost everywhere constant. We denote the corresponding values by $d^{(u)}$ and $d^{(s)}$.

We first consider the cases $k = 1$ and $k = q - 1$. The following result is a slightly more general multidimensional version of the result obtained by Young in the two-dimensional case (see [Y2]).

Proposition 26.1. *For any ergodic hyperbolic measure μ invariant under a $C^{1+\alpha}$-diffeomorphism the limits (26.2) exist almost everywhere and*

(1) *if $k = 1$ then $d^{(u)} = \frac{h_\mu(f)}{k_\mu^{(1)} \lambda_\mu^{(1)}}$;*

(2) *if $k = q - 1$ then $d^{(s)} = -\frac{h_\mu(f)}{k_\mu^{(q)} \lambda_\mu^{(q)}}$.*

The existence of the values $d_\mu^{(s)}(x)$ and $d_\mu^{(u)}(x)$ in the general case was established by Ledrappier and Young in [LY]. We recall that since μ is ergodic $\underline{d}_\mu(x) = \text{const} = \underline{d}(\mu)$ and $\overline{d}_\mu(x) = \text{const} = \overline{d}(\mu)$ almost everywhere (see Section 7).

Proposition 26.2. *For any ergodic hyperbolic measure μ invariant under a $C^{1+\alpha}$-diffeomorphism the limits (26.2) exist almost everywhere and*

(1) $\overline{d}(\mu) \leq d^{(u)} + d^{(s)}$;

(2) $d^{(u)} \geq \frac{h_\mu(f)}{\lambda_\mu^{(1)}}$ *and* $d^{(s)} \geq -\frac{h_\mu(f)}{\lambda_\mu^{(q)}}$.

In [ER], Eckmann and Ruelle discussed the existence of pointwise dimension for hyperbolic invariant measures. We summarize this discussion in the following statement, which was proved by Barreira, Pesin, and Schmeling in [BPS1].

Theorem 26.1. *Let $f: \mathcal{M} \to \mathcal{M}$ be a $C^{1+\alpha}$-diffeomorphism of a compact smooth Riemannian manifold $\mathcal{M}$ and μ a hyperbolic ergodic measure. Then for μ-almost any $x \in \mathcal{M}$,*

$$\underline{d}(\mu) = \overline{d}(\mu) = d^{(u)} + d^{(s)}.$$

Remark.

In [Y2], Young proved this theorem in the two-dimensional case.

Proposition 26.3. *Let f be a $C^{1+\alpha}$-diffeomorphism of a smooth compact surface $\mathcal{M}$ and μ a hyperbolic ergodic measure with Lyapunov exponents $\lambda_\mu^{(1)} > 0 > \lambda_\mu^{(2)}$. Then*

$$\underline{d}(\mu) = \overline{d}(\mu) = h_\mu(f) \left(\frac{1}{\lambda_\mu^{(1)}} - \frac{1}{\lambda_\mu^{(2)}} \right).$$

Proof of the theorem. For the sake of reader's convenience we first consider the special and simpler case of *measures with local semi-product structure*. The proof in the general case will be given later. Since it is technically more complicated it can be omitted in the first reading. We follow [PY].

Without loss of generality we may assume that the holonomy maps $H_{x,y}^{(s)}$ are absolutely continuous with respect to the measures $\mu_x^{(u)}$, $\mu_y^{(u)}$ for any rectangle Π with $\mu(\Pi) > 0$ and μ-almost every $x, y \in \Pi$. We fix such a rectangle Π.

According to Proposition 26.2 it is sufficient to show that $\underline{d}(\mu) \geq d^{(s)} + d^{(u)}$.

Proposition 26.2 (that states the existence of the limits in (26.2)) implies that there exist a closed set $A_1 \subset \Pi$ with $\mu(A_1) > 0$ and a number $r_1 > 0$ satisfying the following condition:

(10) for any $0 < r \leq r_1$ and any $x \in A_1$,

$$\mu_x^{(u)}(B^{(u)}(x,r) \cap \Pi) \leq r^{d^{(u)}-\varepsilon}, \quad \mu_x^{(s)}(B^{(s)}(x,r) \cap \Pi) \leq r^{d^{(s)}-\varepsilon}. \tag{26.3}$$

It follows from the Borel Density Lemma (see Appendix V) that one can find a closed set $A_2 \subset A_1$ with $\mu(A_2) > 0$ and $r_2 > 0$ such that for any $0 < r \leq r_2$ and any $x \in A_2$,

$$\mu(B(x,r)) \leq 2\mu(B(x,r) \cap A_1). \tag{26.4}$$

Fix a point $x_0 \in A_2$ for which $\mu_{x_0}^{(u)}(\mathcal{W}_{\mathrm{loc}}^{(u)}(x_0) \cap A_2) > 0$ and

$$\lim_{r \to 0} \frac{\log \mu(B(x_0,r))}{\log r} = \underline{d}(\mu).$$

We first study the factor measure $\widetilde{\mu}$ induced by the measurable partition $\xi^{(s)}$. Denote by $\pi^{(u)}$ the projection of $B(x_0,r) \cap A_1$ into $\mathcal{W}_{\mathrm{loc}}^{(u)}(x_0)$ given by $\pi^{(u)}(y) = \mathcal{W}_{\mathrm{loc}}^{(u)}(x_0) \cap \mathcal{W}_{\mathrm{loc}}^{(s)}(y)$. For any $x \in B(x_0,r) \cap A_1$ we also have that

$$\pi^{(u)}|\mathcal{W}_{\mathrm{loc}}^{(u)}(x) \cap A_1 = H_{x_0,x}^{(s)}|\mathcal{W}_{\mathrm{loc}}^{(u)}(x) \cap A_1.$$

Without loss of generality we may assume that the holonomy map $H_{x_0,x}^{(s)}$ is absolutely continuous with respect to the conditional measures $\mu_{x_0}^{(u)}$ and $\mu_x^{(u)}$. Therefore, the factor measure $\widetilde{\mu} = \pi_*^{(u)}\mu$ is absolutely continuous with respect to $\mu_{x_0}^{(u)}$:

$$d\widetilde{\mu}(z) = \varphi(z)d\mu_{x_0}^{(u)}(z), \quad z \in B^{(u)}(x_0,r_2) \cap A_1,$$

where $\varphi(z) \geq 0$ is an L^1-function. Since $\mu_{x_0}^{(u)}(\mathcal{W}_{\mathrm{loc}}^{(u)}(x_0) \cap A_2) > 0$, without loss of generality, we may assume that the Radon–Nikodym derivative $d\widetilde{\mu}/d\mu_{x_0}^{(u)}$ is bounded. Therefore, we have for sufficiently small $r > 0$,

$$\widetilde{\mu}(B^{(u)}(x_0,r)) \leq C_4\mu_{x_0}^{(u)}(B^{(u)}(x_0,r)), \tag{26.5}$$

where $C_4 > 0$ is a constant. We now apply the Fubini theorem to the measurable partition $\xi^{(s)}$ and write

$$\mu(B(x_0,r)) = \int_{B(x_0,r)\cap\mathcal{W}_{\mathrm{loc}}^{(u)}(x_0)} \mu_z^{(s)}(\mathcal{W}_{\mathrm{loc}}^{(s)}(z) \cap B(x_0,r))d\widetilde{\mu}(z). \tag{26.6}$$

If the intersection $\mathcal{W}_{\mathrm{loc}}^{(s)}(z) \cap B(x_0,r) \cap A_1$ contains a point x then by Condition (9),

$$\mathcal{W}_{\mathrm{loc}}^{(s)}(z) \cap B(x_0,r) \subset B^{(s)}(x,2Kr)$$

and hence in view of (26.3)

$$\mu_z^{(s)}(\mathcal{W}_{\text{loc}}^{(s)}(z) \cap B(x_0, r)) \le \mu_z^{(s)}(B^{(s)}(x, 2Kr)) \le (2Kr)^{d^{(s)} - \varepsilon}. \qquad (26.7)$$

We also have that

$$\mu_{x_0}^{(u)}(\mathcal{W}_{\text{loc}}^{(u)}(x_0) \cap B(x_0, r)) \le \mu_{x_0}^{(u)}(B^{(u)}(x_0, 2Kr)) \le (2Kr)^{d^{(u)} - \varepsilon}. \qquad (26.8)$$

Conditions (26.5) – (26.8) imply that for sufficiently small r,

$$\begin{aligned}
\mu(B(x_0, r) \cap A_1) &\le \int_{B(x_0,r) \cap \mathcal{W}_{\text{loc}}^{(u)}(x_0)} (2Kr)^{d^{(s)} - \varepsilon} d\widetilde{\mu}(z) \\
&= (2Kr)^{d^{(s)} - \varepsilon} \widetilde{\mu}(B(x_0, r) \cap \mathcal{W}_{\text{loc}}^{(u)}(x_0)) \\
&\le (2Kr)^{d^{(s)} - \varepsilon} \widetilde{\mu}(B^{(u)}(x_0, Kr)) \\
&\le (2Kr)^{d^{(s)} - \varepsilon} C_4 \mu_{x_0}^{(u)}(B^{(u)}(x_0, Kr)) \\
&\le (2Kr)^{d^{(s)} - \varepsilon} C_4 (Kr)^{d^{(u)} - \varepsilon} = C_5 r^{d^{(s)} + d^{(u)} - 2\varepsilon},
\end{aligned}$$

where $C_5 > 0$ is a constant.

Consider a decreasing sequence of positive numbers $\rho_k \to 0$ $(k \to \infty)$ such that $\mu(B(x_0, \rho_k)) \ge \rho_k^{\underline{d}(\mu) - \varepsilon}$ for all k. We can also assume that $\rho_1 < \min\{r_1, r_2, r_3\}$. It follows now from (26.4) that

$$\begin{aligned}
\rho_k^{\underline{d}(\mu) + \varepsilon} &\le \mu(B(x_0, \rho_k)) \le 2\mu(B(x_0, \rho_k) \cap A_1) \\
&\le C_5 \rho_k^{d^{(s)} + d^{(u)} - 2\varepsilon}.
\end{aligned}$$

If $k \to \infty$ this yields

$$\underline{d}(\mu) + \varepsilon \ge d^{(s)} + d^{(u)} - 2\varepsilon.$$

Since ε is arbitrary this implies the desired result.

We now proceed with *measures which do not have local (semi-) product structure*. We will first establish a crucial property of hyperbolic measures: they have *nearly* local product structure. This enables us to apply a slight modification of the above approach to obtain the desired result. In order to highlight the main idea and avoid some complicated technical constructions in the theory of dynamical systems with non-zero Lyapunov exponents we assume that the map f possesses a locally maximal hyperbolic set Λ which supports the measure μ. Although the set Λ has direct product structure hyperbolic measures supported on it, in general, do not have local (semi-) product structure. For the general case see [BPS1].

Consider a Markov partition $\mathcal{R} = \{R_1, \ldots, R_p\}$. In order to simplify notations we set $R_k^l(x) = R_{i_k \ldots i_l}(x)$ (the element of the partition $\bigvee_{i=k}^l f^i \mathcal{R}$ that contains x). We point out the following properties of the Markov partition. Given $0 < \varepsilon < 1$, there exists a set $\Gamma \subset \mathcal{M}$ of measure $\mu(\Gamma) > 1 - \varepsilon/2$, an integer

$n_0 \geq 1$, and a number $C > 1$ such that for every $x \in \Gamma$ and any integer $n \geq n_0$ the following properties hold:

(a) for all integers $k, l \geq 1$ we have

$$C^{-1}e^{-(l+k)(h-\varepsilon)} \leq \mu(R_k^l(x)) \leq Ce^{-(l+k)(h+\varepsilon)}, \qquad (26.9)$$

$$C^{-1}e^{-k(h-\varepsilon)} \leq \mu_x^{(s)}(R_k^0(x)) \leq Ce^{-k(h+\varepsilon)}, \qquad (26.10)$$

$$C^{-1}e^{-l(h-\varepsilon)} \leq \mu_x^{(u)}(R_0^l(x)) \leq Ce^{-l(h+\varepsilon)}, \qquad (26.11)$$

where $h = h_\mu(f)$ (the measure-theoretic entropy of f with respect to μ);
(b)

$$e^{-(d^{(s)}+\varepsilon)n} \leq \mu_x^{(s)}(B^{(s)}(x, e^{-n})) \leq e^{-(d^{(s)}-\varepsilon)n}, \qquad (26.12)$$

$$e^{-(d^{(u)}+\varepsilon)n} \leq \mu_x^{(u)}(B^{(u)}(x, e^{-n})) \leq e^{-(d^{(u)}-\varepsilon)n}; \qquad (26.13)$$

(c) define a to be the integer part of $2(1+\varepsilon)\max\{1/\lambda_1, -1/\lambda_p, 1\}$; then

$$R_{an}^{an}(x) \subset B(x, e^{-n}) \subset R(x), \qquad (26.14)$$

$$R_{an}^0(x) \cap A^{(s)}(x) \subset B^{(s)}(x, e^{-n}) \subset R(x) \cap A^{(s)}(x), \qquad (26.15)$$

$$R_0^{an}(x) \cap A^{(u)}(x) \subset B^{(u)}(x, e^{-n}) \subset R(x) \cap A^{(u)}(x), \qquad (26.16)$$

where the sets $A^{(s)}$ and $A^{(u)}$ are defined by (22.7);
(d) define $Q_n(x)$ to be the set of points $y \in \mathcal{M}$ for which the intersections $R_0^{an}(y) \cap B^{(u)}(x, 2e^{-n})$ and $R_{an}^0(y) \cap B^{(s)}(x, 2e^{-n})$ are not empty; then

$$B(x, e^{-n}) \cap \Gamma \subset Q_n(x) \subset B(x, 4e^{-n}) \qquad (26.17)$$

and $R_{an}^{an}(y) \subset Q_n(x)$ for each $y \in Q_n(x)$;
(e) there exists a positive constant $D = D(\hat{\Gamma}) < 1$ such that for every $k \geq 1$ and $x \in \Gamma$ we have

$$\mu_x^s(R_0^k(x) \cap \Gamma) \geq D, \quad \mu_x^u(R_k^0(x) \cap \Gamma) \geq D;$$

(f) for every $x \in \Gamma$ and $n \geq n_0$ we have

$$R_{an}^{an}(x) \cap A^{(s)}(x) = R_{an}^0(x) \cap A^{(s)}(x), \quad R_{an}^{an}(x) \cap A^{(u)}(x) = R_0^{an}(x) \cap A^{(u)}(x).$$

Property (26.9) shows that the Shannon–McMillan–Breiman theorem holds with respect to the Markov partition $\mathcal{R}$ while properties (26.10) and (26.11) show that "leaf-wise" versions of this theorem hold with respect to the partitions $\mathcal{R}_k^0$ and $\mathcal{R}_0^l$. The inequalities (26.12) and (26.13) are easy consequences of the existence of the stable and unstable pointwise dimensions $d^{(s)}$ and $d^{(u)}$ (see Proposition 26.2). Since the Lyapunov exponents at μ-almost every point are constant the properties (26.14), (26.15), and (26.16) follow from the choice of the number a indicated above. The inclusions (26.17) are based upon the continuous dependence of stable

and unstable manifolds in the $C^{1+\alpha}$ topology on the base point on Λ. Property
(f) follows from the Markov property.

For an arbitrary $C^{1+\alpha}$-diffeomorphism preserving an ergodic hyperbolic measure Ledrappier and Young [LY] constructed a countable measurable partition of
$\mathcal{M}$ of finite entropy which has properties (26.9)–(26.17). In [BPS1], the authors
also showed that this partition satisfies Property (e) and simulates (in a sense)
Property (f). They used this partition to obtain a proof in the general case. We
follow their approach.

It immediately follows from the Borel Density Lemma (see Appendix V) that
one can choose an integer $n_1 \geq n_0$ and a set $\hat{\Gamma} \subset \Gamma$ of measure $\mu(\hat{\Gamma}) > 1 - \varepsilon$ such
that for every $n \geq n_1$ and $x \in \hat{\Gamma}$,

$$\mu(B(x, e^{-n}) \cap \Gamma) \geq \frac{1}{2}\mu(B(x, e^{-n})); \qquad (26.18)$$

$$\mu_x^{(s)}(B^{(s)}(x, e^{-n}) \cap \Gamma) \geq \frac{1}{2}\mu_x^{(s)}(B^{(s)}(x, e^{-n})); \qquad (26.19)$$

$$\mu_x^{(u)}(B^{(u)}(x, e^{-n}) \cap \Gamma) \geq \frac{1}{2}\mu_x^{(u)}(B^{(u)}(x, e^{-n})). \qquad (26.20)$$

Fix $x \in \hat{\Gamma}$ and an integer $n \geq n_1$. We consider the following two classes $\mathfrak{T}(n)$
and $\mathfrak{F}(n)$ of elements of the partition $\mathcal{R}_{an}^{an}$ (we call these elements "rectangles"):

$$\mathfrak{T}(n) = \{R_{an}^{an}(y) \subset R(x) : R_{an}^{an}(y) \cap \Gamma \neq \varnothing\};$$

$$\mathfrak{F}(n) = \{R_{an}^{an}(y) \subset R(x) : R_{an}^0(y) \cap \hat{\Gamma} \neq \varnothing \text{ and } R_0^{an}(y) \cap \hat{\Gamma} \neq \varnothing\}.$$

The rectangles in $\mathfrak{T}(n)$ carry all the measure of the set $R(x) \cap \Gamma$. Obviously, the
rectangles in $\mathfrak{T}(n)$ that intersect $\hat{\Gamma}$ belong to $\mathfrak{F}(n)$. If these were the only ones in
$\mathfrak{F}(n)$, the measure $\mu|R(x) \cap \Gamma$ would have the local direct product structure at the
"level" n and its pointwise dimension could be estimated as above. In the general
case, the rectangles in the class $\mathfrak{F}(n)$ are obtained from the rectangles in $\mathfrak{T}(n)$
(that intersect $\hat{\Gamma}$) by "filling in" the gaps in the product structure. See Figure
21 where we show the rectangle $R(x)$ which is partitioned by "small" rectangles
$R_{an}^{an}(y)$. The black rectangles comprise the collection $\mathfrak{T}(n)$. By adding the gray
rectangles one gets the direct product structure on the level n.

We wish to compare the number of rectangles in $\mathfrak{T}(n)$ and $\mathfrak{F}(n)$ intersecting
a given set. This will allow us to evaluate the deviation of the measure μ from
the direct product structure at each level n. Our main observation is that for
"typical" points $y \in \hat{\Gamma}$ the number of rectangles from the class $\mathfrak{T}(n)$ intersecting
$A^{(s)}(y)$ (respectively $A^{(u)}(y)$) is "asymptotically" the same up to a factor that
grows at most subexponentially with n. However, in general, the distribution of
these rectangles along $A^{(s)}(y)$ (respectively $A^{(u)}(y)$) may "shift" when one moves
from point to point. This causes a deviation from the direct product structure.
We will use a simple combinatorial argument to show that this deviation grows
at most subexponentially with n.

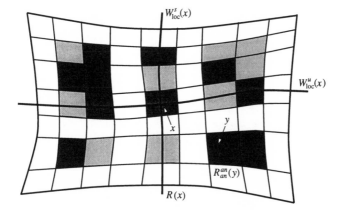

Figure 21. FILLING IN THE PRODUCT STRUCTURE.

To effect this, for each set $A \subset R(x)$, we define numbers

$$N(n, A) = \text{card } \{R \in \mathfrak{T}(n) : R \cap A \neq \varnothing\},$$

$$N^{(s)}(n, y, A) = \text{card } \{R \in \mathfrak{T}(n) : R \cap A^{(s)}(y) \cap \Gamma \cap A \neq \varnothing\},$$

$$N^{(u)}(n, y, A) = \text{card } \{R \in \mathfrak{T}(n) : R \cap A^{(u)}(y) \cap \Gamma \cap A \neq \varnothing\},$$

$$\hat{N}^{(s)}(n, y, A) = \text{card } \{R \in \mathfrak{F}(n) : R \cap A^{(s)}(y) \cap A \neq \varnothing\},$$

$$\hat{N}^{(u)}(n, y, A) = \text{card } \{R \in \mathfrak{F}(n) : R \cap A^{(u)}(y) \cap A \neq \varnothing\}.$$

Note that $N(n, R(x))$ is the cardinality of the set $\mathfrak{T}(n)$, and $N^{(s)}(n, y, R(x))$ (respectively $N^{(u)}(n, y, R(x))$) is the number of rectangles in $\mathfrak{T}(n)$ that intersect Γ and the stable (respectively unstable) local manifold at y. The product $\hat{N}^{(s)}(n, y, R(x)) \times \hat{N}^{(u)}(n, y, R(x))$ is the cardinality of the set $\mathfrak{F}(n)$ for "typical" points $y \in R(x)$ (which comprise a set of "big" measure).

We first obtain some estimates of the growth of the above numbers.

Lemma 1. *For each $y \in R(x) \cap \Gamma$ and integer $n \geq n_0$ we have*

$$N^{(s)}(n, y, Q_n(y)) \leq \mu_y^{(s)}(B^{(s)}(y, 4e^{-n})) \cdot Ce^{an(h+\varepsilon)} N^{(u)}(n, y, Q_n(y))$$
$$\leq \mu_y^{(u)}(B^{(u)}(y, 4e^{-n})) \cdot Ce^{an(h+\varepsilon)}.$$

Proof of the lemma. We have

$$\mu_y^{(s)}(B^{(s)}(y, 4e^{-n})) \geq \mu_y^{(s)}(Q_n(y)) \geq N^{(s)}(n, y, Q_n(y))$$
$$\times \min\{\mu_y^{(s)}(R) : R \in \mathfrak{T}(n) \text{ and } R \cap A^{(s)}(y) \cap \Gamma \cap Q_n(y) \neq \varnothing\}.$$

Let $z \in R \cap A^{(s)}(y) \cap \Gamma \cap Q_n(y)$ for some $R \in \mathfrak{T}(n)$. By Property (f) we obtain $\mu_y^{(s)}(R) = \mu_y^{(s)}(R_{an}^0(z)) = \mu_z^{(s)}(R_{an}^0(z))$. The first inequality follows now from (26.10). The proof of the second inequality is similar. ∎

Lemma 2. *For each $y \in R(x) \cap \hat{\Gamma}$ and integer $n \geq n_1$ we have*

$$N(n, Q_n(y)) \geq \mu(B(y, e^{-n})) \cdot 2Ce^{-2an(h+\varepsilon)}.$$

Proof of the lemma. It follows from (26.17) and (26.18) that

$$\frac{1}{2}\mu(B(y, e^{-n})) \leq \mu(B(y, e^{-n}) \cap \Gamma) \leq \mu(Q_n(y) \cap \Gamma)$$

$$\leq N(n, Q_n(y)) \cdot \max\{\mu(R) : R \in \mathfrak{T}(n) \text{ and } R \cap Q_n(y) \neq \varnothing\}.$$

The desired inequality follows from (26.9). ∎

Lemma 3. *For μ-almost every $y \in R(x) \cap \hat{\Gamma}$ there is an integer $n_2(y) \geq n_1$ such that for each $n \geq n_2(y)$ we have*

$$N(n + 2, Q_{n+2}(y)) \leq \hat{N}^{(s)}(n, y, Q_n(y)) \cdot \hat{N}^{(u)}(n, y, Q_n(y)) \cdot 2C^2 e^{4a(h+\varepsilon)} e^{4an\varepsilon}.$$

Proof of the lemma. Since $\hat{\Gamma} \subset \Gamma$, by (26.17) and the Borel Density Lemma (applied to the set $A = \hat{\Gamma}$; see Appendix V), for μ-almost every $y \in \hat{\Gamma}$ there is an integer $n_2(y) \geq n_1$ such that for all $n \geq n_2(y)$,

$$2\mu(Q_n(y) \cap \Gamma) \geq 2\mu(B(y, e^{-n}) \cap \Gamma) \geq \mu(B(y, e^{-n}))$$

$$\geq \mu(B(y, 4e^{-n-2})) \geq \mu(Q_{n+2}(y)). \qquad (26.21)$$

For any $m \geq n_2(y)$, by (26.9) and Property (d), we have

$$\mu(Q_m(y)) = \sum_{R_{am}^{am}(z) \subset Q_m(y)} \mu(R_{am}^{am}(z)) \geq N(m, Q_m(y)) \cdot C^{-1} e^{-2am(h+\varepsilon)}.$$

Similarly, for every $n \geq n_2(y)$ we obtain

$$\mu(Q_n(y) \cap \hat{\Gamma}) = \sum_{R_{an}^{an}(z) \subset Q_n(y)} \mu(R_{an}^{an}(z) \cap \hat{\Gamma}) \leq N_n \cdot C e^{-2an(h-\varepsilon)},$$

where N_n is the number of rectangles $R_{an}^{an}(z) \in \mathfrak{T}(n)$ that have non-empty intersection with $\hat{\Gamma}$. Set $m = n + 2$. The last two inequalities together with (26.21) imply that

$$N(n + 2, Q_{n+2}(y)) \leq N_n \cdot 2C^2 e^{4a(h+\varepsilon)+4an\varepsilon}. \qquad (26.22)$$

On the other hand, since $y \in \hat{\Gamma}$ the intersections $R_0^{an}(y) \cap A^{(u)}(y) \cap \hat{\Gamma}$ and $R_{an}^0(y) \cap A^{(s)}(y) \cap \hat{\Gamma}$ are non-empty.

Consider a rectangle $R_{an}^{an}(v) \subset Q_n(y)$ that has non-empty intersection with $\hat{\Gamma}$. Then the rectangles $R_{an}^0(v) \cap R_0^{an}(y)$ and $R_{an}^0(y) \cap R_0^{an}(v)$ belong to $\mathfrak{F}(n)$ and intersect respectively the stable and unstable local manifolds at y. Hence, one can associate to any rectangle $R_{an}^{an}(v) \subset Q_n(y)$ in $\mathfrak{T}(n)$, that has non-empty intersection with $\hat{\Gamma}$ the pair of rectangles $(R^{(s)} = R_{an}^0(v) \cap R_0^{an}(y), R^{(u)} = R_{an}^0(y) \cap R_0^{an}(v))$ such that $R^{(s)}, R^{(u)} \in \mathfrak{F}(n)$, $R^{(s)} \cap A^{(s)}(y) \cap Q_n(y) \neq \varnothing$, and $R^{(u)} \cap A^{(u)}(y) \cap Q_n(y) \neq \varnothing$. Clearly this correspondence is injective. Therefore,

$$\hat{N}^{(s)}(n, y, Q_n(y)) \cdot \hat{N}^{(u)}(n, y, Q_n(y)) \geq N_n.$$

The desired inequality follows from (26.22). ∎

Lemma 4. *For each $x \in \hat{\Gamma}$ and integer $n \geq n_1$ we have*

$$\hat{N}^{(s)}(n, x, R(x)) \leq D^{-1} C^2 e^{an(h+3\varepsilon)},$$

$$\hat{N}^{(u)}(n, x, R(x)) \leq D^{-1} C^2 e^{an(h+3\varepsilon)}.$$

Proof of the lemma. Since the partition $\mathcal{R}$ is countable one can find points y_i such that the union of the rectangles $R_0^{an}(y_i)$ is the set $R(x)$ and these rectangles are mutually disjoint. Without loss of generality we can assume that $y_i \in \hat{\Gamma}$ whenever $R_0^{an}(y_i) \cap \hat{\Gamma} \neq \varnothing$. We have

$$
\begin{aligned}
N(n, R(x)) &\geq \sum_i N^{(s)}(n, y_i, R_0^{an}(y_i)) \\
&\geq \sum_{i:R_0^{an}(y_i) \cap \hat{\Gamma} \neq \varnothing} N^{(s)}(n, y_i, R_0^{an}(y_i)).
\end{aligned}
\tag{26.23}
$$

By Properties (e), (f), and (26.10) we obtain that

$$
\begin{aligned}
N^{(s)}(n, y_i, R_0^{an}(y_i)) &\geq \frac{\mu_{y_i}^{(s)}(R_0^{an}(y_i) \cap \Gamma)}{\max\{\mu_z^{(s)}(R_{an}^{an}(z)) : z \in A^{(s)}(y_i) \cap R(x) \cap \Gamma \neq \varnothing\}} \\
&= \frac{D}{\max\{\mu_z^{(s)}(R_{an}^0(z)) : z \in A^{(s)}(y_i) \cap R(x) \cap \Gamma \neq \varnothing\}} \\
&\geq DC^{-1} e^{an(h-\varepsilon)}.
\end{aligned}
\tag{26.24}
$$

Similarly, (26.9) implies that

$$
N(n, R(x)) \leq \frac{\mu(R(x))}{\min\{\mu(R_{an}^{an}(z)) : z \in R(x) \cap \Gamma\}} \leq C e^{2anh + 2an\varepsilon}.
\tag{26.25}
$$

We now observe that

$$
\hat{N}^{(u)}(n, x, R(x)) = \text{card } \{i : R_0^{an}(y_i) \cap \hat{\Gamma} \neq \varnothing\}.
\tag{26.26}
$$

Putting (26.23)–(26.26) together we conclude that

$$
\begin{aligned}
C e^{2an(h+\varepsilon)} \geq N(n, R(x)) &\geq \sum_{i:R_0^{an}(y_i) \cap \hat{\Gamma} \neq \varnothing} N^{(s)}(n, y_i, R_0^{an}(y_i)) \\
&\geq \hat{N}^{(u)}(n, x, R(x)) \cdot DC^{-1} e^{an(h-\varepsilon)}.
\end{aligned}
$$

This yields $\hat{N}^{(u)}(n, x, R(x)) \leq D^{-1} C^2 e^{an(h+3\varepsilon)}$. The other inequality can be proved in a similar way. $\blacksquare$

We emphasize that the procedure of "filling in" rectangles to obtain the class $\mathfrak{F}(n)$ may substantially increase the number of rectangles in neighborhoods of some points. However, the next lemma shows that this procedure does not add too many rectangles at almost every point. In other words the procedure is "locally homogeneous" on a set of "big" measure.

Lemma 5. *For μ-almost every $y \in R(x) \cap \hat{\Gamma}$ we have*

$$\varlimsup_{n \to \infty} \frac{\hat{N}^{(s)}(n, y, Q_n(y))}{N^{(s)}(n, y, Q_n(y))} e^{-7an\varepsilon} < 1, \quad \varlimsup_{n \to \infty} \frac{\hat{N}^{(u)}(n, y, Q_n(y))}{N^{(u)}(n, y, Q_n(y))} e^{-7an\varepsilon} < 1.$$

Proof of the lemma. By (26.20) for each $n \geq n_1$ and $y \in \hat{\Gamma}$,

$$\mu_y^{(s)}(Q_n(y)) \geq \mu_y^{(s)}(B^{(s)}(y, e^{-n}) \cap \Gamma) \geq \frac{1}{2}\mu_y^{(s)}(B^{(s)}(y, e^{-n})).$$

Since $R_{an}^{an}(z) \subset R_{an}^0(z)$ for every z we obtain by virtue of (26.10) and (26.12) that

$$N^{(s)}(n, y, Q_n(y)) \geq \frac{\mu_y^{(s)}(Q_n(y))}{\max\{\mu_z^{(s)}(R_{an}^{an}(z)) : z \in A^{(s)}(y) \cap R(x) \cap \Gamma \neq \varnothing\}}$$

$$\geq \frac{1}{2} \frac{\mu_y^{(s)}(B^{(s)}(y, e^{-n}))}{\max\{\mu_z^{(s)}(R_{an}^0(z)) : z \in A^{(s)}(y) \cap R(x) \cap \Gamma \neq \varnothing\}}$$

$$\geq \frac{1}{2C} \frac{e^{-(d^{(s)}+\varepsilon)n}}{e^{-an(h-\varepsilon)}}.$$

Consider the set

$$F = \left\{ y \in \hat{\Gamma} : \varlimsup_{n \to \infty} \frac{\hat{N}^{(s)}(n, y, Q_n(y))}{N^{(s)}(n, y, Q_n(y))} e^{-7an\varepsilon} \geq 1 \right\}.$$

For each $y \in F$ there exists an increasing sequence $\{m_j\}_{j=1}^{\infty} = \{m_j(y)\}_{j=1}^{\infty}$ of positive integers such that

$$\hat{N}^{(s)}(m_j, y, Q_{m_j}(y)) \geq \frac{1}{2} N^{(s)}(m_j, y, Q_{m_j}(y)) e^{7am_j\varepsilon}$$

$$\geq \frac{1}{4C} e^{-d^{(s)}m_j + am_j h + 5am_j\varepsilon} \qquad (26.27)$$

for all j (note that $a > 1$).

We wish to show that $\mu(F) = 0$. Assuming the contrary consider the set $F' \subset F$ of points $y \in F$ for which the following limit exists

$$\lim_{r \to 0} \frac{\log \mu_y^{(s)}(B^{(s)}(y, r))}{\log r} = d^{(s)}.$$

Clearly, $\mu(F') = \mu(F) > 0$. Then one can find $y \in F$ such that

$$\mu_y^{(s)}(F) = \mu_y^{(s)}(F') = \mu_y^{(s)}(F' \cap R(y) \cap A^{(s)}(y)) > 0.$$

It follows from Frostman's lemma that

$$\dim_H(F' \cap A^{(s)}(y)) = d^{(s)}. \tag{26.28}$$

Consider the collection of balls

$$\mathcal{B} = \{B(z, e^{-m_j(z)}) : z \in F' \cap A^{(s)}(y), \, j = 1, 2, \dots\}.$$

By the Besicovitch Covering Lemma (see Appendix V) one can find a countable subcover $\mathcal{C} \subset \mathcal{B}$ of $F' \cap A^{(s)}(y)$ of arbitrarily small diameter and finite multiplicity ρ (which depends only on $\dim \mathcal{M}$). This means that for any $L > 0$ one can choose a sequence of points $\{z_i \in F' \cap A^{(s)}(y)\}_{i=1}^{\infty}$ and a sequence of integers $\{t_i\}_{i=1}^{\infty}$, where $t_i \in \{m_j(z_i)\}_{j=1}^{\infty}$ and $t_i > L$ for each i, such that the collection of balls $\mathcal{C} = \{B(z_i, e^{-t_i}) : i = 1, 2, \dots\}$ comprises a cover of $F' \cap A^{(s)}(y)$ whose multiplicity does not exceed ρ. We write $Q(i) = Q_{t_i}(z_i)$.

The Hausdorff sum corresponding to this cover is

$$\sum_{B \in \mathcal{C}} (\operatorname{diam} B)^{d^{(s)} - \varepsilon} = \sum_{i=1}^{\infty} e^{-t_i(d^{(s)} - \varepsilon)}.$$

By (26.27) we obtain

$$\sum_{i=1}^{\infty} e^{-t_i(d^{(s)} - \varepsilon)} \leq \sum_{i=1}^{\infty} \hat{N}^{(s)}(t_i, z_i, Q(i)) \cdot 4Ce^{-at_i h - 4at_i \varepsilon}$$

$$\leq 4C \sum_{q=1}^{\infty} e^{-aqh - 4aq\varepsilon} \sum_{i:t_i=q} \hat{N}^{(s)}(q, z_i, Q(i)).$$

Since the multiplicity of the subcover $\mathcal{C}$ is at most ρ each set $Q(i)$ appears in the sum $\sum_{i:t_i=q} \hat{N}^{(s)}(q, z_i, Q(i))$ at most ρ times. Hence,

$$\sum_{i:t_i=q} \hat{N}^{(s)}(q, z_i, Q(i)) \leq \rho \hat{N}^{(s)}(q, y, R(y)).$$

It follows from Lemma 4 that

$$\sum_{B \in \mathcal{C}} (\operatorname{diam} B)^{d^{(s)} - \varepsilon} \leq 4C \sum_{q=1}^{\infty} e^{-aq(h + 4\varepsilon)} \rho \hat{N}^{(s)}(q, y, R(y))$$

$$\leq 4 \frac{D}{C^3} \rho \sum_{q=1}^{\infty} e^{-aq(h + 4\varepsilon - h - 3\varepsilon)} = 4 \frac{D}{C^3} \rho \sum_{q=1}^{\infty} e^{-aq\varepsilon} < \infty.$$

Since the number L can be chosen arbitrarily large (and so are the numbers t_i) it follows that $\dim_H(F' \cap A^{(s)}(y)) \leq d^{(s)} - \varepsilon < d^{(s)}$. This contradicts (26.28). Hence, $\mu(F) = 0$. This yields the first inequality. The proof of the second inequality is similar. ∎

Lemmas 3, 4, and 5 show that the deviation of the distribution of rectangles in $\mathfrak{T}$ from direct product structure is subexponentially small. The rest of the proof simulates the case of measures which have local (semi-) product structure considered above.

By Lemmas 2 and 3 for μ-almost every $y \in R(x) \cap \hat{\Gamma}$ and $n \geq n_2(y)$ we obtain

$$\mu(B(y, e^{-n-2})) \leq \hat{N}^{(s)}(n, y, Q_n(y)) \times \hat{N}^{(u)}(n, y, Q_n(y))$$
$$\times 4C^3 e^{4a(h+\varepsilon)} e^{-2anh+6an\varepsilon}.$$

By Lemma 5 for μ-almost every $y \in R(x) \cap \hat{\Gamma}$ there exists an integer $n_3(y) \geq n_2(y)$ such that for all $n \geq n_3(y)$,

$$\hat{N}^{(s)}(n, y, Q_n(y)) < N^{(s)}(n, y, Q_n(y)) e^{7an\varepsilon},$$
$$\hat{N}^{(u)}(n, y, Q_n(y)) < N^{(u)}(n, y, Q_n(y)) e^{7an\varepsilon}.$$

This implies that

$$\mu(B(y, e^{-n-2})) \leq N^{(s)}(n, y, Q_n(y)) \times N^{(u)}(n, y, Q_n(y))$$
$$\times 4C^3 e^{4a(h+\varepsilon)} e^{-2anh+20an\varepsilon}.$$

Applying Lemma 1 we obtain

$$\mu(B(y, e^{-n-2})) \leq \mu_y^{(s)}(B^{(s)}(y, 4e^{-n})) \times \mu_y^{(u)}(B^{(u)}(y, 4e^{-n}))$$
$$\times 4C^5 e^{4a(h+\varepsilon)} e^{22an\varepsilon}.$$

This implies that

$$\varliminf_{n \to \infty} \frac{\log \mu(B(y, e^{-n}))}{-n} \geq d^{(s)} + d^{(u)} - 22a\varepsilon$$

for μ-almost every $y \in \hat{\Gamma}$. Since $\mu(\hat{\Gamma}) > 1 - \varepsilon$ and $\varepsilon > 0$ is arbitrary we conclude that

$$\underline{d} = \varliminf_{r \to 0} \frac{\log \mu(B(y, r))}{\log r} = \varliminf_{n \to \infty} \frac{\log \mu(B(y, e^{-n}))}{-n} \geq d^{(s)} + d^{(u)}$$

for μ-almost every $y \in \mathcal{M}$. This completes the proof. ∎

Let Λ be a hyperbolic set for a $C^{1+\alpha}$-diffeomorphism f on a compact smooth manifold $\mathcal{M}$. Assume that $f|\Lambda$ is topologically transitive. The set Λ is called a **hyperbolic attractor** if there exists an open set U such that $\Lambda \subset U$ and $\overline{f(U)} \subset U$. Clearly, $\Lambda = \bigcap_{n \geq 0} f^n(U)$. One can show that if Λ is a hyperbolic attractor then $\mathcal{W}_{\mathrm{loc}}^{(u)}(x) \subset \Lambda$ for every $x \in \Lambda$ (see, for example, [KH]). Consider the unique equilibrium measure $\mu = \mu_\varphi$ on Λ corresponding to the function $\varphi = J^{(u)}(x)$. It is known that μ_φ is the SRB-measure (see [Bo2]). Obviously, $d^{(u)} = \dim \mathcal{W}_{\mathrm{loc}}^{(u)}(x) \stackrel{\mathrm{def}}{=} m$ for every $x \in \Lambda$. By Proposition 26.2 and the entropy formula (see [LY]) we have that

$$d^{(s)} \geq -\frac{h_\mu(f)}{\lambda_\mu^{(q)}} = -\frac{\sum_{j=1}^{k} k_\mu^{(j)} \lambda_\mu^{(j)}}{\lambda_\mu^{(q)}}.$$

This implies that

$$\dim_H \Lambda \geq \dim_H \mu \geq m - \frac{\sum_{j=1}^{k} k_\mu^{(j)} \lambda_\mu^{(j)}}{\lambda_\mu^{(q)}}.$$

Appendix V

Some Useful Information

1. Outer Measures [Fe].

Let (X, ρ) be a complete metric space and m a σ-sub-additive outer measure on X, i.e., a set function which satisfies the following properties:

(1) $m(\varnothing) = 0$;
(2) $m(Z_1) \leq m(Z_2)$ if $Z_1 \subset Z_2 \subset X$;
(3) $m(\bigcup_{i \geq 0} Z_i) \leq \sum_{i \geq 0} m(Z_i)$, where $Z_i \subset X, i = 0, 1, 2, \ldots$.

A set $E \subset X$ is called measurable (with respect to m or simply m-measurable) if for any $A \subset X$,

$$m(A) = m(A \cap E) + m(A \setminus E).$$

The collection $\mathfrak{A}$ of all m-measurable sets can be shown to be a σ-field and the restriction of m to $\mathfrak{A}$ to be a σ-additive measure (which we will denote by the same symbol m).

An outer measure m is called

(1) Borel if all Borel sets are m-measurable;
(2) metric if

$$m(E \cup F) = m(E) + m(F)$$

for any positively separated sets E and F (i.e., $\rho(E, F) = \inf\{\rho(x, y) : x \in E, y \in F\} > 0$);
(3) regular if for any $A \subset X$ there exists an m-measurable set E containing A for which $m(A) = m(E)$. One can prove that any metric outer measure is Borel.

2. Borel Density Lemma [Gu].

We state the result known in the general measure theory as the Borel Density Lemma. We present it in the form which best suits to our purposes.

Borel Density Lemma. *Let $A \subset X$ be a measurable set of positive measure. Then for μ-almost every $x \in A$,*

$$\lim_{r \to 0} \frac{\mu(B(x, r) \cap A)}{\mu(B(x, r))} = 1.$$

Furthermore, if $\mu(A) > 0$ then for each $\delta > 0$ there is a set $\Delta \subset A$ with $\mu(\Delta) > \mu(A) - \delta$ and a number $r_0 > 0$ such that for all $x \in \Delta$ and $0 < r < r_0$,

$$\mu(B(x, r) \cap A) \geq \frac{1}{2}\mu(B(x, r)).$$

3. Covering Results [F4], [Fe].

In the general measure theory there is a number of "covering statements" which describe how to obtain an "optimal" cover from a given one. We describe two of them which we use in the book. Consider the Euclidean space $\mathbb{R}^m$ endowed with a metric ρ which is equivalent to the standard metric.

Vitali Covering Lemma. *Let $\mathcal{A}$ be a collection of balls contained in some bounded region of $\mathbb{R}^m$. Then there is a (finite or countable) disjoint subcollection $\mathcal{B} = \{B_i\} \subset \mathcal{A}$ such that*

$$\bigcup_{B \in \mathcal{A}} B \subset \bigcup_i \tilde{B}_i,$$

where $\tilde{B}_i$ is the closed ball concentric with B_i and of four times the radius.

Besicovitch Covering Lemma. *Let $Z \subset \mathbb{R}^m$ be a set and $r \colon Z \to \mathbb{R}^+$ a bounded function. Then the cover $\mathcal{G} = \{B(x, r(x)) : x \in Z\}$ contains a finite or countable subcover of finite multiplicity which depends only on m.*

As an immediate consequence of the Besicovitch Covering Lemma we obtain that for any set $Z \subset \mathbb{R}^m$ and $\varepsilon > 0$ there exists a cover of Z by balls of radius ε of finite multiplicity which depends only on m.

4. Cohomologous Functions [R1].

Two functions φ_1 and φ_2 on a compact metric space X are called cohomologous if there exist a Hölder continuous function $\eta \colon X \to \mathbb{R}$ and a constant C such that

$$\varphi_1 - \varphi_2 = \eta - \eta \circ f + C.$$

In this case we write $\varphi_1 \sim \varphi_2$. If the above equality holds with $C = 0$ the functions are called strictly cohomologous. We recall some well-known properties of cohomologous functions:

(1) if $\varphi_1 \sim \varphi_2$ then for every $x \in X$,

$$\lim_{n \to \infty} \frac{1}{n} \sum_{k=0}^{n-1} [\varphi_1(x) - \varphi_2(x)] = C;$$

(2) $\varphi_1 \sim \varphi_2$ if and only if equilibrium measures of φ_1 and φ_2 on X coincide;
(3) if the functions φ_1 and φ_2 are strictly cohomologous then $P_X(\varphi_1) = P_X(\varphi_2)$.

5. Legendre Transform [Ar].

We remind the reader of the notion of a Legendre transform pair of functions. Let h be a strictly convex C^2-function on an interval I, i.e., $h''(x) > 0$ for all $x \in I$. The Legendre transform of h is the differentiable function g of a new variable p defined by

$$g(p) = \max_{x \in I}(p\,x - h(x)). \tag{A5.1}$$

One can show that: 1) g is strictly convex; 2) the Legendre transform is involutive; 3) strictly convex functions h and g form a Legendre transform pair if and only if $g(\alpha) = h(q) + q\alpha$, where $\alpha(q) = -h'(q)$ and $q = g'(\alpha)$.

Bibliography

[AKM]   R. L. Adler, A. G. Konheim, M. H. McAndrew: Topological entropy. *Trans. Amer. Math. Soc.* **114** (1965), 309–319.

[AY]   J. C. Alexander, J. A. Yorke: Fat Baker's Transformations. *Ergod. Theory and Dyn. Syst.* **4** (1984), 1–23.

[Ar]   V. I. Arnold: *Mathematical Methods of Classical Mechanics.* Springer-Verlag, Berlin–New York, 1978.

[B]   V. Bakhtin: Properties of Entropy and Hausdorff Measure. *Vestnik. Moscow. Gos. Univ. Ser. I, Mat. Mekh. (in Russian)*, **5** (1982), 25–29.

[Ba]   M. Barnsley, R. Devaney, B. Mandelbrot, H. O. Peitgen, D. Saupe, R. Voss: *The Science of Fractal Images.* Springer-Verlag, Berlin–New York, 1988.

[Bar1] L. Barreira: Cantor Sets with Complicated Geometry and Modeled by General Symbolic Dynamics. *Random and Computational Dynamics*, **3** (1995), 213–239.

[Bar2] L. Barreira: A Non-additive Thermodynamic Formalism and Applications to Dimension Theory of Hyperbolic Dynamical Systems. *Ergod. Theory and Dyn. Syst.* **16** (1996), 871–928.

[BPS1] L. Barreira, Ya. Pesin, J. Schmeling: On the Pointwise Dimension of Hyperbolic Measures — A Proof of the Eckmann-Ruelle Conjecture. *Electronic Research Announcements*, **2:1** (1996), 69–72.

[BPS2] L. Barreira, Ya. Pesin, J. Schmeling: On a General Concept of Multifractality: Multifractal Spectra for Dimensions, Entropies, and Lyapunov Exponents. Multifractal Rigidity. *Chaos*, **7:1** (1997), 27–38.

[BPS3] L. Barreira, Ya. Pesin, J. Schmeling: Multifractal Spectra and Multifractal Rigidity for Horseshoes. *J. of Dynamical and Control Systems*, **3:1** (1997), 33–49.

[BS]   L. Barreira, J. Schmeling: Sets of "Non-typical" Points Have Full Hausdorff Dimension and Full Topological Entropy. Preprint, IST, Lisbon, Portugal; WIAS, Berlin, Germany.

[Bi]   P. Billingsley: *Probability and Measure.* John Wiley & Sons, New York–London–Sydney, 1986.

[BGH]   F. Blanchard, E. Glasner, B. Host: A Variation on the Variational Principle and Applications to Entropy Pairs. *Ergod. Theory and Dyn. Syst.* **17** (1997), 29–43.

[Bot]   H. G. Bothe: The Hausdorff Dimension of Certain Solenoids. *Ergod. Theory and Dyn. Syst.* **15** (1995), 449–474.

[Bo1]   R. Bowen: Topological Entropy for Non-compact Sets. *Trans. Amer. Math. Soc.* **49** (1973), 125–136.

296

[Bo2]   R. Bowen: Equilibrium States and the Ergodic Theory of Anosov Diffeomorphisms. Lecture Notes in Mathematics, Springer-Verlag, Berlin–New York.

[Bo3]   R. Bowen: Hausdorff Dimension of Quasi-circles. *Publ. Math. IHES*, **50** (1979), 259–273.

[BK]   M. Brin, A. Katok: On Local Entropy. Lecture Notes in Mathematics, Springer-Verlag, Berlin–New York.

[C]   C. Carathéodory: Über das Lineare Mass. *Göttingen Nachr.* (1914), 406–426.

[CG]   L. Carleson, T. Gamelin: *Complex Dynamics.* Springer-Verlag, Berlin–New York, 1993.

[CM]   R. Cawley, R. D. Mauldin: Multifractal Decompositions of Moran Fractals. *Advances in Math.* **92** (1992), 196–236.

[CE]   P. Collet, J.-P. Eckmann: Iterated Maps on the Interval as Dynamical Systems. *Progress in Physics.* Birkhäuser, Boston, 1980.

[CLP]   P. Collet, J. L. Lebowitz, A. Porzio: The Dimension Spectrum of Some Dynamical Systems. *J. Stat. Phys.* **47** (1987), 609–644.

[Cu]   C. D. Cutler: Connecting Ergodicity and Dimension in Dynamical Systems. *Ergod. Theory and Dyn. Syst.* **10** (1990), 451–462.

[EP]   J. P. Eckmann, I. Procaccia: Fluctuations of Dynamical Scaling Indices in Nonlinear Systems. *Phys. Rev. A* **34** (1986), 659–661.

[ER]   J. P. Eckmann, D. Ruelle: Ergodic Theory of Chaos and Strange Attractors. 3, *Rev. Mod. Phys.* **57** (1985), 617–656.

[EM]   G. A. Edgar, R. D. Mauldin: Multifractal Decompositions of Digraph Recursive Fractals. *Proc. London Math. Soc.* **65:3** (1992), 604–628.

[Er1]   P. Erdös: On a Family of Symmetric Bernoulli Convolutions. *Amer. J. Math.* **61** (1939), 974–976.

[Er2]   P. Erdös: On the Smoothness Properties of a Family of Symmetric Bernoulli Convolutions. *Amer. J. Math.* **62** (1940), 180–186.

[F1]   K. Falconer: Hausdorff Dimension of Some Fractals and Attractors of Overlapping Construction. *J. Stat. Phys.* **47** (1987), 123–132.

[F2]   K. Falconer: The Hausdorff Dimension of Self-affine Fractals. *Math. Proc. Camb. Phil. Soc.* **103** (1988), 339–350.

[F3]   K. Falconer: A subadditive thermodynamic formalism for mixing repellers. *J. Phys. A: Math. Gen.* **21** (1988), L737–L742.

[F4]   K. Falconer: Dimension and Measures of Quasi Self-similar Sets. *Proc. Amer. Math. Soc.* **106** (1989), 543–554.

[F5]   K. Falconer: *Fractal Geometry, Mathematical Foundations and Applications.* John Wiley & Sons, New York–London–Sydney, 1990.

[Fe]   H. Federer: *Geometric Measure Theory.* Springer-Verlag, Berlin–New York, 1969.

[Fr]   O. Frostman: Potential d'Équilibre et Capacité des Ensembles avec Quelques Applications à la Théorie des Fonctions. *Meddel. Lunds Univ. Math. Sem.* **3** (1935), 1–118.

[Fu]   H. Furstenberg: Disjointness in Ergodic Theory, Minimal Sets, and a Problem in Diophantine Approximation. *Mathematical Systems Theory*, **1** (1967), 1–49.

[GaP]   D. Gatzouras, Y. Peres: Invariant Measures of Full Dimension for Some Expanding Maps. Preprint, Univ. of California, Berkeley CA, USA.

[G]   P. Grassberger: Generalized Dimension of Strange Attractors. *Phys. Lett. A* **97:6** (1983), 227–230.

[GP]   P. Grassberger, I. Procaccia: Characterization of Strange Attractors. *Phys. Rev. Lett.* **84** (1983), 346–349.

[GHP]   P. Grassberger, I. Procaccia, G. Hentschel: On the Characterization of Chaotic Motions. *Lect. Notes Physics*, **179** (1983), 212–221.

[GOY]   C. Grebogi, E. Ott, J. Yorke: Unstable Periodic Orbits and the Dimension of Multifractal Chaotic Attractors. *Phys. Rev. A* **37:5** (1988), 1711–1724.

[GY]   M. Guysinsky, S. Yaskolko: Coincidence of Various Dimensions Associated With Metrics and Measures on Metric Spaces. *Discrete and Continuous Dynamical Systems*, (1997).

[Gu]   M. de Guzmán: *Differentiation of Integrals in* $\mathbb{R}^n$. Lecture Notes in Mathematics, vol. 481. Springer-Verlag, Berlin–New York, 1975.

[HJKPS]   T. C. Halsey, M. Jensen, L. Kadanoff, I. Procaccia, B. Shraiman: Fractal Measures and Their Singularities: The Characterization of Strange Sets. *Phys. Rev. A* **33:2** (1986), 1141–1151.

[Ha]   B. Hasselblatt: Regularity of the Anosov Splitting and of Horospheric Foliations. *Ergod. Theory and Dyn. Syst.* **14** (1994), 645–666.

[H]   F. Hausdorff: Dimension und Ausseres Mass. *Math. Ann.* **79** (1919), 157–179.

[HP]   H. G. E. Hentschel, I. Procaccia: The Infinite Number of Generalized Dimensions of Fractals and Strange Attractors. *Physica D*, **8** (1983), 435–444.

[Il]   Y. Il'yashenko: Weakly Contracting Systems and Attractors of Galerkin Approximations of Navier–Stokes Equations on the Two-Dimensional Torus. *Selecta Math. Sov.* **11:3** (1992), 203–239.

[K]   A. Katok: Bernoulli Diffeomorphisms on Surfaces. *Ann. of Math.* **110:3** (1979), 529–547.

[JW]   B. Jessen, A. Wintner: Distribution Functions and the Riemann Zeta Function. *Trans. Amer. Math. Soc.* **38** (1938), 48–88.

[KH]   A. Katok, B. Hasselblatt: *Introduction to the Modern Theory of Dynamical Systems*. Encyclopedia of Mathematics and its Applications,, vol. 54. Cambridge University Press, London–New York, 1995.

[Ki]   Y. Kifer: Characteristic Exponents of Dynamical Systems in Metric Spaces. *Ergod. Theory and Dyn. Syst.* **3** (1983), 119–127.

[Kr]   S. N. Krushkal': *Quasi-conformal Mappings and Riemann Surfaces*. John Wiley & Sons, New York–London–Sydney, 1979.

[LG]   S. P. Lalley, D. Gatzouras: Hausdorff and Box Dimensions of Certain Self-affine Fractals. *Indiana Univ. Math. J.* **41** (1992), 533–568.

[LN]   Ka-Sing Lau, Sze-Man Ngai: Multifractal Measures and a Weak Separation Condition. Preprint, Univ. of Pittsburgh, Pittsburgh PA, USA.

[L]   F. Ledrappier: Dimension of Invariant Measures. *Proceedings of the Conference on Ergodic Theory and Related Topics.* vol. 11, pp. 116–124. Teubner, 1986.

[LM]   F. Ledrappier, M. Misiurewicz: Dimension of Invariant Measures for Maps with Exponent Zero. *Ergod. Theory and Dyn. Syst.* **5** (1985), 595–610.

[LP]   F. Ledrappier, A. Porzio: Dimension Formula for bernoulli Convolutions. *J. Stat. Phys.* **76** (1994), 1307–1327.

[LY]   F. Ledrappier, L.-S. Young: The Metric Entropy of Diffeomorphisms. Part II. *Ann. of Math.* **122** (1985), 540–574.

[Lo]   A. Lopes: The Dimension Spectrum of the Maximal Measure. *SIAM J. Math. Analysis*, **20** (1989), 1243–1254.

[M1]   B. Mandelbrot: *The fractal Geometry of Nature*. W. H. Freeman & Company, New York, 1983.

[M2]   B. Mandelbrot: *Fractals and Multifractals*. Springer-Verlag, Berlin–New York, 1991.

[M]   R. Mañé: *Ergodic Theory and Differentiable Dynamics*. Springer-Verlag, Berlin–New York, 1987.

[Ma1]   A. Manning: A Relation between Lyapunov Exponents, Hausdorff Dimension, and Entropy. *Ergod. Theory and Dyn. Syst.* **1** (1981), 451–459.

[Ma2]   A. Manning: The Dimension of the Maximal Measure for a Polynomial Map. *Ann. of Math.* **119** (1984), 425–430.

[Mas]   B. Maskit: *Kleinian Groups*. Springer Verlag, Berlin–New York, 1988.

[MW]   R. Mauldin, S. Williams: Hausdorff Dimension in Graph Directed Constructions. *Trans. Amer. Math. Soc.* **309** (1988), 811–829.

[MM]   H. McCluskey, A. Manning: Hausdorff Dimension for Horseshoes. *Ergod. Theory and Dyn. Syst.* **3** (1983), 251–260.

[Mu]   C. McMullen: The Hausdorff Dimension of General Sierpiński Carpets. *Nagoya Math. J.* **96** (1984), 1–9.

[Mi]   M. Misiurewicz: Attracting Cantor Set of Positive Measure for a C^{∞}-map of an Interval. *Ergod. Theory and Dyn. Syst.* **2** (1982), 405–415.

[Mo]   P. Moran: Additive Functions of Intervals and Hausdorff Dimension. *Math. Proc. Camb. Phil. Soc.* **42** (1946), 15–23.

[O]   L. Olsen: A Multifractal Formalism. *Advances in Math.* **116** (1995), 82–196.

[PV]   J. Palis, M. Viana: On the Continuity of Hausdorff Dimension and Limit Capacity of Horseshoes. Lecture Notes in Mathematics, Springer-Verlag, Berlin–New York.

[Pa]   W. Parry: Symbolic Dynamics and Transformations of the Unit Interval. *Trans. Amer. Math. Soc.* **122** (1966), 368–378.

[PaPo]   W. Parry, M. Pollicott: Zeta Functions and the Periodic Orbit Structures of Hyperbolic Dynamics. *Astérisque*, **187–189** (1990).

[P1]   Ya. Pesin: Lyapunov Characteristic Exponents and Smooth Ergodic Theory. *Russian Math. Surveys*, **32** (1977), 55–114.

[P2]   Ya. Pesin: Dimension Type Characteristics for Invariant Sets of Dynamical Systems. *Russian Math. Surveys*, **43:4** (1988), 111–151.

[P3]   Ya. Pesin: On Rigorous Mathematical Definition of Correlation Dimension and Generalized Spectrum for Dimensions. *J. Stat. Phys.* **7:3–4** (1993), 529–547.

[PP]   Ya. Pesin, B. Pitskel': Topological Pressure and the Variational Principle for Non-compact Sets. *Functional Anal. and Its Applications*, **18:4** (1984), 50–63.

[PT]   Ya. Pesin, A. Tempelman: Correlation Dimension of Measures Invariant Under Group Actions. *Random and Computational Dynamics*, **3:3** (1995), 137–156.

[PW1]   Ya. Pesin, H. Weiss: On the Dimension of Deterministic and Random Cantor-like Sets, Symbolic Dynamics, and the Eckmann–Ruelle Conjecture. *Comm. Math. Phys.* **182:1** (1996), 105–153.

[PW2]   Ya. Pesin, H. Weiss: A Multifractal Analysis of Equilibrium Measures for Conformal Expanding Maps and Markov Moran Geometric Constructions. *J. Stat. Phys.* **86:1-2** (1997), 233–275.

[PW3]   Ya. Pesin, H. Weiss: The Multifractal Analysis of Gibbs Measures: Motivation, Mathematical Foundation, and Examples. *Chaos*, **7:1** (1997), 89–106.

[PY]   Ya. Pesin, Ch. Yue: Hausdorff Dimension of Measures with Non-zero Lyapunov Exponents and Local Product Structure. Preprint, Penn State Univ., University Park PA, USA.

[PoW]   M. Pollicott, H. Weiss: The Dimensions of Some Self-affine Limit Sets in the Plane and Hyperbolic Sets. *J. Stat. Phys.* (1993), 841–866.

[PS]   L. Pontryagin, L. Shnirel'man: *On a Metric Property of Dimension.* Appendix to the Russian translation of the book "Dimension Theory" by W. Hurewicz, H. Walhman, Princeton Univ. Press, Princeton, 1941.

[PU]   F. Przytycki, M. Urbański: On Hausdorff Dimension of Some Fractal Sets. *Studia Mathematica*, **93** (1989), 155–186.

[PUZ]   F. Przytycki, M. Urbański, A. Zdunik: Harmonic, Gibbs, and Hausdorff Measures on Repellers for Holomorphic Maps, I. *Ann. of Math.* **130** (1989), 1–40.

[Ra]   D. Rand: The Singularity Spectrum for Cookie-Cutters. *Ergod. Theory and Dyn. Syst.* **9** (1989), 527–541.

[R]   A. Rényi: Dimension, Entropy and Information. *Trans. 2nd Prague Conf. on Information Theory, Statistical Decision Functions, and Randone Processes*, (1957), 545–556.

[Ri]   R. Riedi: An Improved Multifractal Formalism and Self-similar Measures. *J. Math. Anal. Appl.* **189** 2 (1995), 462–490.

[Ro]   C. Rogers: *Hausdorff Measures.* Cambridge Univ. Press, London–New York, 1970.

[R1]   D. Ruelle: *Thermodynamic Formalism.* Addison-Wesley, Reading, MA, 1978.

[R2]   D. Ruelle: Repellers for Real Analytic Maps. *Ergod. Theory and Dyn. Syst.* **2** (1982), 99–107.

[S]   R. Salem: Algebraic Numbers and Fourier Transformations. *Heath Math. Monographs.* 1962.

[SY]   T. Sauer, J. Yorke: Are the Dimensions of a Set and Its Image Equal under Typical Smooth Functions?. *Ergod. Theory and Dyn. Syst.* (1997).

[Sch]   J. Schmeling: On the Completeness of Multifractal Spectra. Preprint, Free Univ., Berlin, Germany.

300

[Sc]   M. Schröeder: *Fractals, Chaos, Power, Laws.* W. H. Freeman & Company, New York, 1991.

[Sh]   M. Shereshevsky: A Complement to Young's Theorem on Measure Dimension: The Difference between Lower and Upper Pointwise Dimensions. *Nonlinearity,* **4** (1991), 15–25.

[Shi]   M. Shishikura: The Boundary of the Mandelbrot Set Has Hausdorff Dimension Two. *Astérisque,* **7** (1994), 389–405.

[Sim1]   K. Simon: Hausdorff Dimension for Non-invertible Maps. *Ergod. Theory and Dyn. Syst.* **13** (1993), 199–212.

[Sim2]   K. Simon: The Hausdorff Dimension of the General Smale–Williams Solenoid. *Proc. Amer. Math. Soc.* (1997).

[Si]   D. Simpelaere: Dimension Spectrum of Axiom A Diffeomorphisms. II. Gibbs Measures. *J. Stat. Phys.* **76** (1994), 1359–1375.

[Sm]   S. Smale: Diffeomorphisms with Many Periodic Points. In *In Differential and Combinatorial Topology,* pp. 63–80. Princeton Univ. Press, Princeton, NJ, 1965.

[So]   B. Solomyak: On the Random Series $\sum +\lambda^n$ (An Erdös Problem). *Ann. of Math.* **142** (1995), 1–15.

[St]   S. Stella: On Hausdorff Dimension of Recurrent Net Fractals. *Proc. Amer. Math. Soc.* **116** (1992), 389–400.

[T1]   F. Takens: Detecting Strange Attractors in Turbulence. Lecture Notes in Mathematics, Springer-Verlag, Berlin–New York.

[T2]   F. Takens: Limit Capacity and Hausdorff Dimension of Dynamically Defined Cantor Sets. Lecture Notes in Mathematics, Springer-Verlag, Berlin–New York.

[Tel]   T. Tél: Dynamical Spectrum and Thermodynamic Functions of Strange Sets From an Eigenvalue Problem. *Phys. Rev. A* **36:5** (1987), 2507–2510.

[V1]   S. Vaienti: Generalized Spectra for Dimensions of Strange Sets. *J. Phys. A* **21** (1988), 2313–2320.

[V2]   S. Vaienti: Dynamical Integral Transform on Fractal Sets and the Computation of Entropy. *Physica D,* **63** (1993), 282–298.

[W]   P. Walters: A Variational Principle for the Pressure of Continuous Transformations. *Amer. J. Math.* **97** (1971), 937–971.

[We]   H. Weiss: The Lyapunov and Dimension Spectra of Equilibrium Measures for Conformal Expanding Maps. Preprint, Penn State Univ., University Park PA, USA.

[Y1]   L.-S. Young: Capacity of Attractors. *Ergod. Theory and Dyn. Syst.* **1** (1981), 381–388.

[Y2]   L.-S. Young: Dimension, Entropy, and Lyapunov Exponents. *Ergod. Theory and Dyn. Syst.* **2** (1982), 109–124.

[Z]   A. Zdunik: Harmonic Measure Versus Hausdorff Measures on Repellers for Holomorphic Maps. *Trans. Amer. Math. Soc.* **2** (1991), 633–652.

[W]   A. Wintner: On Convergent Poisson Convolutions. *Amer. J. Math.* **57** (1935), 827–838.

Index